TEXTILES FOR ADVANCED APPLICATIONS

Edited by **Bipin Kumar** and **Suman Thakur**

Textiles for Advanced Applications
http://dx.doi.org/10.5772/66015
Edited by Bipin Kumar and Suman Thakur

Contributors

Murat Eyvaz, Serkan Arslan, Ercan Gürbulak, Ebubekir Yüksel, Izabela Luiza Ciesielska-Wrobel, Ivana Schwarz, Stana Kovacevic, Jacqueline Domjanić, Snježana Brnada, Urška Rozman, Sonja Šostar Turk, Emil Pal, Vida Gönc, PhD Daniela Zavec Pavlinić, Peng Wang, Xavier Legrand, Damien Soulat, Felecia Davis, Nesrin Karaduman, Yekta Karaduman, Hüseyin Özdemir, Gökce Özdemir, Adrian Buhu, Liliana Buhu, Alenka Pavko Čuden, Darja Rant, Nazan Kalebek, Gabriela Telipan, Beatrice Moasa, Elena Helerea, Razvan Scarlat, Eftalea Carpus, Gheorghe Enache, Mutlu Kurban, Osman Babaarslan, İsmail Hakkı Çağatay, Suman Thakur, Adriana Ungureanu

Published by InTech
Janeza Trdine 9, 51000 Rijeka, Croatia

Publishing Process Manager Dajana Pemac
Technical Editor SPi Global
Cover InTech Design team

First published September, 2017
Printed in Croatia
Legal deposit, Croatia: National and University Library in Zagreb

Additional hard copies can be obtained from orders@intechopen.com

Textiles for Advanced Applications, Edited by Bipin Kumar and Suman Thakur
p. cm.
Print ISBN 978-953-51-3500-5
Online ISBN 978-953-51-3501-2

For EU product safety concerns:
IN TECH d.o.o., Prolaz Marije Krucifikse Kozulić 3, 51000 Rijeka, Croatia,
info@intechopen.com or visit our website at intechopen.com.

Meet the editors

Dr. Bipin Kumar graduated from Indian Institute of Technology Delhi, India, with a PhD degree in Textile Engineering in 2013. Currently, he is appointed as an assistant professor at IIT Delhi, India. Prior to joining IIT Delhi, he was working as RAP (2016–2017) at the Hong Kong Polytechnic University, Hong Kong. He also served as a Fulbright Scholar (2014–2016) at the University of California, Davis, USA. His main research focuses on smart fibrous/polymeric materials and related fabric structures. He has over 30 journal publications, 2 patents, 2 authored books, 10 book chapters, and over 20 conference proceedings. He holds several editorial and professional memberships including ACP, Techtextil North America, ICC, JEFF, FTEE, and CTFTTE. Currently, he is engaged in R&D of functional textiles for several engineering and medical applications.

Dr. Suman Thakur, currently a postdoctoral fellow at Hong Kong Polytechnic University, obtained his Ph.D. in smart polymer nanocomposites from the Department of Chemical Sciences, Tezpur University, India. He had also worked as a postdoctoral researcher at Chonbuk National University, South Korea in 2015. His current research interest includes the synthesis of bio-based smart polymers, nanomaterials including graphene based nanomaterials/nanohybrids and polymer nanocomposites for their different advanced applications such as shape memory materials, self-healing polymer, photocatalysis, etc.

Contents

Preface

Textile has been associated with human lives dating far back in history around 6500 BC. The most obvious function of a textile in clothing, from the very beginning to the present day, is to improve the comfort of the wearer by protecting him/her from various elements of external factors. Initially, textiles were primarily used for clothing to provide comfort and protection. Besides the application for clothing, textile has been used for the purpose of decorating the interior, as a dominant part of textile-based products in the market (application in vehicles, construction, agriculture, health care, defense and security, power and environmental technologies, and many more). This world of conventional textiles (clothing and furnishing) has begun to change in the past two centuries, as new cultural influences meet technological innovations. Researches and processing technologies have been resulting in development of smart and advanced textiles that can serve functional purposes of products well beyond their traditional roles of clothing and fashion. Textile today is being progressively considered as a useful engineering material in diverse applications, solving challenging problems related to health, living, mobility, production, environment, energy, etc.

Textile materials and structures are intrinsically strong, while remaining flexible, permeable, moldable, and drapable. Despite these unique properties, many more new and advanced features are now expected from a textile. Thanks to the numerous innovations in the past, competitive market needs, and technological developments, today the potential of textiles has multiplied by manifolds, and their application can be seen in several domains. Buildtech area includes textiles or composite materials used in the construction of buildings and structures for various applications, such as facade foundation systems, concrete reinforcement, insulations, interior construction, visual protection, noise prevention, protection against the sun and building safety, air conditioning, and many more. Geotech includes textile products most commonly used in geotechnical applications, because of their separation, reinforcement, filtration, drainage, and protection functions. Indutech includes technical textile products used in the manufacturing sector like filters, conveyor fabrics, industrial belts and hose, etc. Meditech products are textile materials used in hygiene, health and personal care, as well as surgical applications. Mobiltech segment of technical textiles is used in the construction of all sorts of transport for human or cargo. Ecotech segment refers to the use of technical textiles in environmental engineering, primarily for environmental protection. The primary segment is landfill waste management, which refers to the use of geosynthetic products to secure landfills against leakage of municipal or hazardous waste. Protech is a collection of textile materials and products used to protect the body during exposure to hazards in the working environment. Protech includes clothing and accessories that provide protection against harmful chemical and extreme temperature conditions and low visibility and garment for ballistic protection (space suits, air-conditioned clothing, armor, motorcycle leathers, etc.).

This book presents a global view of the development and applications of technical textiles with the description of materials, structures, properties, characterizations, functions and relevant production technologies, case studies, challenges, and opportunities. More emphasis is placed on the principles of textile science and technology to provide solutions to several engineering problems. All chapter topics are exclusive and selectively chosen and designed, and they are extensively explored by different authors having specific knowledge in each area. Chapter 1 gives an overview of the development of textiles and their applications throughout history. Chapters 2 and 3 suggest some specific examples of the materials, structures, and technologies used for the development of textiles. More emphasis is given in the textile for composite applications, which is currently a hot topic across the world (Chapters 4 to 7). Thereafter, the potential applications of textile in protection and filtration are mentioned (Chapters 8 to 11). Some topics on smart textiles are also covered; smart textiles is the next generation of textiles intended for fashion, furnishing, and technical textile applications, which are designed and manufactured to sense and respond in a predefined manner to environmental stimuli. The last chapter reveals the practical and theoretical examples of smart textiles from an economic perspective concerning their integration in different value chains, commercial or innovative.

Bipin Kumar
Assistant Professor
Indian Institute of Technology Delhi
India

Section 1

Introduction to Textiles as Advanced Applications

Chapter 1

Textile Application: From Need to Imagination

Ivana Schwarz and Stana Kovačević

Additional information is available at the end of the chapter

http://dx.doi.org/10.5772/intechopen.68376

Abstract

To understand textiles and their application today, as well as future applications, it is necessary to comprehend the development of textiles and their applications throughout history. The gradual development of textile production processes and the use of different materials, influenced the development and application of materials themselves. Numerous innovations made since Industrial revolution, events in technology development and international competition have shaped the industry and continue to affect the textile production even today. Nowadays, textiles can be divided into two main sectors according to their application: conventional textiles (textiles for fashion clothing) and technical textiles with numerous applications for nearly all society needs. Due to market needs and technology development, the interference of all areas of science occurs, resulting in amazing innovations that follow existing trends and set future trends in terms of interactivity, digital and electronic functionality, social and environmental awareness, esthetics, etc. These are the reasons for great freedom, development prospects, expression of creativity and thus of innovativeness. The need for them today is greater and more important than ever before in the history of textiles and their application.

Keywords: textile history, development, materials, technical textiles, future trends

1. Introduction

Textiles have been important in human history and reflect the materials available to a civilization as well as to the technologies that had been mastered. From the ancient times to the present day, methods of textile production have continually evolved, and the choices of textiles available have influenced how people carried their possession, clothed themselves, and decorated their surroundings. The social significance of finished products reflects their culture. Archeological findings reveal many secrets about the history of textiles, their application and development, as well as art, and numerous historical documents. Knowledge of such materials remains inferential, since textile deteriorates quickly compared to stone, bone, shell, and metal artifacts.

2. Textiles in prehistoric age

Historical facts give strong reasons for believing that humans began wearing clothes 100,000–50,000 years ago. The evidence supporting such beliefs is genetic analysis that indicates the fact that the human body louse, which lives in clothing, may have diverged from the head louse 170,000 years ago. These estimates predate the first known human exodus from Africa [1, 2].

Our knowledge of ancient textiles has expanded in the recent times thanks to modern technological developments. Possible sewing needles have been dated to around 40,000 years ago and the earliest dyed flax fibers have been found in a prehistoric cave in the Republic of Georgia and date back to 36,000 BC, which suggests that textile-like materials were made even in prehistoric times [3].

The earliest evidence of weaving comes from the impressions of textiles, basketry, and nets on little pieces of hard clay, dating from 27,000 BC and found in Dolni Vestonice in the Czech Republic. At a slightly later date (25,000 BC), the Venus figurines were depicted with clothing (**Figure 1**). Evidence exists of flax cultivation from 8000 BC in the Near East [4, 5].

It is believed that the first actual textile, as opposed to skins sewn together, was probably felt. Many cultures have legends about the origins of felt making. A Sumerian legend claims that the secret of felt making was discovered by Urn Amman of Lagash. Another early textile method dated from 6500 BC is Nålebinding, a fabric creation technique predating both knitting and crochet [7].

The earliest known woven textiles of the Near East may be fabrics used to wrap the dead, excavated at a Neolithic site at Çatalhöyük in Anatolia, carbonized in a fire and radiocarbon dated to 6000 BC.

In Europe, Near East, and North Africa, from prehistoric times to early Middle Ages, two loom types dominated the textile production: the warp-weighted and the two-beam loom. The width of woven fabric is determined by the length of the cloth beam and could be 2–3 m wide. Woven clothing was very often made from full loom widths draped, tied, or pinned in place.

Neolithic period (around 5500 BC) is characterized by Ancient Egypt from where evidence of the flax fabric production originate. Documentation of domesticated wild flax cultivation, likely imported from the Levant, exists from 6000 BC. Other fibers like rush, reed, palm, and papyrus were used alone or with linen, to make rope and other textiles. Production and use of wool in that period are not significantly documented. The inhabitants of the Indus Valley Civilization used cotton for making clothing as early as 5000–4000 BC [5, 6, 8].

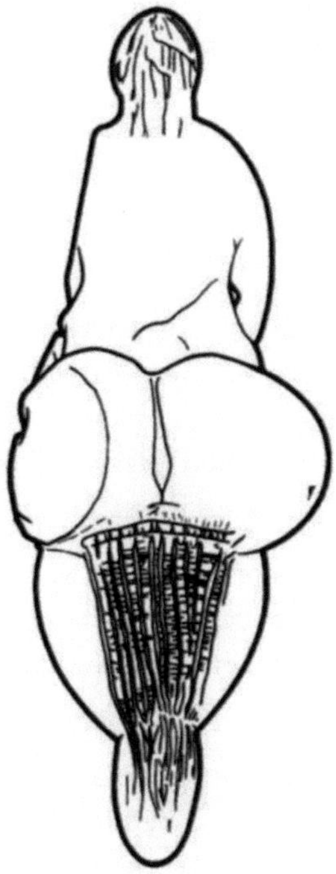

Figure 1. Paleolithic Venus figure wearing a skirt of twisted strings [6].

Around 3000 BC, the breeding of sheep with woolly fleece, rather than hair, occurs. At that time, different spinning techniques were used (drop spindle, hand-to-hand spinning, rolling on the thigh). The yarn was also spliced, and the horizontal ground loom was used for weaving process. From the period of New Kingdom, a vertical two-beam loom was introduced, probably from Asia [9].

3. Textile in ancient history

Ancient History is divided according to cultures whose traces have been kept until today (3500 BC–fifth century AD). The most important civilizations are Ancient Mesopotamia, Ancient Egypt, Ancient China, Ancient India, Ancient Greece, and Ancient Rome. Textiles and its development had a major impact on the development of its application and thus on the development other society and culture of these civilizations.

Time between 5000 and 3000 BC was a period from which the earliest evidence of silk production in China date back, found at the sites of Yang Shao culture in Xia, Shanxi. Another, with the remains, rich sites of Hemudu culture in Yuyao, Zhejiang, reveals fragments of primitive loom, dated to about 4000 BC, while in a Liangzhu culture site at Qianshanyang in Huzhou, Zhejiang, were found silk scarps, dating back to 2700 BC. Other valuable fragments have been recovered from royal tombs (1600–1046 BC) of the Shang Dynasty. Clothing of the elite was made of silk in vivid primary colors [10].

Linen bandages were used in the burial custom of mummification, and such linen, woven with 540 threads per inch, was found on an Egyptian mummy (2500 BC). Art depicts Egyptian men wearing linen kilts and women in narrow dresses in various forms of shirts and jackets, often of sheer-pleated fabric and goddess Isis holding a shuttle (**Figure 2**) [5].

The earliest evidence of weaving in Japan is associated with the Jōmon period. Cloth fragments made of bark fibers and hemp fibers were discovered. Some pottery pattern imprints depict fine mat designs, proving their weaving techniques, and also show clothing with patterns that are embroidered or painted arched designs. Since bone needles were also found, it is assumed that they wore dresses that were sewn together [12].

The classical Filipino clothing varied according to cost and current fashions and so indicated social standing. In ascending order of value, textiles were made of abaca, abaca decorated with colored cotton thread, cotton, cotton decorated with silk thread, silk, imported print stuff, and an elegant abaca woven of selected fibers almost as thin as silk.

Fabric in Ancient Greece was woven on a warp-weighted loom. The first extant image of weaving in western art comes from a Terracotta lekythos (**Figure 3**). The vase 550–530 BC depicts two women weaving at an upright loom. Wool was the preferred fabric in Ancient Greek

Figure 2. The Mural "Relief of Thutmose I," showing men's and women's clothing, around 1500 BC [11].

Figure 3. Terracotta lekythos (oil flask), display weaving, Greek [13].

clothing, although linen, hemp, and small amounts of expensive imported silk and cotton were also worn. Ancient Greeks and Romans developed an enormous trade in textile (300 BC). Silk became the luxury cloth in Rome, while around 65 BC, cotton awnings were used [5, 9].

3.1. Silk road: the ancient textile trade route

The Silk road was network of a series of ancient trade and cultural transmission routes, through regions of the Asian continent, linking East and West, China with Mediterranean Sea. The trade route was initiated around 114 BC, although earlier trade across the continents had already existed. Trade on the Silk road, where the exchange of luxury textiles was predominant, was a significant factor in the development of the great civilizations of China, Egypt, Mesopotamia, Persia, the Indian subcontinent, and Rome and helped to lay the foundations for the modern world. For the West World, silk was the most important merchandise that passed the Silk road. The development of silk weaving can be traced back to 2000 BC. The production of silk in sufficient quantities and for export developed only completion of "time rival empire" in the third century BC. At the time, silk was an extremely rare material in the West, like purple and glass counted in luxury goods of the Roman Empire. During the Middle Ages, simple clothing was favored by the majority of people, while only the richest could afford finer materials such as silks and linens, even at the modest prices (**Figure 4**). At the time of the Augustan peace, the Roman upper class expressed an increased "appetite" for silk, spices, and jewels, because they wanted to emulate the rich lifestyle of the Far East [10, 14].

Figure 4. Silk woven fabric from China, the Western Han Dynasty, second century BC [15].

4. Textiles in the medieval period

The elite of early medieval Europe imported silk and cotton from the Byzantine and later Muslim, as well as bleached linen, dyed wool. Besides weaving, embroidery was very widespread. Lower classes wore local or homespun wool, very often undyed, trimmed with bands of decoration (various embroideries, tablet-woven bands, or colorful borders) woven into the fabric in the loom. Evidence exists from 200 AD of earliest woodblock printing (flowers in three colors on silk) from China, while the oldest samples of cloth printed by Woodblock printing from Egypt date from 600 AD. Finely decorated examples of cotton socks made by true knitting using continuous thread appear in Egypt around 1000 AD [9].

At the same time, a great expansion of wool industry was present in England. Henry I sponsored the first woolen cloth guild and relocated skilled Flemish weavers to English villages to increase production. In 1153 AD, the first annual cloth fair was held in England. England became a European center of textile production.

During thirteenth century, great progress in the dyeing and wool processing has been made, which was the most important material for outerwear of that time. For the clothing that was in direct contact with skin, the linen that could be laundered and bleached in the sun, was increasingly used. Raw cotton was imported from Egypt and elsewhere, and was used for many applications such as padding and quilting, as well as cloths such as buckram and fustian.

Valuable knowledge of fine textiles like light silks was brought to Western Europe from Levant by the Crusaders. Silk that was imported in Northern Europe, was very expensive luxury material (like exclusive woven brocades from Italy), which could afford only the well-off. Fashionable Italian silks of this period featured repeating patterns of roundels and animals, deriving from Ottoman silk-weaving centers in Bursa and Yuan Dynasty from China [16].

5. Modern age: flourishing of textiles

The Renaissance, which marked a break with the Middle Ages, is one of the most creative periods in history. It is one of the largest movements in the culture of Western Europe, which has led to a reversal in all segments of the Arts and Sciences, including textiles.

Further development and progress in the process of dyeing and tailoring, in Western Europe during the fourteenth century, accelerated the expansion of fashion and drastically changed worldviews and ways of thinking. In subsequent centuries, clothing and draperies became increasingly elaborate, but still retaining the manufacturing methods.

Raw materials such as flax and hemp were highly represented in the fabrics production during the Renaissance Europe, but wool has still remained a dominant material. Wool fabrics were available in a wide variety of qualities and processing, from rough undyed, to fine dense broadcloth with a velvety nap, dyed in rich colors (red, green, gold, blue). High-value broadcloth was the most important export product and the backbone of the British economy.

By the fifteenth century in the Mediterranean, silk fabric was established as a luxury product and it was extensively used in the Italian fashion to dress high-class citizens (silk velvets with silver-gilt wefts) (**Figure 5**).

The prosperity increase during the fifteenth century influenced the textile industry, where the middle class, following the fashion of elite class, began wearing more complex clothes and materials.

During the sixteenth and seventeenth century in Europe, the great flowering of needle lace occurred. Geometric reticella deriving from cutwork was elaborated into true needle lace, which reflected the scrolling floral designs popular for embroidery (**Figure 6**). Lacemaking centers were established in France to reduce the outflow of cash to Italy [16, 18].

In 1589, William Lee invents stocking frame, the first and hand-operated weft-knitting machine. During the seventeenth century, England passed numerous laws that for example forbade English textile craftsmen to emigrate to America, forbade English colonies in America from trading wool materials, required all persons to be buried in woolen cloth (because more cloth

Figure 5. Velvet silk with silver thread (fifteenth century), Italian, Venice [17].

Figure 6. Italian needle lace, seventeenth century [19].

was being produced than could be sold). All the secrets of weaving crafts were strictly kept in family circles. Until 1842, there was a law, which allows death sentence in the case of loom exports out of the country.

In the late seventeenth and early eighteenth centuries a great number of innovations and patents emerged which greatly enhanced the production and use of textiles. The flying shuttle, patented in 1733 by John Kay, doubled the weaving productivity, which led to even greater imbalance increase between spinning and weaving processes. The flying shuttle became widely used after the invention of drop box in 1760 (by John's son Robert). Wool spinning with even greater thickness was enabled with patenting of roller spinning frame and the flyer-and-bobbin system, by Lewis Paul, who in 1748 along with Daniel Bourn also patented carding machines. Lewis's invention was later developed and improved by Richard Arkwright in his water frame and Samuel Crompton in his spinning mule. James Hargreaves invented the spinning jenny, which he patented in 1770, while the spinning frame or water frame was developed and patented in 1769 by Richard Arkwright. Samuel Crompton's Spinning Mule, introduced in 1779 was a combination of the spinning jenny and the water frame, which was able to produce finer thread than hand spinning and at a lower cost. Mule spun thread was of suitable strength to be used as warp, and finally allowed Britain to produce good quality calico cloth. Edmund Cartwright developed a vertical power loom, which he patented in 1785 and in 1786, he patented a two-man operated loom, which was more conventional. 1804 was the year when Joseph Marie Jacquard presented the first mechanical weaving loom, which used punched cards in a continuous sequence to enable a single weaver to produce complex patterned fabrics, and thus simplified the production. This is an early example of precomputer technology [5].

Since about 1760 to sometime between 1820 and 1840 was the period of transition to new manufacturing processes, known as Industrial Revolution. Textiles were the dominant industry of the Industrial Revolution in terms of employment, value of output and capital invested; the textile industry was also the first to use modern production methods. The industrial revolution has enabled the application of waterwheels and steam engines for mechanization of textile production. As a result, small cottage-based production was switched to mass production based on assembly line organization. Opposite the fabric, clothing production was still made by hand.

Mechanized textile production spread from Great Britain to continental Europe in the early nineteenth century, with important centers of textiles in Belgium and later in France. Since then industrialization has spread throughout much of the world. A large number of power looms were beginning to be installed in the factories in America (1816).

New advances of the Industrial Revolution such as steamboats, canals, and railroads, have significantly influenced the changes in the transportation sector, in the form of cost reduction, which also contributed to the price reduction of finished textile products. Therefore, textile has become more accessible and hence cheaper, despite a dislocated but industrial production, as opposed to more expensive hand-made goods produced locally. Such triggered production influenced the development of the national market, which has from 1810 till 1840 tripled the output, causing a slow neglect of hand weaving.

Most of the employees in the textile factories were women, who began working for various reasons, but all leading toward gaining a sense of independence and growth as a personal goal. This can be considered as the beginning of women's emancipation.

In the early 1880s, Joseph Swan invented the first synthetic fiber, drawn from a cellulose liquid. In 1889, Hilaire de Chardonnet developed the first manufactured fiber—artificial silk (Chardonnet silk). The first successful process of producing artificial fibers was developed in 1894 and was called Viscose, while the commercial viscose was first produced by the UK Company Courtaulds Fibers in 1905 by the name Rayon [20–23].

In period between 1865 and 1948, textile industry has expanded to such an extent that the most prestigious educational institutions such as UC Davis, University of Nebraska-Lincoln, Iowa State University, and many others, have established various departments of textiles and clothing. Textiles have even entered high school and high school libraries, with various books collections on the history of clothing and textiles.

6. Contemporary history: the birth of the textile future

After the World War I comes the period of unstoppable development. Textile industry is one of the fastest growing industries, whose production has been altered almost beyond recognition by industrialization and the introduction of modern manufacturing techniques with automation, advanced textile processing techniques, and applications.

In 1935, Wallace C. Carothers developed the first synthetic fiber— Nylon, while polyester, acrylic and other artificial fibers were introduced between 1940 and 1950.

Knitting machine controlled by computer that produced fabrics with highly complex patterns at tremendous speed was presented in 1970. In early 1980s, first robots were introduced into the textile industry [23].

Electronics, automation, and robotics became the new trends, which resulted in more productivity with less labor and operational costs. By late 1980s, textile mills used high-speed looms with darts (many tiny shuttles) and with a jet of water or air carrying the filling through the warp up to 1000 times a minute, while industrial looms today incorporate air-jets to weave at speeds over 2000 picks per minute.

The 1980s was the period of digitalization and the CAD/CAM systems (computer-aided design and computer-aided manufacturing) have also impacted the textile industry. This tool, offers benefits such as improved product design, increase in productivity, higher utilization, better quality control and enabling companies the retain their profitability and competitiveness (**Figure** 7). Machine designs became increasingly sophisticated and precise, enabling innovations in specialty fabrics.

With the development of nanotechnology, and in 1980s with the full development of the sol-gel and electrospinning techniques, the use of nanoparticles and nanofibers to produce specialized nanofabric became a huge subject of interest. In the coming decades, an increase in global funding accelerated nanofabric research studies. It was believed that the future of fabrics would be nanotechnology-based.

The textile industry is growing rapidly with increasing demands in the global market. The invention of modern technologies has benefited the textile industry in automating long and complex textile processes. This resulted in improvement in speed, quality, and cost of textile manufacturing. Textile processing includes many processes like preparatory processes, desizing, scouring, bleaching, dyeing, printing, and finishing. There are also innumerable machines in the market for basic functions such as spinning to the most advanced functions of decorating textiles like embroidery. Now, it is easy to carry out most of these processes on textile processing machinery. The results are reduced labor costs, efforts, and time for production.

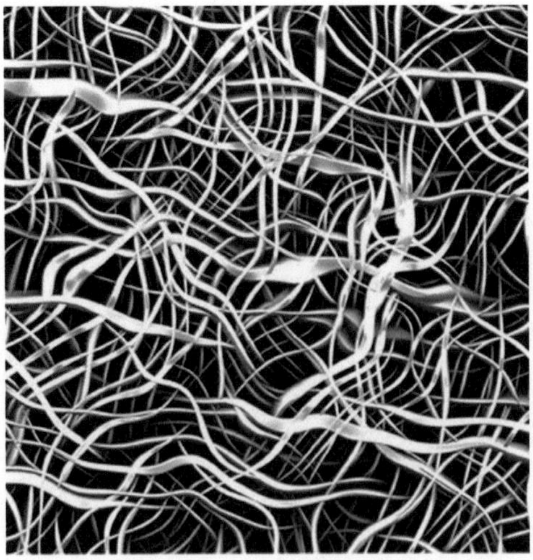

Figure 7. Textile engineers are developing high-tech textile products through computer-aided design [24].

In this period, activities in the textile industry extended far beyond clothing and home furnishings. Former conventional textiles received a companion on the path of incredible textile developments that generated a completely new world named "technical textiles."

7. Textile applications: from needs to needs (from the earlier needs to today's needs)

Looking at the distant history, the emergence of textiles and their uses succeeded leather. First textiles were crude and simple, but over time, the technology advanced and textiles became more beautiful and comfortable to wear and for other applications. Throughout years, we use textiles for different purposes, improve them, and make them more attractive. Some are fashionable, some fall out of fashion, some come back, and some are just useful. The history of textiles is as entangled as textiles themselves.

Throughout history, textiles have been used for different applications. The origins of textile application can be traced back to ancient human history, as well as the development of textile application. It is considered that the first application of textiles was for clothing, which is exclusively human trait. It is also a feature of most societies, in social, economic, cultural, and traditional sense. Although the initial reasons for wearing clothes were not identical to present ones, anthropologists believe that the animal skins and vegetation were used as coverings for body protection against external conditions. There is another view that the original coverings were used for the purpose of magic, decoration, cult, or prestige, and later founded the application of practicality [25].

The most obvious function of clothing, from the very beginnings to the present day, is to improve the comfort of the wearer by protecting the wearer from various elements of external factors. It performed a range of social and cultural functions, such as individual, occupational, gender, and religion differentiation as well as social status and standard of modesty. These features have changed through historical periods and regions but have retained their essence to the present day (**Figure 8**).

Throughout history, clothing has been made from a very wide variety of materials, ranging from leather and furs to woven materials, to elaborate and exotic natural and synthetic fabrics, which were enabled by the development of textile raw materials, production, and application. Different cultures have evolved various ways of creating clothes out of cloth. One

Figure 8. Clothing in history, showing (from top) Egyptians, Ancient Greeks, Romans, Byzantines, Franks and thirteenth through fifteenth century Europeans [26].

approach simply involves draping the cloth. Another approach involves cutting and sewing the cloth, but using every bit of the cloth rectangle in constructing the clothing [16].

In thousands of years that humans have spent constructing clothing, they have created an astonishing array of styles, many of which have been reconstructed from surviving garments, photos, paintings, mosaics, as well as from written descriptions. Costume history serves as a source of inspiration to current fashion designers, as well as a topic of professional interest to costumers constructing for theater or films and historical reenactment.

Throughout history, people have created the textiles and clothing inspired by the historical remains of different objects, mosaics, written documents, and photographs, complementing them with new knowledge, capabilities, and current situations. Even today, the history of textile is an inexhaustible source of inspiration for fashion designers and artistic costume designers.

Mechanization of the textile industry made many varieties of cloth widely available at affordable prices. Styles have changed and the availability of synthetic fabrics has changed the definition of trendy and stylish. Nowadays, clothing may also function as a form of adornment and an expression of personal taste or style [16, 18].

Different techniques of making textiles were gradually developed and improved. Production methods and techniques are closely related to the application of the product for which the material is intended.

Felting is the oldest technique, which involves pressing a mat of fibers together in a liquid to create a tangled, flat material. A liquid, such as soapy water, is usually added to lubricate the fibers and to open up the microscopic scales on strands of wool.

Weaving is a textile production method, which implies the interweaving of two thread systems, vertical system, called warp, and a horizontal system, called weft. The process of weaving is carried out on the loom of different types. There is also a manual way of making fabrics, hand weaving, and the manufactured products are unique and expensive.

Knitting involves interlacing loops of yarn, which are formed on a knitting needle, together in a line. Knitting has several active loops at one time on the knitting needle waiting to interlock with another loop.

Lacing is a technique that uses fine woven fabrics with open holes throughout the piece, through which yarns are interlaced, resulting in a layer known as a pile, which is prominent in the manufacture of carpets and velvet. Lace can be produced either by hand or by machine.

Nonwoven textiles are manufactured by the bonding of fibers to make fabric. Bonding may be thermal or mechanical, or adhesives can be used.

Special fabrics like carpets, rugs, velvet, velour, and velveteen are made by interlacing a secondary yarn through woven fabric, creating a tufted layer known as a nap or pile.

Braiding is a technique that implies interlacing minimum three separate yarns or fabric strips, forming a complex structure in flat or tubular form, with exceptional strength properties.

Laminating and coating are techniques by which textile materials obtain properties that were not originally its features. The difference between two is in application method: coatings are applied in their preparatory state (mostly in liquid form), while lamination process requires the prepreparation of a laminate membrane which is applied to the fabric. Due to great possibilities of added properties, application of such materials, within technical textile field, is also very wide.

Embroidery is a technique of decorating finished textile materials and one of the few methods that have not changed throughout history. However, it is important to note that the technical achievements and a high standard of production from the distant past, today can rarely be achieved.

3D textile materials imply flexible textile products with significantly prominent third dimension compared to the other two planar dimensions. 3D textiles manufacturing methods include braiding, weft or warp knitting, weaving, stitched assemblies, as well as combined proprietary technologies.

Besides application for clothing, from way back in the history, textile has been used for the purpose of decorating the interior, as dominant part of textile-based products in the market. The world of conventional textile began to change, as new cultural influences meet technological innovations. Researches and processing technologies have been resulting in development of fabrics that can serve functional purposes of products well beyond their traditional roles of clothing and furnishing (application in vehicles, construction, agriculture, health care, defense and security, power and environmental technologies, and many more) [27].

The original feature of clothing, body protection, evolved due to technological developments and application needs, in extreme but necessary aspects of protection (protective clothing). However, the distinction between conventional and protective clothing is not always clear-cut, since clothes designed to be fashionable often have protective value and clothes designed for function often consider fashion in their design.

Protective clothing is just one segment of the large field of nonconventional textiles, called technical textiles, which has been primarily developed and produced for its performance or functionality, to meet the exacting specified high-performance requirements of end-use, and not for its appearance or esthetics, unlike conventional textile. Such a brief definition and description of technical textiles leaves a significantly large space for explanation of technical textiles, especially due to the increasing number of products that contain a combination of distinct properties, decorative appearance and function in equal measure (fireproof furniture, breathable footwear for recreation). Technical textiles cover one of the most dynamic and most widespread areas of textiles, materials, processes, products, and applications. They change so fast that they could not be specifically defined and documented and be able to describe and classify their goal. They often require the in-depth material knowledge, engineering capacities, high manufacturing, and quality control standards, as well as close development of collaboration with end-users. As a result of cooperation between different industrial sectors and areas of application, users of those areas are extremely important in the process of innovating technical textiles, because with their functional, technical, and performance requirements, influence and define textile-based products [27].

Technical textiles, besides earlier mentioned protective clothing, include textile structures for a wide range of application, their usage and associated products.

Agrotech includes technical textile products used in agriculture, horticulture (including floriculture), fisheries, and forestry. Applications for technical textiles in agriculture include all activities concerned with the growing and harvesting of crops and animals. They are used for crop protection fabrics against sun and rain, further for fishing ropes, nets, tarpaulins, horticulture twines, shade fabrics, etc.

Buildtech area includes textiles or composite materials used in the construction of buildings and structures for various applications, such as facade foundation systems, concrete reinforcement, insulations, interior construction, visual protection, noise prevention, protection against the sun and building safety, air conditioning, and many more.

Clothtech is part of technical textiles, which mainly comprises textile components used for specific functional applications in garments, shoes, and bags. These components are largely hidden e.g., interlinings in shirts, sewing threads, shoelaces, labels, hook, and loop fasteners.

Geotech includes textile products most commonly used in geotechnical applications, because of their separation, reinforcement, filtration, drainage, and protection functions. This category of products includes permeable fabrics or synthetic materials, wovens or nonwovens, which can be used with geotechnical engineering material. Application areas include civil engineering (roads and pavements, slope stabilization and embankment protection, tunnels, rail-track bed stabilization, ground stabilization, drainage, etc.), marine engineering (soil erosion control and embankment protection, breakwaters), and environmental engineering (landfills and waste management).

Hometech is field of technical textiles, which includes products for household, primarily for interior decoration and furniture, carpeting, floor and wall coverings, cushion materials, textile reinforced structures, filters, sun protection products, and others. These products can create comfortable, practical, hygienic, and beautiful solutions for modern living. Recent developments in the home furnishings industry include the creation of nonwovens that kill dust mites in bedding, repel dirt, and contain antimicrobial qualities.

Indutech includes technical textile products used in the manufacturing sector like filters, conveyor fabrics, industrial belts and hose, abrasive products, etc.

Meditech products are textile materials used in hygiene, health, and personal care as well as surgical applications. The Meditech products are available in woven, knitted and nonwoven forms based on the area of application (surgical gowns, drapes, dressings, masks, caps, sutures, medical implants, sanitary napkins, diapers). Their use is based on a number of typical basic textile properties like softness and lightness, flexibility, absorption, filtering.

Mobiltech segment of technical textiles is used in the construction of all sorts of transport for human or cargo. Products belonging to this area can be divided into two categories—visible components (seat covers, carpets, seat belts, airbags) and hidden components (noise vibration, harness components, cables, tyre cords, liners). They provide good performance and safety, reduce the weight of the car, enhance the comfort and esthetics and provide advanced insulation, fire retardancy and resistance to water, fuels, extremes of temperature, and abrasion.

Ecotech segment refers to the use of technical textiles in environmental engineering, primarily for environmental protection. The primary segment is landfill waste management, which refers to the use of geosynthetic products to secure landfills against leakage of municipal or hazardous waste. Other areas include secondary protection in chemical/oil industries. For this purpose, recycled schemes, products for oil spill treatment, and insulation products are used.

Packtech includes several flexible packaging materials used for industrial, agricultural, consumer, and other goods. It ranges from synthetic bags used for industrial packaging to jute sacks used for packing food grains. It also provides innovative packing solutions such as oxygen scavenging, packaging for foodstuffs and antistatic packaging for computer equipment. With the advantage of moisture proof quality, they are used for packing moisture sensitive goods.

Protech, as already mentioned, is a collection of textile materials and products used to protect the body during exposure to hazards in the working environment. Protech includes clothing and accessories that provides protection against harmful chemical and extreme temperature conditions, low visibility and garment for ballistic protection (space suits, air conditioned clothing, armor, motorcycle leathers, high-visibility clothing, clothing against heat and radiation for fire fighters, against molten metals for welders, etc.).

Sportech segment includes technical textile products used in sports and leisure such as shoes, sports equipment, flying and sailing sports, climbing, angling, cycling, winter and summer sports, and indoor sports and other [28].

Textiles can be made from many materials of different origin: natural (animal, plant and mineral) and synthetic fibers. Simultaneously with the discovery and development of synthetic fibers and their fast growth production, new technologies and techniques were developed to manufacture technical textiles. The main processes used for technical textiles are knitting, weaving, braiding, nonwoven, tufting and others, whereby nonwoven technology predominates.

8. Textile fibers: shaped nature for developed application

Natural fibers are used since the earliest days of human civilization. The most used plant natural fibers, throughout history and today, are cotton, flax and hemp, although sisal, jute, kenaf, and coconut are also widely used, while the most used animal fibers are silk, wool, angora, mohair, and alpaca. Mineral fibers are fibers obtained from mineral sources and may be used in their naturally occurring form or after slight modifications. Their use was more present in recent times than throughout history.

After World War II, there was an enormous rise in the production of synthetic fibers, and the use of natural fibers significantly decreased. In recent years, strengthening of environmental awareness has contributed to the revival and increase of the use of natural fibers in the textile, building, plastics, and automotive industries. In those existing application areas, further improvements will be achieved in the performance of existing products that will expand the areas of application and at the same time affect the reduction of environmental burdens.

Wool is one of the earliest used fibers for textile materials and due to its very good heat insulation properties it was used in very cold and very hot climates. Wool is a multifunctional fiber with a range of diameters and with ability to absorb and release moisture, which makes it very comfortable. Therefore is wool suitable for various applications (clothing, household fabrics, and technical textiles), but nevertheless, two-thirds of the total produced raw material is used in the manufacture of garments. In combination with other fibers, wool adds drape, and crease resistance. One-third of produced fibers is used in hometech sector for household textiles because of its inherent resistance to flame and heat, making it one of the safest in this field. In industry it is mostly applied as a thermal and acoustic insulation.

Silk and its manufacturing process were held as the greatest secret under the threat of death penalty in China. However, nothing can stay secret forever and when it was discovered, cultivation and silk processing spread rapidly in other countries. Because of its beauty and comfort properties it is mainly used for high-fashion clothes, lingerie, and underwear. It is also used for technical textile as household textiles (upholstery, wall coverings, carpets) and as medical textiles (surgical sutures, biodegradable micro tubes, molded inserts for bones, teeth reconstruction).

Another material, which is one of the earliest textile products, is flax (linen). It was highly regarded from the times of the earliest civilizations until today for its quality and beauty. Flax is highly respected fiber for its retention properties of cooling during hot days (symbol of breezy summer elegance), and for that reason its application is the largest in the production of clothing (over 70%). Like other natural fibers, its application is significant in the field of household textiles (bed linen, furnishing fabrics, interior decoration accessories), but also as reinforcement and fillers in automotive industry and other consumer products.

Cotton has been part of human history for 7000 years. Today, it is practically an irreplaceable material for clothing (60% of cotton production) and many other applications, but over time, it shaped the history of many countries and peoples as it does now. Cotton is also used to make home furnishings (draperies, bedspreads, and window blinds), and it is the most commonly used fiber in table and bed linen. It is made into specialty materials suitable for a great variety of applications: fire-proof apparel, cotton wool, compresses, gauze bandages, sanitary towels, and cotton swabs. Industrial products containing cotton include book bindings, industrial thread, and tarpaulins.

Hemp has been used for centuries to make rope, canvas, and paper. Long hemp fibers can be spun and woven to make crisp, linen-like fabric used in clothing, home furnishing textiles, and floor coverings. Blending with cotton, linen, silk, and wool gives hemp a softer feel, while adding resistance and durability to the product. Hemp fiber is raw material with a valuable characteristic suitable for multiple industrial applications (insulation products and composites) [29, 30].

The invention and development of synthetic fibers is related to the first half of the twentieth century. The production of synthetic fibers accounts for about half of all fiber usage, with applications in every field of textile technology. Synthetic fibers and fabrics are all made from a type of polymer, but they each have unique properties and characteristics making them useful for specific applications. Among a number of different synthetic fibers, the most valuable and dominant fibers are four of them—nylon, polyolefin, acrylic, and polyester (approximately 98% of all synthetic fiber production, with polyester alone accounting for around 60%). The main advantages of synthetic fibers are strength, thermoplasticity, abrasion resistance, low absorbency, durability, resistant to moths and fungi, easy maintenance, availability, and inexpensiveness.

Nylon (1931) is the fiber of great properties: durable, strong, resists stains, hides soil, resists mildew and bacteria, prevents static, resistant to abrasion and wrinkling, does not absorb water, and it dries quickly. Nylon can be used in carpet, and high-filament nylon yarns are often blended with spandex and used in athletic apparel, swimwear, and hosiery.

Glass fiber (1938) is used in the production of technical textiles for industrial, automotive application, home insulation, specialty papers in battery separators and filtration, as reinforcement of composite materials (flame-retardant and protective fabric, soundproof, fireproof, and insulating fabrics). Glass fibers are woven and coated with Teflon to produce beta cloth, a virtually fireproof fabric that replaced nylon in the outer layer of United States space suits since 1968.

Metal fibers (1946) have a variety of uses, including the elimination and prevention of static charge build-up, conducting electricity to transmit information, conduction of heat, adding metallic properties to clothing for the purpose of fashion, production of cloth-of-gold and jewelry.

Polyolefin fibers (1949) have great physical-mechanical properties—strength, resistance to abrasion, and its advantages includes resistant to stains, sunlight, odor and chemicals, mildew, rot, and weather. They are fast drying and have a high wick-ability making them useful for spill cleanup.

Acrylic fibers (1950) are unique among synthetic fibers because they have an uneven surface. This fiber is called artificial wool because it has the warmth and softness of wool but does not absorb water. It is often used as cold weather fiber for blankets and sweaters.

Polyethylene fiber (1950) with high molecular weight (HMWP) is one of the world's strongest and lightest fibers. Polyethylene fiber is pound-for-pound 10 times stronger than steel. The material floats, resists chemicals and water, and exhibits superior fiber-to-fiber abrasion. Polyethylene fibers are used in police and military ballistic vests, helmets and armored vehicles, sailcloth, fishing lines and lifting slings, cut-resistant gloves, and a wide range of safety apparel.

Polyester (1953) is the most important synthetic fiber. It is versatile and has low raw material and production costs. Polyester is resistant to abrasion, has the ability to spring back into shape, does not absorb water, and dries quickly. It is utilized in all types of clothing, home furnishings, and as a reinforcing fiber in tires, belts, and hoses. New insulating polyester fiber-fills are used in high-performance outdoor wear.

Spandex (1958) is a lightweight manufactured material that can be stretched over 500% without breaking. It is a soft fabric that is resistant to abrasion and can resist body oils, perspiration, and detergents.

Carbon fiber (1958) is a polymer, sometimes known as graphite fiber, which on top of being strong and light, has high stiffness, tensile strength, chemical resistance and thermal expansion, and can be mixed with other materials. The material is used to produce high-quality devices and can be used for composites in many industries such as aerospace, automotive, military, and recreational applications [31, 32].

A number of those fibers are used in conventional textile manufacturing, but the growing textile industry has resulted in growing demands for various textile processing and applications, and thus fibers themselves. Among other innovations, textile engineers developed and still develop high-tech fibers that are made from synthetic fibers by the modification process, i.e., are mostly regenerated with high physical and chemical properties [33].

There is a long list of synthetic fibers, which are available today and frequently used in every new product. Today, a majority of technical textiles are manufactured from a variety of synthetic fibers based on the desired properties of the end product. Nevertheless, some raw materials hold a good market share for the technical textile industry: cotton (Buildtech, Clothtech, Packtech, Mobiltech, Meditech, Sportech, Hometech, Indutech), jute (Buildtech, Geotech, Packtech, Agrotech, Sportech), silk (Clothtech, Meditech, Sportech), and coir (Hometech, Geotech) holds a good market share as a key raw material for technical textiles.

9. Yesterday's fiction-today's reality-tomorrow's history

The majority of textile-based products are manufactured long before they are sold to the end user. Product developers, manufacturers, and distributors collectively try to guesstimate actual demand and exact customer preferences. The various unpredictable factors can negatively influence and result in destruction of economic value of the product.

Mass production that dominated for many years was able to satisfy the basic market needs in the field of clothing, home furnishing and some standard industrial uses. However, the development of technology and the emergence of different needs, make changes from standard production process to customization process of textile manufacturing. This process involves many segments like logistics, communication, financial transactions, IT, and software providers for product development, production organization, and supply chain management. Therefore, the involvement and cooperation of experts from all fields of science and industry is crucial. The fact, that every customer is different and individual for himself, with its own wishes and needs, is now accepted. Uncompromisingly optimal satisfaction of user specific requirements is increasingly required feature of the finished product. Textile plays a major role in user personal identification and expression. Therefore, textiles represent ideal products for customization and personalization [27].

Big investments in the development of new products are present with the objective of innovation, maintaining manufacturer market position, and creating new potential areas of business. Innovations enter all areas of textile with the aim of improving properties and tracking of trends for the needs of the market. The development of textiles is not only determined by the end-use needs, but also by the development of science and technology in general, which is implemented on the textile base—fiber or through treatment of finished materials. As already mentioned, interference of all scientific areas is present, resulting in the deletion of sharp boundaries between areas, leading to great freedom, creativity, and innovation. The possibility of expressing creativity and the need for innovation is bigger today and more important than ever before in the history of textiles.

Remarkable innovations in textiles were achieved in the past. Examples are innovations in: aerospace textiles (aircraft, space suits, space shuttles, space transportation), medical textiles that provide bioreceptive and biocompatible materials (artificial heart valve), sport textiles which provide light weight material with safety features like breathability, waterproofness, conductivity, durability, flexibility, comfort, eco-friendliness, digital printing, etc.

Textiles science and technology constantly evolve increasingly integrating other fields of science in order to achieve better and more advanced features. Some examples of recent development in the field of technical fibers are: hollow fibers, blended fibers, ultra-microfibers, ultra-lightweight, and high-stretch synthetic fibers; while a few examples of fabrics are: breathable artificial fabrics, thin and light reflective fabrics, metallic textiles, exquisite fabrics, phase change materials, 3D structured fabrics, etc.

The latest innovation in textile sector imply: development in nanopolymer technology (nanopolymer coatings that impart properties to material like increasing effectiveness, decreasing maintenance time and cost) (**Figure 9**), nanocoatings for stainless materials, sensing T-shirt (with textile pressure sensors to increase the comfort and effectiveness of spinal braces), edema stocking (monitors and measures changes in leg volume), EQ-top seismic wallpaper (composite of strong glass fibers to create durable and elastic panels) (**Figure 10**), new life polyester yarn (made from 100% postconsumer recycled plastic bottles, mechanical processed), mushroom material (home-compostable bio-plastic), and many others.

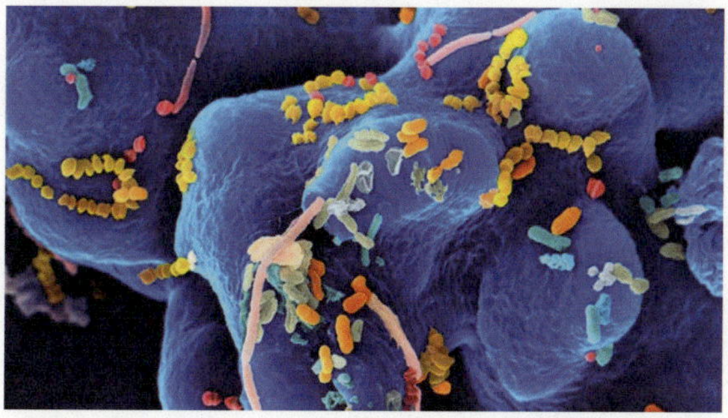

Figure 9. Nanotechnology used to achieve antimicrobial performance in textiles [34].

Figure 10. EQ-Top seismic wallpaper can save lives in an earthquake [35].

Understanding the basic textile structure provides a foundation for development of materials with different qualities or properties that by upgrading and adjusting, according to desired properties, results in interactive textile structures. Textile structures have become an important basis for high-tech innovations that affect the use, design, and esthetics of textile products. Innovation achievement in conventional materials (woven, warp knitted, weft knitted fabrics) is possible by applying specific structures and raw materials, as well as integration of various elements, in order to achieve the properties of added values (social-ecological awareness, comfort properties, sound and light insulation, various aspects of protection, regulation of body temperature, vibration muscle control). The development of science and technology contributed to the advanced way of thinking and realizing needs, resulting in innovations such as 3D textiles, based on emphasizing the relationship between three dimensions (x, y, z), with almost limitless variety of application solutions. Such textile may be braided, woven, knitted (warp and weft knitted), nonwoven, stitched, embroidered or made by some new special forms (**Figure 11**). Diverse uses include almost all aspects of technical textiles and even fashion [36].

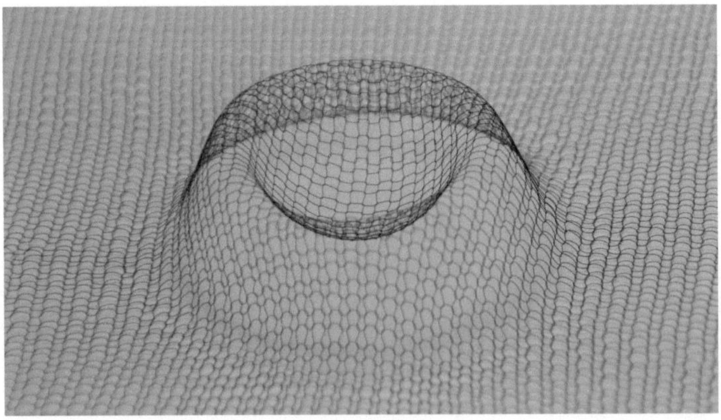

Figure 11. Thermally activated 3D textile with shape memory [37].

Spacer fabrics are very exciting group of 3D textiles, which offer some of the most innovative solutions of today. Their construction offers the unique possibility to combine various properties, functions, and fibers, all within one 3D structure. Those structures are truly multifunctional and provides a number of exceptional properties: high strength, stability, structural integrity and compression elasticity, breathability, and air permeability, adjustable vapor transport and temperature regulation, cushioning, insulation, light weight, age resistance, recyclable, ability to form complex shapes, design flexibility, and much more (**Figure 12**).

There is one term attracting particular attention and that is Smart textiles. It represents the next generation of textiles intended for fashion, furnishing, and technical textile applications, which are designed and manufactured to sense and respond in a predefined manner to environmental stimuli. The degree of smartness varies and it is possible to enhance the intelligence further by including the technologies that provides the increased functionality. Materials integrated into smart textiles include optical fibers, metals and conductive polymers (**Figure 13**). The market is primarily driven by the increasing usage of nanotechnology in these fabrics, giving them special functionality (**Figure 14**). It is also driven by the rising demand of wearable technology, which is one of the most important factors at a global level market. Due to the miniaturization of electronic components, it has been possible to integrate electronic components (sensors, actuators, control unit) into textiles, making them as a part of wearable textiles (e-textiles).

These textiles have numerous potential applications, such as the ability to communicate with other devices, conduct energy, transform into other materials and protect the wearer from environmental hazards (**Figure 15**). Smart textiles are also applied in field of Medtech and Sportech (fabrics that: release medication or moisturizer into the skin, control the vibration of muscles during athletic activities, and regulate body temperature). In the fashion segment, smart textiles fulfills esthetic requirements, like color change, light up in patterns, display pictures, or videos (**Figure 16**). From the last mentioned, it is evident that the area of electronics

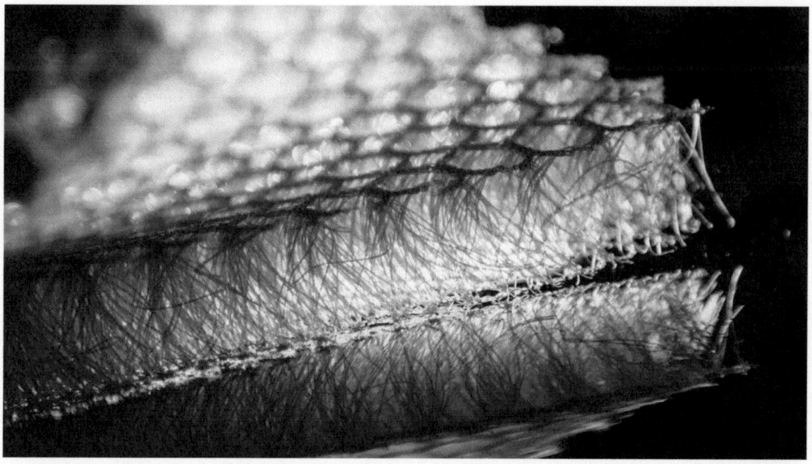

Figure 12. 3D spacer fabric [38].

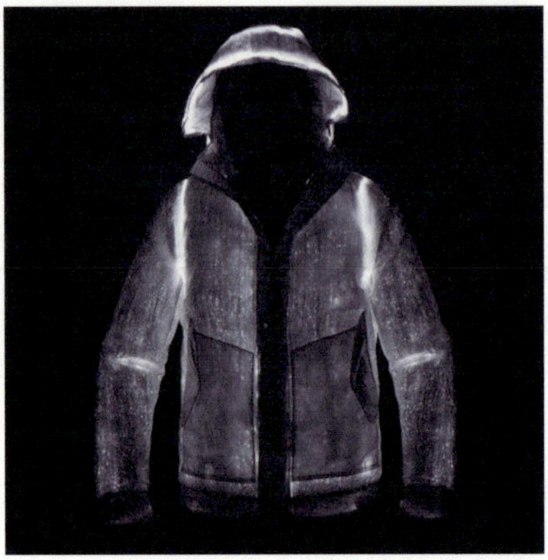

Figure 13. LED jacket, fabric woven from optical fibers within nylon and steel fabric [39].

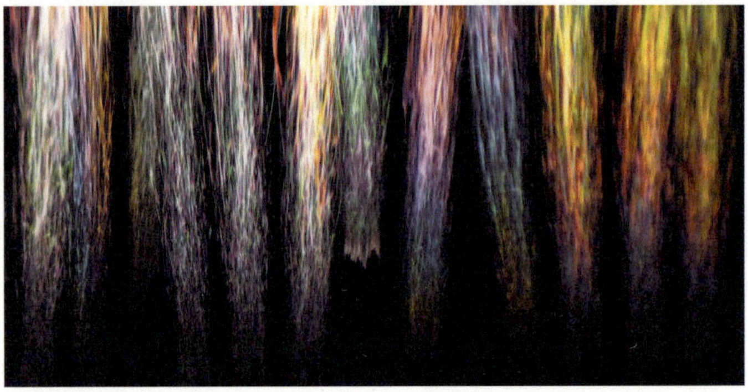

Figure 14. The nanofiber composite reinforcement veil [40].

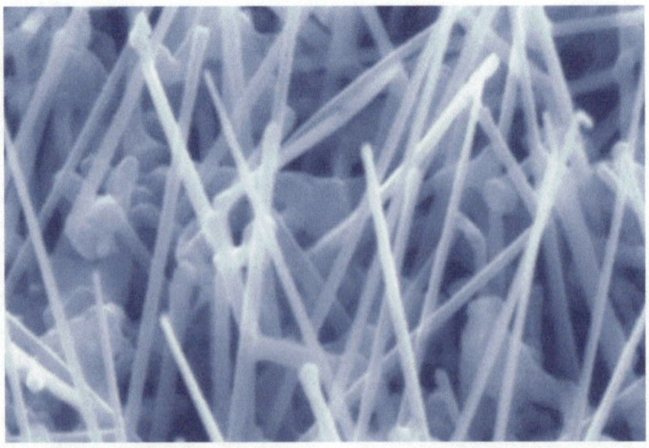

Figure 15. The prototype systematic fiber-based electrochemical micro-super-capacitor is composed of flexible electrode, fine plastic wire and fiber electrode made of Kevlar—alternative to batteries and rechargeable batteries for energy storage [42].

Figure 16. The world's first stretchable and conformable thin-film transistor (TFT) driven LED display laminated into textiles developed by Holst Centre, Imec and CSMT [43].

(smart textiles) enters into the area of conventional textiles for clothes (fashion textile), based on which it can be concluded that there are no longer clear boundaries and separation between these areas. Advanced technologies enter into all textile segments [41].

The increasing number of research and development in recent years indicates one of the major opportunities in the global textile market, enabling the existing technology to be improved and delivering new innovative products in the future. Some of them will be definitely in the field of sports and military industry as fabrics that help regulate body temperature, reduce wind resistance, and control muscle vibration; in the health and beauty industry as drug-releasing medical textiles and fabrics with moisturizing, perfuming, and antiaging properties. There are also indications that future textiles will be self-repairing (with microcapsules containing glue-like substance, which is released if the garment snags, filling in the gaps and hardens), esthetically smart (with illumination properties and properties of collecting and utilization of energy from the environment by harnessing vibration, sound or heat and react to input), interactive (digital functionality, interaction of fabric with an external device) and three-dimensional (with applications in the medicine and aerospace industry). The development of future yarns is oriented towards direct embedding of electronics into yarn and size reduction in diameter (0.2 mm), and also to the exploitation of natural resources (Ingeo– fiber from fermented corn starches, silk-like fiber derived from spoiled milk). The trend of combination and correlation between natural and synthetic sources is about pushing the limits, reacting to the challenge, and delivering the desired level of performance. The emphasis in future trends of textiles is also based on material structure, comfort, and protection properties and strong social and environmental awareness [44].

The trend of the future is to do something different, something new. It is all about pushing the boundaries. A rebellious and freethinking mood is an inspiration for this textile trend, which is, not quite breaking the rules, but more about bending them. There is a spirited thinking that something does not have to be perfect, but still should deliver performance. In this way, one is looking away from traditional perfect surface applications to something more unruly and unique. It is essential to look to innovative fiber content through to textures, finishes, and printing—multiple performances and multiple applications, in delivering apparel with an edge and more importantly, next generation attitude. Rethinking old traditions and delivering new concepts, exploring what we have and how we can reengineer it to create the next generation material is the way the future of textile should develop [45, 46].

The development of technology, and thus the development of textiles moves at a speed that is hard to track and even harder to understand. Its possibilities cross the boundaries of reality and pass into a fictional world, so beautiful, so free and without limitation.

Author details

Ivana Schwarz* and Stana Kovačević

*Address all correspondence to: ivana.schwarz@ttf.hr

University of Zagreb, Faculty of Textile Technology, Zagreb, Croatia

References

[1] Travis J. The Naked Truth? Lice hint at a recent origin of clothing. Science News. 2003; **164**(8):118

[2] Reed DL, et al. Genetic analysis of lice supports direct contact between modern and archaic humans. PLoS Biology. 2004;**2**(11):340

[3] Kvavadze E, et al. 30,000-year-old wild flax fibers. Science. 2009;**325**(5946):1359

[4] Balter M. Clothes make the (hu)man. Science. 2009;**325**(5946):1329

[5] Kovačević S, Schwarz I. Weaving complex patterns—From weaving looms to weaving machines. In: C. Volosencu, editor. Cutting Edge Research in Technologies. England: In Tech; 2015. pp. 93-111.

[6] Barber EJW. Prehistoric Textiles: The Development of Cloth in the Neolithic and Bronze Ages. Princeton, USA: Princeton University Press; 1990. p. 504

[7] Schoeser M. World Textiles: A Concise History. London, UK: Thames & Hudson Ltd.; 2003. p. 224

[8] Jenkins DT. The Cambridge History of Western Textiles. Cambridge, UK: Cambridge University Press; 2003. p. 1191

[9] Harris J. Textiles, 5,000 years: an international history and illustrated survey. New York, USA: H.N. Abrams Ltd.; 1993. p. 320

[10] Watt JCY, Wardwell AE. When Silk was Gold: Central Asian and Chinese Textiles. New York, USA: Metropolitan Museum of Art; 1997. p. 238

[11] A History of Textiles in Egypt. [Internet]. 2017. Available from: https://www.zum.de/whkmla/sp/1011/ignoramus/igno2.html#II [Accessed: 03 January, 2017]

[12] Slade T. Japanese fashion: A cultural history. Oxford, UK: Berg Publishers; 2009. p. 192

[13] The Met. [Internet]. 2017. Available from: http://www.metmuseum.org/art/collection/search/253348) [Accessed: 05 January, 2017]

[14] Elisseeff V. The Silk Roads: Highways of Culture and Commerce. Oxford, New York: Berghahn Books; 1998. p. 332

[15] Silk. [Internet]. 2017. Available from: http://www.wikiwand.com/en/Silk [Accessed: 04 January, 2017]

[16] Anderson Black J, Garland M. A History of Fashion. New York: Orbis; 1980. p. 304

[17] The Met. Heilbrunn Timeline of Art History, Length of velvet. [Internet]. 2017. Available from: http://www.metmuseum.org/toah/works-of-art/12.49.8/ [Accessed: 17 January, 2017]

[18] Gordon B. Textiles: The Whole Story: Uses, Meanings, Significance. London, UK: Thames & Hudson Ltd.; 2011. p. 304

[19] The Met. Heilbrunn Timeline of Art History, Gros point needle lace border. [Internet]. 2017. Available from: http://www.metmuseum.org/toah/works-of-art/33.90.35/ [Accessed: 21 January, 2017]

[20] Collier AM. A Handbook of Textiles. New York: Pergamon Press; 1970. p. 258

[21] Ashton T. The Industrial Revolution (1760-1830). New York, USA: Oxford University Press; 1998. p. 162

[22] Dublin T. Transforming Women's Work: New England Lives in the Industrial Revolution. New York: Cornell University Press; 1995. p. 324

[23] McDonald F. Textiles: A History. UK: Remember When; 2011. p. 304

[24] Varrasi J. Transforming the Textile Industry. [Internet]. 2012. Available from: https://www.asme.org/engineering-topics/articles/manufacturing-processing/transforming-the-textile-industry [Accessed: 14 January, 2017]

[25] Gilligan I. The prehistoric development of clothing: Archaeological implications of a thermal model. Journal of Archaeological Method and Theory. 2010;**17**(1):15-80

[26] Beach CB, McMurry FM. The New Student's Reference Work. Chicago: F. E. Compton and company; 1909. p. 570

[27] EURATEX. European technology platform for the future of textiles and clothing—A vision for 2020. [Internet]. 2004. Available from: http://www.certh.gr/dat/141D2148/file.pdf [Accessed: 18 January, 2017]

[28] Horrocks AR, Anand SC. Handbook of Technical Textiles: Technical Textile Applications. Cambridge, UK: Woodhead Publishing; 2016. p. 431

[29] Gordon Cook J. Handbook of Textile Fibres: Natural Fibres. USA: McGraw Hill Professional; 1984. p. 208

[30] Sreenivasa Murthy HV. Introduction to Textile Fibres. New Delhi, India: Woodhead Publishing India; 2015. p. 250

[31] Lewin M. Handbook of Fiber Chemistry. Boca Raton, US: CRC Press; 2006. p. 1056

[32] McIntyre JE. Synthetic Fibres: Nylon, Polyester, Acrylic, Polyolefin. Cambridge, UK: Woodhead Publishing; 2005. p. 300

[33] Alagirusamy R, Das A. Technical Textile Yarns. Netherlands: Elsevier; 2010. p. 632

[34] Advanced Textiles Source. [Internet]. 2015. Available from: http://advancedtextilessource.com/ [Accessed: 08 January, 2017]

[35] New EQ-Top Seismic Wallpaper Can Save Lives in an Earthquake [Internet]. 2017. Available from: http://inhabitat.com/new-eq-top-seismic-wallpaper-can-save-lives-in-an-earthquake/ [Accessed: 18 January, 2017]

[36] Chen X. Advances in 3D Textiles. Cambridge, UK: Woodhead Publishing; 2015. p. 404

[37] Hello Materials Blog. [Internet]. 2017. Available from: https://hellomaterialsblog.com/ [Accessed: 19 January, 2017]

[38] Springs Creative. [Internet]. 2017. Available from: http://springscreative.com/about/ [Accessed: 15 February, 2017]

[39] NEMEN LED. [Internet]. 2017. Available from: http://www.nemen.it/led/ [Accessed: 19 January, 2017]

[40] Revolution Fibres. [Internet]. 2017. Available from: www.revolutionfibres.com [Accessed: 21 January, 2017]

[41] Tao X. Smart fibres, fabrics and clothing. Cambridge, UK: Woodhead Publishing; 2001. 336 p

[42] T-Shirt Replaces Battery: Fiber-Based Electrochemical Micro-Supercapacitor. [Internet]. 2011. Available from: https://phys.org/news/2011-01-t-shirt-battery-fiber-based-electrochemical-micro-supercapacitor.html [Accessed: 09 January, 2017]

[43] Turning Clothing into Information Displays. [Internet]. 2015. Available from: http://www.nanowerk.com/nanotechnology-news/newsid=41199.php [Accessed: 20 January, 2017]

[44] Briggs-Goode A, Townsend K. Textile Design: Principles, Advances and Applications. Cambridge, UK: Woodhead Publishing; 2011. p. 408

[45] Quinn B. Textile Visionaries: Innovation and Sustainability in Textile Design. UK: Laurence King Publishing Ltd; 2013. p. 312

[46] Pailes-Friedman R. Smart textiles for designers: Inventing the future of fabrics. UK: Laurence King Publishing; 2016. p. 192

Section 2

Materials, Structures and Technologies for Advanced Textiles

Chapter 2

Woven Fabrics for Technical and Industrial Products

Adrian Buhu and Liliana Buhu

Additional information is available at the end of the chapter

http://dx.doi.org/10.5772/intechopen.68989

Abstract

Textile products are classified into products for clothing, household, and technical textiles. Products for clothing and household goods such as curtains, textile wallpapers, fabrics, furniture, carpets, and so on can be easily defined. Textiles that do not fit into these categories may be considered technical textiles. Technical textiles are products designed to perform a specific function. In this category are the woven fabrics presented in this chapter, such as webbings or woven fabrics used to produce reinforcing elements of composite materials.

Keywords: technical textiles, tyre cords, weaving, webbing, composites

1. Introduction

For millennia, humans have been using fibers and textiles. The most common product is clothing, which is also the most important in terms of amount of production. Textiles were and still are being used for medical applications, such as a wound dressing made from silk in Roman times. Today, parts of organs, blood vessels, and ligaments are produced using textile structures. Without fiber-reinforced composites, modern aircraft production would not be possible, and in the house and road-building industries, fibers and textiles are increasingly being used (**Figure 1**). Another category of textile products is represented by filters, which are made from textile structures using a wide range of textile materials ranging from polymers (like polyester) to steel.

For this wide range of products, fibers and textiles are used for three reasons: their mechanical properties, such as tenacity, elongation, shrinkage, and E-modulus that can be adjusted; their unique high ratio of surface to mass; and their variable porosity. Depending on their field of application, textile products have to fulfill the specific requirements like:

1. Aesthetic properties:
   - handle;
   - optical appearance and look;
   - color; and
   - susceptibility to dirt.
2. Physiological properties for wear:
   - skin-friendly wear properties;
   - air permeability;
   - water resistance; and
   - moisture take-up.
3. Physical properties:
   - strength;
   - elongation;
   - wrinkle resistance; and
   - abrasion resistance.
4. Chemical/biological properties:
   - resistance to chemical cleaning agents;
   - resistance to microorganisms and pests;
   - fastness against light, sweat, and friction; and
   - water fastness.

Depending on the specific fields of application, these requirements are of greater or lesser importance. For example, the aesthetic properties can be very important for jeans wear, depending on the type of use (as street wear or for work). The physiological properties for wear and the physical properties are even more important because they influence the personal comfort of the user. In addition, the consumer expects good durability. Chemical properties, however, are somewhat less important, in Ref. [1]. For carpets and rugs, the evaluation of properties is completely different. The aesthetic parameters are very important as well as the purchase price. The physiological character is rather unimportant because it does not influence the physical comfort of an individual. Good physical properties are essential because a carpet has to be long lasting in wear and look. The chemical properties must be distinguished between lightfastness, which is important, and resistance against chemical cleaning, which is desirable but of minor importance.

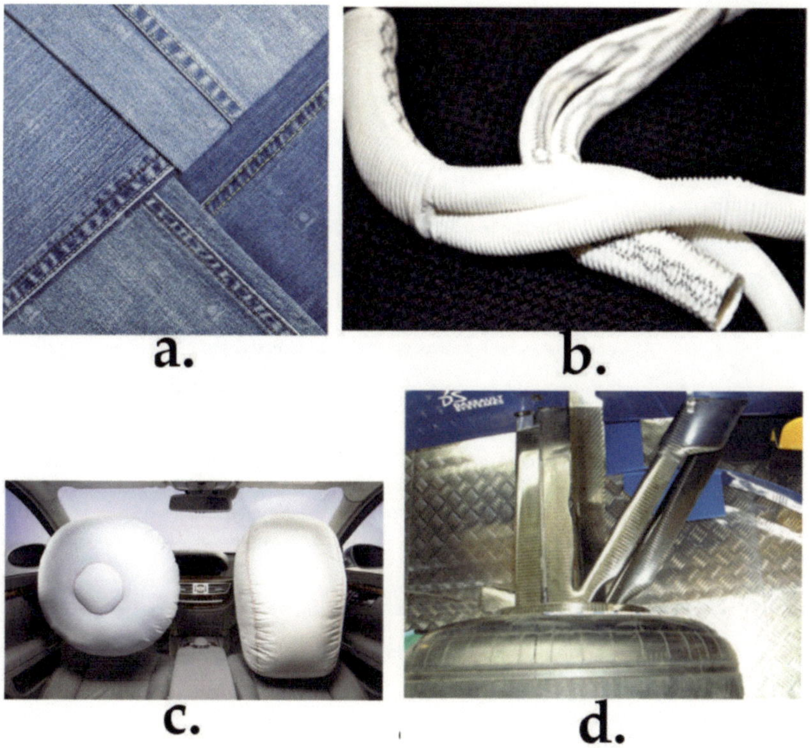

Figure 1. Textile products: (a) denim; (b) medical textiles; (c) airbag; and (d) wheel suspension made from CFC.

The airbag is not affected by fashionable influences; therefore, aesthetic aspects may be neglected. The physiological properties are less important because no direct body contact occurs in normal use. The airbag is a safety product; therefore, the physical properties are essential (it has to be temperature resistant). These aspects determine which raw fiber materials need to be used, and the price must be considered. Thus, when designing a product, it is important to choose materials that fulfill its specific requirements.

2. Woven structures for technical applications

Textile products are classified into products for clothing, household, and technical textiles. Clothing and household products (curtains, wallpaper, textiles, upholstery materials, carpets, and floor coverings) seem to be easily defined. It can be concluded that all other textile products constitute technical textiles group, but this definition cannot be accepted easily. For example, pressurized suits for astronauts, cold water-resistant suits for divers, and protective clothing for steelmakers cannot enter the field of clothing, which are technical textiles. A definition of technical textiles could be: "Technical textiles are all those products that are designed especially to meet their functionality," in Ref. [1]. The terms that define technical textiles are presented in **Table 1**.

Agrotech	Horticulture, landscaping, agriculture, forestry, and livestock breeding
Buildtech	Membranes, light and heavy construction, industrial construction and temporary interior design, hydraulic engineering
Clothtech	Clothing, footwear
Hometech	Upholstery, interior decoration, carpets, floor coverings
Geotech	Construction of underground galleries, drains and waste deposits, mining: protective nets, scaffolding, textiles for preventing erosion, shoreline and sandbars reinforcement
Indutech	Filtration, cleaning, mechanical engineering, chemical and electrical industry, composite, belts, conveyor belts, abrasive discs
Medtech	Health, hygiene, underwear, workwear, veins, dialysis, implants, and medicinal thread
Mobiltech	Bicycles, cars, motorcycles, trains, buses, ships, vehicles, aviation and aerospace, hot air balloons, aeroplanes, kites, airbags, seat belts, seat covers, upholstery, automotive interiors, carpets, upholstery, door networks heart, fabric awnings, toothed belts, pipes, fittings clutch and brake, sealants, composites, armor for vehicles
Ecotech	Environmental protection, recycling, storage
Packtech	Packing, armoring, cord nets, tapes
Protech	Protection of staff and the environment, insulation, water retention, body armor, vests warning, soundproofing, protecting buildings
Sporttech	Sport and leisure, functional sports clothing, sports equipment, textile membrane for surfboards, and sailing and gliding

Table 1. Technical textiles classification.

Based on this definition, one can identify a variety of products including textiles. For example, plastics reinforced with fibers fulfill the criteria that other materials hardly can achieve: low density, stiff adjustable attenuation, good thermal expansion that reduces the direction of the fibers, increased stability to vibration for longer due to insertion of fibers along the lines of force, increased chemical resistance, and a high capacity for absorbing energy from destruction. It can be used for different types of textile structures and the matrix can be used for thermosets or thermoplastics, as shown in **Figure 2a**, [2, 3].

Another example is the conveyer belts that are used, for example, in the paper industry or in the construction and food industry, the baking belts to transport different materials. They are usually made of one or two layers, strengthening material being carried out by an elastomeric bonding and between them an intermediate layer of reinforcement. Textile insertions can be made of polyester or polyamide placed in the longitudinal direction, on a low elongation, or they can be placed in transverse direction on a low elongation and a good draping. Higher layers, intermediate and binding, are elastomers (rubber, synthetic rubber, PVC, polyester, silicone), and the top layer is fixed by vulcanization. Safety and protection belts used in mines must be made of flame retardant synthetic materials, (**Figure 2b**), [4].

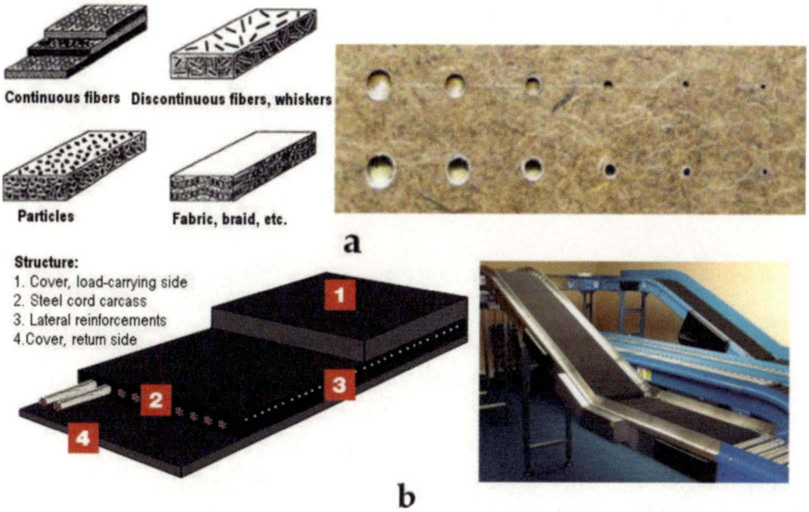

Figure 2. Technical textiles products. a) plastics reinforced with fibers; b) conveyer belts with textile insertions.

2.1. Tyre cords fabrics

A product that includes fabrics is tyres, which support a large and dynamic load and must be flexible. A rubber matrix defines the shape of the tyre functions and acts as a protective elastic layer, and the resistance layer embedded absorbs the forces exerted on the tyre. The main requirements that must meet this layer of resistance are dimensional stability (high Young's modulus, low shrinkage), fatigue resistance, and resistance to adhesion to rubber and rubber chemical products. Depending on the intended use, there are different types of tyres with different structures (for bicycles, motorcycles, cars, aeroplanes, trucks, etc.) [1].

The carrier part of tyre is carcass, which is composed of different layers of fabrics which are wrapped around the bead. Tape made of woven nylon or steel stabilizes the rolling track. The rolling track is made of a nonabrasive rubber with high adhesive capacity. Inside the tyre is inserted a layer of rubber. In the case of diagonal tyres, the fabric layers are arranged diagonally (at an angle between 30 and 45°) from one bead to the other, and these tyres are radial. Radial strips of fabric are diagonally arranged to each other (at an angle from 20 to 25°). These tapes guarantee stability and tyres are made generally of steel fabrics because they are superior textile tapes. For several years, using a tape of twisted polyamide yarns over these bands decreases the danger of separation edges tape and increases the capacity running at high speeds. The tyre housing is made of viscose fabrics, polyester, or polyamide and sometimes by aramid. Viscose yarns adhere well to rubber and are used for tyres speed in Europe. Polyamide is used in particular diagonal tyres because the rubber has a good grip and high resistance to fatigue. Polyester has gained a market share of radial tyres because it has an excellent price/performance ratio (**Figure 3**), [5, 6].

The purpose of obtaining technology yarns cord is to increase the fatigue strength of the material under compression. This is done by twisting after spinning when there are cord

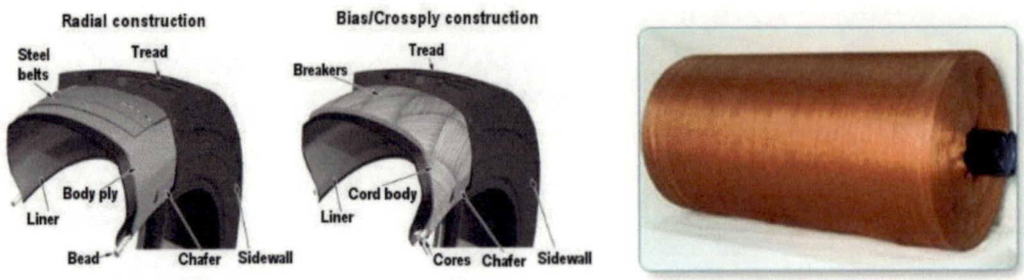

Figure 3. Technical textiles used for tyre cord: structure of tyre and tyre cord made by nylon.

yarns, a process that consists of two distinct phases of pretwisting and twisting. The first phase gives torsion to untwisted filament yarns in normal sense Z. Then twisting two yarns in the S sense is such to eliminate previous twisting on a ring twisting machine and the winding bobbin is cylindrical. These bobbins are fed from a double-twist twisting machine. The pretwisting process step is dropped in the process of direct cabling. One bobbin with untwisted filament yarns is situated in the stationary twist pot of the bobbin (inside thread), and a second bobbin is fixed in the railing above the machine (exterior/outside thread). The inside thread is led directly to the outside thread in an axial direction upward to the joining point. The outside thread is guided from the bottom up into the rotating part of the bobbin. At the storage disk, it appears again, forms a balloon of thread, and covers the inside yarn above the twisting pot. The two yarns must be wrapped with tension equal in the point of contact to prevent a yarn with length variations that affect tensile strength and fatigue.

After twisting the cord yarns, they will be woven. The warp and weft systems of normal fabrics have close densities, but in terms of deformation and vulcanization behavior of these features are not suitable. Therefore, cord fabrics of yarns have different structures and densities. Tyre cord fabrics will have 10 picks/10 cm, and cotton yarn structure is open. The weft does not have any effect upon the casing; it constitutes a yarn support to facilitate following operations.

The fabric cord was impregnated to increase adhesion with rubber and to change its contraction modulus for a better dimensional stability. Impregnating machine has a length of 100 m and a height of five floors. The tyre cord fabric is pulled off the woven fabric and passed through an impregnating bath with a resorcinol-formaldehyde-latex solution that improves the adhesiveness on the rubber. This impregnating is sufficient for viscose and polyamide, but polyester and aramid have to be pretreated additionally with bonding agents based on epoxy resin. After the drying zone, the fabric passes through a heat-setting zone in which it is subjected to a defined temperature and tension treatment for the adjustment of modulus and shrinkage. According to the material used, it may be necessary to add a normalization zone for the compensation of inside tensions or a second impregnating bath with an additional drying zone. After passing the machines, the fabric is coiled on a cylinder. The impregnated fabrics are coated with a thin rubber layer by calendering. In the last processing step before the actual tyre production, the fabric is cut under a certain angle in tapes with a desired width.

2.2. Textile webbing

The webbings are narrow fabrics that are distinguished by the type of yarns that are produced, variation in tensile strength, and width. Webbings are produced on narrow looms. Overall, narrow fabrics are compared to ropes because they are used primarily for harnesses. Because these are versatile fabrics, these have various industrial applications used in on the military field and the automotive industry. Typically, the webbings obtained are compact or tubular having different applications and functions. If the ropes are thick, webbings are lightweight products. The raw materials used are made up of different types of polyester, polyamide, or polypropylene. Cotton webbings are not only used mainly in clothing but also in other commercial applications. The webbings can be obtained in a variety of structures, colors, and prints. Manufacturers can produce reflective straps for safety applications (**Figure 4**).

Typical applications of webbings may be associated with the following industries:

- seat belts and harnesses—automotive industry;
- equipment for hiking, climbing harnesses, and backpacks—consumer sports equipment;
- safety and signaling strips—hospitals and medical industry;
- upholstery (support for chairs, etc.)—the production of furniture; and
- uniforms and accessories for different professions—police, fire, and military.

The webbings are known as compact and flat webbings, having different thicknesses and are distinguished by being flat and having different uses such as safety belts. They are likely to be lightweight to intense tear that tend to affect the surface of the product. Flat webbings are too rigid to be used in applications requiring knotting, and for this reason, it is used for products that are high on sewing, for example, the backpack straps. Tubular webbings are thicker and more durable than the compact and are composed of two fabrics. They are used in applications requiring knotting (as a rope to lift) and support higher tensions. For destinations such as climbing, it is recommended to use tubular webbing woven into a continuous loop.

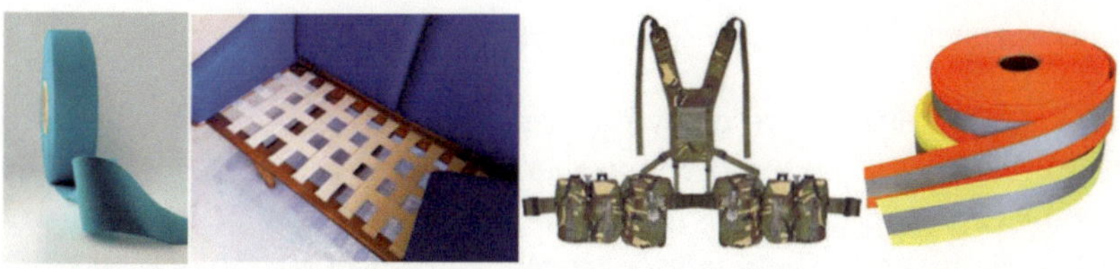

Figure 4. Applications of textile webbings.

2.2.1. Raw materials for webbing

Polyamide webbings are elastic products of high strength and are used to produce. In the wet environment, polypropylene elongates by 2% and it is recommendable that the fabric cannot be exposed to water for a long time. Polyamide webbings tend to absorb water which leads to mold growth if they are not properly maintained.

Polyester webbings are durable and have an appearance similar to that of polyamide webbings. The products are recommended for applications that require lifting heavy weights. Also, these webbings do not absorb a lot of water, being resistant to mold. Common uses are for racing harnesses and seat belts.

Polypropylene webbings are used for the products used for the environment activities. Some of these products include mesh for windows, webbing, and bags. In terms of physicochemical properties are comparable to those of polyamide, but the polypropylene products are lighter than similar polyamide. It is used to produce materials which are resistant to water and UV radiation. Products have different thickness and have a low resistance to abrasion and suitable for work that requires medium resistance.

Based on the properties of these fabrics, when are used for seat belts, harnesses, and other safety products, a periodic check should be carried out. This is required to see if the equipment security requirements have been modified. The webbings used as harnesses and belts in motor racing industry lose their elasticity and break when they are used frequently or exposed to oil or heat. It is recommended to replace them after a period of 2–5 years or sooner if used frequently, such as seat belts or seats.

Another problem raised by webbing is maintenance. As a general rule, they must be kept clean and dry, even if polypropylene is waterproof. It is recommended to wash with mild detergent because these materials are painted while the colors fade or even disappear if subjected to certain environmental conditions or cleaning.

Textile products must satisfy certain quality requirements, depending on their destination. These requirements refer, but are not limited, to the type of the raw material, the fineness of yarn, the density of the fabric structure, the fabric weight, breaking strength and elongation at break, and color and finish. One of the requirements for leather goods webbing is to respect certain values for breaking force. For this reason, it is necessary that the yarns be selected as appropriate, with appropriate breaking force. Characteristics of the polypropylene yarns used for warp and weft are shown in **Table 2**.

2.2.2. Mechanical characteristics of leather goods webbings

Samples subjected to testing were obtained from the black polypropylene yarns, having a width of 30 mm. All fabrics had the same number of yarns in the warp, respectively 104 yarns, but instead were varied in the following characteristics: density of the weft yarns and the warp and weft yarn fineness. These fabrics are made on short technology: the direct warping of yarns on narrow warping machines, warp installation in weaving machine, and weaving on a weaving narrow loom.

Characteristics		Values	
Raw material	–	polypropylene	
Color	–	black	
Count	Dtex	550	1100
Tenacity, min.	cN/dtex	3	2.5
Elongation	%	20 ± 3	
Breaking force, mean	cN	1610	2665

Table 2. Warp and weft yarns characteristics.

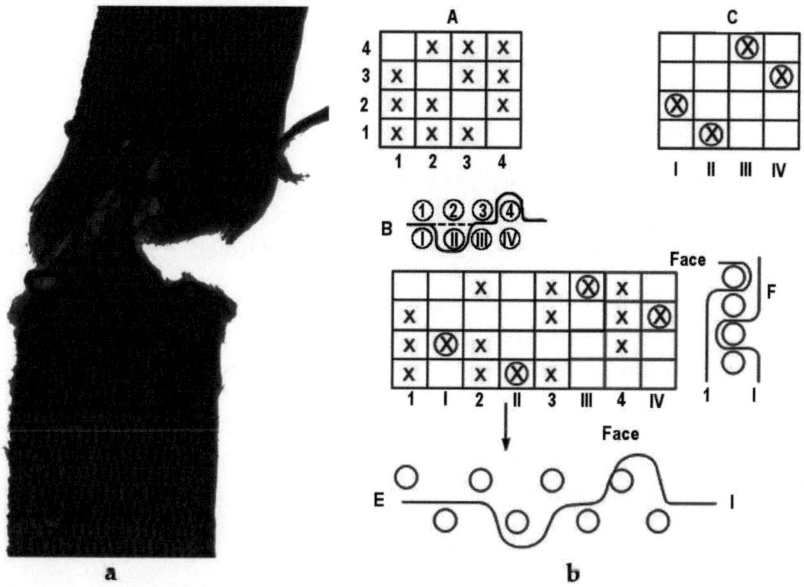

Figure 5. Webbing for leather goods: (a) aspect and (b) structure [7].

Variants	Tt_{wa} (dtex)	Tt_{we} (dtex)	Fr_{wa} (N)	Fr_{we} (N)	D_{we} (fire/cm)	F_{maxt} (kN)	a_{wa} (%)
1	550	1100	161	266.5	8	2.13	26.11%
2	550	550	161	161	8	2.35	22.56%
3	550	550	161	161	10	1.95	18.69%
4	550	1100	161	266.5	10	1.65	26.65%
5	550	1100	161	266.5	6	1.86	18.26%
6	550	550	161	161	6	1.87	17.51%

Table 3. Values for breaking force and breaking elongation.

Woven fabric structure was backed by the warp; the structure and aspect is presented in **Figure 5**.

The test was performed on a universal testing machine WDW 50E of mechanical type, with computer-aided control and maximum load capacity of 50 kN and the distance between grips was set at 300 mm. The values obtained for the variants tested are shown in **Table 3**.

Tensile tests were performed to breakage of test pieces, until recording time to the maximum load value was considered as $F_{maxt,}$ that were the first warp yarn breakages. Load-elongation curves of the leather webbings had a similar look for all variants analyzed webbings.

Shown below are two effort-elongation curves of these fabrics for specific two of the six tested variants, presented in **Figures 6** and **7**.

Analyzing the following figures aspect, it is found that this curve shows three distinct zones. Of these, the central area presents a linear evolution, i.e. the force applied to the sample and the corresponding elongation varies directly proportional.

- an initial area, of relatively slow growth (hyperbolic), up to a loading level between 8 and 10% of maximum force F_{max};
- an area of relatively abrupt increase, linear with a slope greater, up to a force of almost 95% of maximum force F_{max}, respectively; limits of these zones are between 8 and 95%. In this area, the fabric behaves almost proportional having a relatively high modulus of elasticity;
- an area with a small length, in which the breakage occurs. In this area, two situations are revealed: in the first situation, breakage occurs gradually due to break on all the threads, and the second situation is where breakage occurs suddenly.

Processing of data results has led at extracting meaningful data from this set of values. Values for the breaking force of tapes for handbags were grouped according to two criteria: the strength of the weft and warp breakage. These criteria are grouped according to the values shown in **Table 4**.

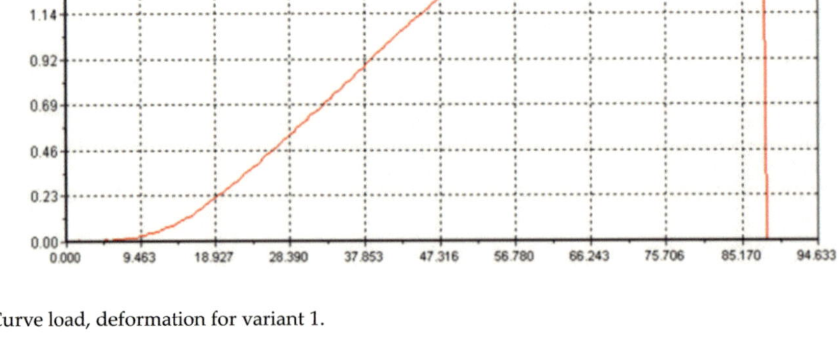

Figure 6. Curve load, deformation for variant 1.

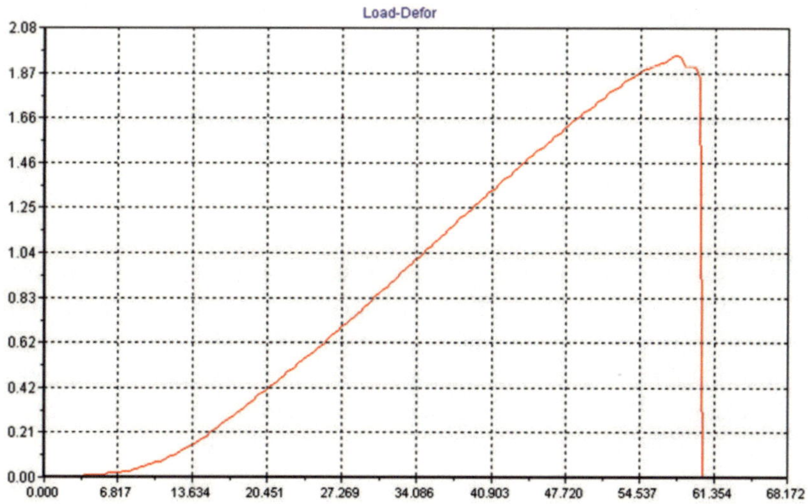

Figure 7. Curve load, deformation for variant 3.

Db, (ends/cm)	Fru (cN)	Mean values of Fru (kN)	au (%)
6		**2.80**	**18.7**
	161	1.87	19.0
	266.5	3.74	18.5
8		**3.09**	**23.8**
	161	2.24	22.9
	266.5	3.94	24.7
10		**2.74**	**29.2**
	161	1.80	24.1
	266.5	3.69	34.3

Table 4. Medium values for breaking force and elongation.

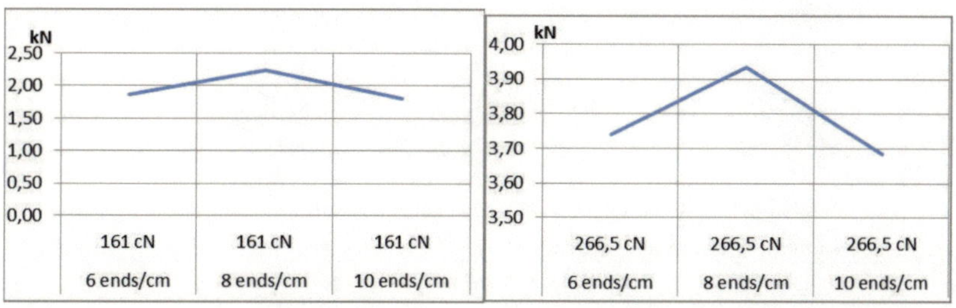

Figure 8. The mean values of webbings breaking force.

The values of webbings breaking force are comprised between 1.8 and 2.4 kN for warp yarns with breaking force of 161 cN and between 3.69 and 3.74 kN for warp yarns of breaking force of 266.5 cN. The bold values in Table 4 are the average values for fabrics of the same density in the weft directions, but with different yarns in the direction of the warp. These values are not significant in the performed analysis because these are not referred at the same fabric. So if the warp yarns are used with greater resistance, resistance webbing in proportion will increase. (**Figure 8**).

The maximum values for both warps appear at weaves with the weft density average value (8 ends/cm). Even if woven webbing with a greater density, 10 ends/cm, is more compact and appears more resistant, in reality, the higher density caused the biggest undulation and crushing of yarns in fabric. The effect is manifested through subsequent breakage of the warp yarns to the breakage of the strap. The breaking of the webbing was occurred suddenly, for the fabrics with weft density with 8 ends/cm and those of the webbing with weft density of 6 yarns/cm.

In the case of variant elongation, the values are shown in **Table 4**, and the change in elongation in the weft according to the breaking force and the wefts is represented in **Figure 9**.

The values of elongation of webbing are comprised between 19 and 24.1% for the warp yarns with breaking force of 161 cN and between 18.5 and 34.3% for warp yarns with breaking force of 266.5 cN. If the weft yarn density increases, the elongation at break increases, as shown in **Figure 9**. The weft density influences the degree of crimping of the warp yarns, respectively, its increase, and so the elongation will be greater due to the higher recovery of crimping.

Observation of effort-elongation curves leads to the conclusion that between breaking force and elongation at break is a relationship of proportionality.

The checking of the correlation of experimental data packet, Data Analysis, using Microsoft Excel and the result is presented in **Table 5**. The bold values in the table represent significant correlation coefficients.

From this table, it is observed that between the breaking force of the webbing and breaking forces of the weft and warp yarns, there is a certain dependency (coefficient of 0.914, −0.301), and elongation at break is influenced by the breaking force by the warp yarns and the weft density (0.348, respectively 0.654).

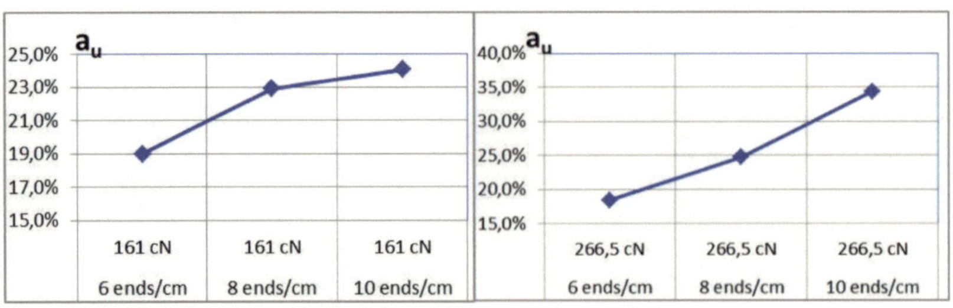

Figure 9. The mean values of elongation of the webbing.

	Fru	Frb	Db	Frt	au
Fru	1				
Frb	0	1			
Db	0	0	1		
Frt	**0.914408**	−0.30089	−0.02464	1	
au	0.347696	0.294629	**0.654416**	0.215268	1

Table 5. Matrix of correlation coefficients.

SUMMARY OUTPUT

Regression Statistics	
Multiple R	0,962956
R Square	0,927285
Adjusted R Square	0,900017
Standard Error	0,328369
Observations	12

ANOVA

	df	SS	MS	F	Significance F
Regression	3	11,000	3,667	34,006	0,000
Residual	8	0,863	0,108		
Total	11	11,863			

	Coefficients	Standard Error	t Stat	P-value	Lower 95%	Upper 95%	Lower 95,0%	Upper 95,0%
Intercept	0,526	0,721	0,729	0,487	-1,137	2,188	-1,137	2,188
Fru	0,017	0,002	9,591	0,000	0,013	0,021	0,013	0,021
Frb	-0,006	0,002	-3,156	0,013	-0,010	-0,002	-0,010	-0,002
Db	-0,015	0,058	-0,258	0,803	-0,149	0,119	-0,149	0,119

Figure 10. Regression analysis of the breaking force of the webbing.

Regression analysis of the experimental values was performed with regression package. This package was applied to the original experimental data; results are presented in **Figure 10**.

It is observed that experimental data are correlated (coefficient 0.927), and the model is especially significant. Testing the significance of the coefficients leads to the following observation: the breaking force of the webbing is influenced by the breaking force of the warp yarns and the weft yarns. It reforms the statistical analysis and removes the weft density.

The new values resulting from the statistical analysis are better than previous (higher correlation coefficient, 0.91) and the coefficients are significant. On the basis of the output data of the regression, the equation is shown in Eq. (1):

$$F_{rt} = 0.406 + 0.017 \cdot F_{rwa} - 0.006 \cdot F_{rwe} \quad (1)$$

In Eq. (1), breaking force of the webbing (F_{rt}) depends on the breaking force of the warp (F_{rwa}) and the weft (F_{rwe}) yarns. The increase by 1 cN of breaking force of the warp yarns will increase, to 0.423 kN, the breaking force of the webbing. Breaking force of weft does not significantly influence the strength of the webbing as its growth causes a reduction of total breaking force, as shown in Eq. (1).

Of the regression analysis on the influence of mechanical characteristics of yarns on elongation of webbing, it is noticed that in this case, the correlation coefficient is lower (0.428) which requires the consideration and other factors.

Based on these results the following conclusions may come off:

- the breaking force of the webbing is proportional to the breaking force of the warp yarns because they all participate in the strength of the fabric;
- the weft yarns do not influence the breaking strength since it has been found that to using resistance yarns, usually thicker warp yarns break more easily;
- fabric structure affects the strength by creating an equilibrium between the crimping of yarns and the number of bonding points in which frictional forces appear;
- webbing testing requires the use of grips to take up the tensile force along the fabric and to do fixing without crushing the fabric portion between the clamps.
- tightening pressed of webbings from filament yarns produces jaw-crushing yarns and breaking occurs near them. This affects the veracity of values.

2.3. Textiles used in composites

In the last two decades, the uses of textile structures made from high-performance fibers are finding increasing applications in composites. High-performance textile structures may be defined as materials that are highly engineered fibrous structures having high specific strength, high specific modules, and designed to perform at high temperature and high pressure (loads) under corrosive and extreme environmental conditions. Significant developments have taken place in fibers, matrix polymers, and composite manufacturing techniques. Composites that are a part of industrial textiles have a significant role in many applications especially in automobiles and aerospace applications [8].

Composite materials reinforced with woven fabrics, braids, and knits are becoming increasingly popular in various structural applications from automotive, aerospace, furniture, and so on. Processing techniques of materials to obtain composites include technologies to obtain reinforcement layers, stratification technology, transfer resins in textile layers through molding, molding with vacuum/pressure, and autoclaving fabric (reinforcement structures) for impregnation products with properties of thermosetting and compression/preforming molding of thermoplastic and thermosetting composites [9].

2.3.1. 2D woven structures in composites

Preimpregnated fabrics play an important role in the technology of composite structures because they can perform various structures from different natural materials (fibers and textile fabrics, glass fibers, aramid fibers, carbon, and mixed structures). Advantages of woven reinforced composites are reduced cost, improved machinability, and, in particular, the use of a wide range of textile structures. Woven reinforced polymeric composite materials have broad applications in the structure of aeroplanes and ships, having good stability and easy machinability. Inserting textile elements in composite structures aims to:

- improve the mechanical behavior of the composite material, the advantageous orientation of the textile insertion relative to the direction of mechanical stress; and
- improve the resistance of bonding areas of the pieces.

The stiffness and strength of textile-reinforced composites depend on the characteristics of the yarns, the matrix properties, and structure parameters of the insertion (the thickness of the fabric, warp and weft density, tensile strength in the direction of the two yarn systems, and the structure of the insertion). The thickness depends on the density and fineness of the fabric yarns, while the structure determines how the warp and weft yarns interact. Typical structures of plies which can be used in the composites are presented in **Figure 11**.

Depending on the type of loads that are subject to insertions, they may have the same strength and stiffness in both directions (warp and weft). If the request is important only in one direction (e.g., warp direction) fabrics can be obtained with a large number of yarns in that direction and less in the opposite system (weft). These fabrics are called unidirectional because they offer high strength and stiffness in one direction. Unidirectional composites have high machinability.

Compared to unidirectional composites and nonwovens, composites that use fabrics like the reinforcement system are more resistant to impact and have uniform properties in all directions [10].

Textile structures used as insertions in composites can be obtained by different methods of binding/joining of textile materials (fibers, fiber preforms, yarns, etc.) such as weaving, knitting, and braiding. They are different textile structures used as reinforcement, such as fabrics, braids, and knitted ones (**Figure 12**).

The choice of a particular type of textile fabric used as reinforcement elements for the production of composites depends on the capacity of multiaxial reinforcement and between the layers, namely the ability to obtain different forms of spatial composites. Depending on how composites processing and requirements, certain structures of reinforcement elements may be adopted.

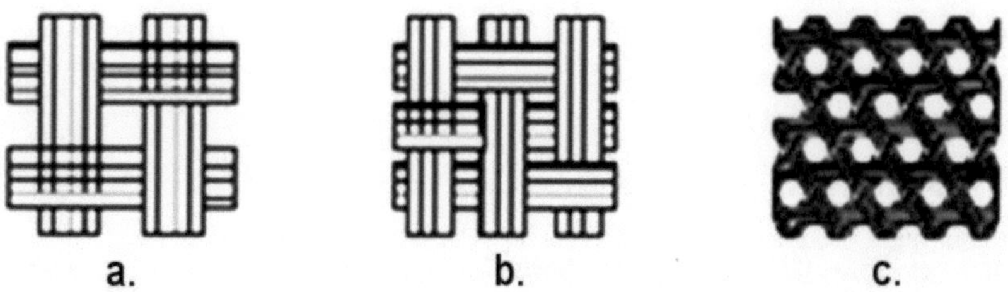

Figure 11. Different types of textile structures: (a) plain woven fabric; (b) twill fabric; and (c) braid.

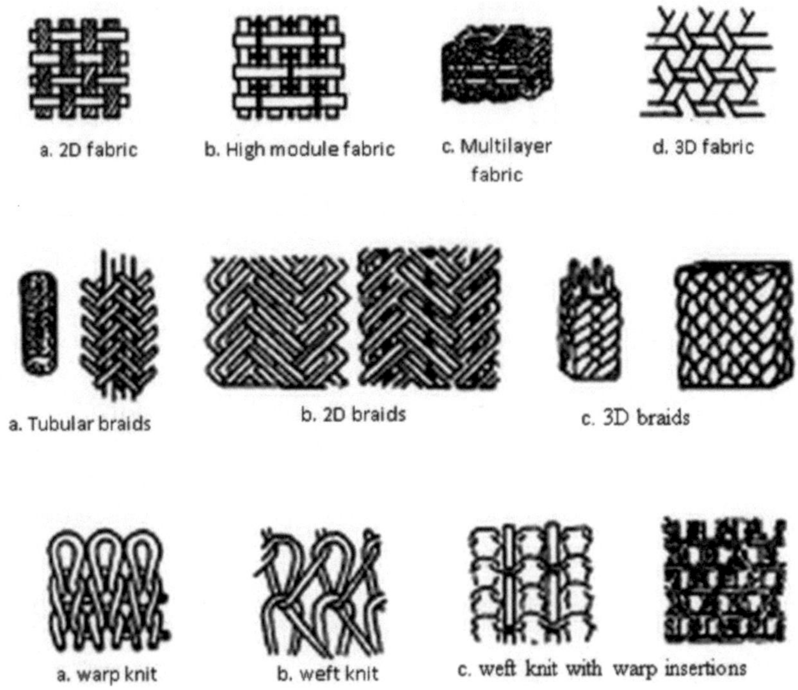

Figure 12. Textile structures used like insertions.

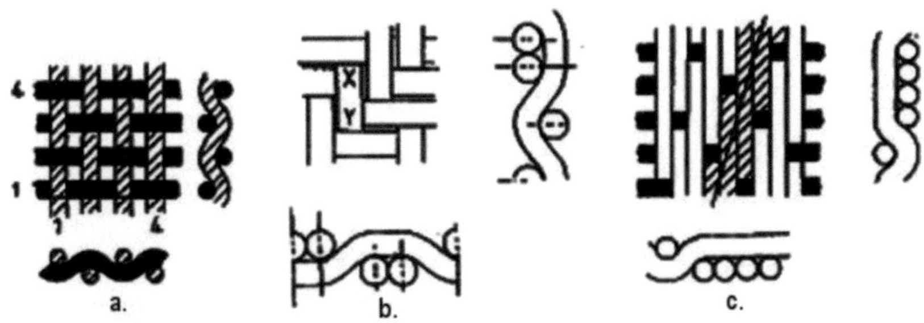

Figure 13. Fabrics fundamental ties: (a) plain; (b) twill; and (c) atlas.

The armor drawing used for connecting yarn systems participating in the fabric affords numerous combinations of woven. Bi-dimensional fabrics (plane) obtained by combining the two systems of yarn (warp and weft) disposed mutually at an angle of 90° by repeating the cell structural or topological model.

Regarding fabric structures, fundamental ties are distinguished, as presented in **Figure 13**, which entail all other types of ties between yarn systems participating in the fabric: plain weave, twill, and satin. The differences between the three types of ties are the number of binding points and lengths of the segments associated with each yarn system that binds the opposite system.

A plain weave fabric is characterized by:

- the most frequent crimping the warp and weft yarns in the ratio tie, which leads to a maximum shrinkage of yarn systems to weaving;
- fabric surface is monotonous, and the reflected light is diffuse;
- if the density of warp yarns is equal to the density of weft yarns, and the count of the yarns of the two systems is same, the effect will be the same on both sides of the fabric and there is no dominant system, and the fabric has the same behavior to mechanical load of the two yarn systems;
- it has high resistance to wear and friction applications; and
- plain weave gives the fabric a high level of structural integrity and greater expandability due to high frequency crimping.

Satin fabric is characterized by:

- smooth, shiny uniform appearance due to distribution points binding;
- uneven distribution of yarn systems on the front and back fabric leads to a dominating system (weft on the face and warp of the back face of the fabric);
- Increasing the system effect, i.e. the length of the warp segment in the weave repeat causes a decrease in the compactness and mechanical properties of the fabric;
- allowing a better transfer of the structure of yarns and fabrics for resisting a translation module efficiently due to low binding yarns of opposing systems, allowing better mobility between them.

Braided reed or cane is used for a long time to obtain pieces of furniture (chairs, tables, etc.) and triaxial ties between systems are used to build the structure. There was a concern for construction machinery to carry out such structures naturally [11].

2.3.2. 3. D *woven structures in composites*

Due to their flat surfaces, modeled fabrics can be profiled. In this case, the fabric must be materials with high-capacity stretching. Obtaining the reinforcing fabric to be used as textile insertions in composites with three-dimensional geometry requires:

- A series of complex weaving technologies, such as weaving with binding between the two fabric layers of 2D and 3D weaving surface (fabric structures comprising semi double, double);
- Methods for the design of links that generate smooth fabrics to cover certain areas of the composite and methods to predict and prevent various defects such as creasing, folding, and tearing [12, 13].

At the same time, textile fabrics used as insertions have benefits as good machinability and a corresponding draping. In terms of mechanical properties, textile-reinforced composites are advantageous over unidirectional laminate composites because they have no reinforcements oriented in the thickness direction. On the other, the textiles are characterized by 3D architecture due to the connections between the yarns that are part of different systems and crossing the different layers of the insertion [14].

By technologies of weaving, braiding, and knitting, it can produce bi- or 3D structures. The orientations of yarns, their distribution in the insertion structure, and number of yarns from preform thickness determine the type of structure (bi or three-dimensional). A bi-dimensional structure assumes the existence on the thickness of the insertion of two or three yarns, which are oriented in the x-y plane.

A 3D structure is obtained by using three or more yarns and the thickness of the yarns go through the structure in all three directions.

Characteristic of triaxial fabric structures is the hexagonal orientation of yarn systems, participating in the structure, as shown in **Figure 14**, which leads to elevated shear strength of the fabric.

3D woven structures are obtained mainly by using multiple warp and weft systems. In this connection, multilayers or double structures were used for making packaging, laces, textile tapes, and carpets. By using these types of ties, it is possible to produce a solid orthogonal panel(**Figure 15a**); solid panels with variable thickness (**Figure 15b**); panel coreless structures like beams(**Figure 15c**); or similar structure types of lattice girders (**Figure 15d**).

The inconvenience of plane reinforcing elements, respectively, low resistance in the diagonal direction, is removed by replacing with multilayer fabric made by using the triaxial weaving technology.

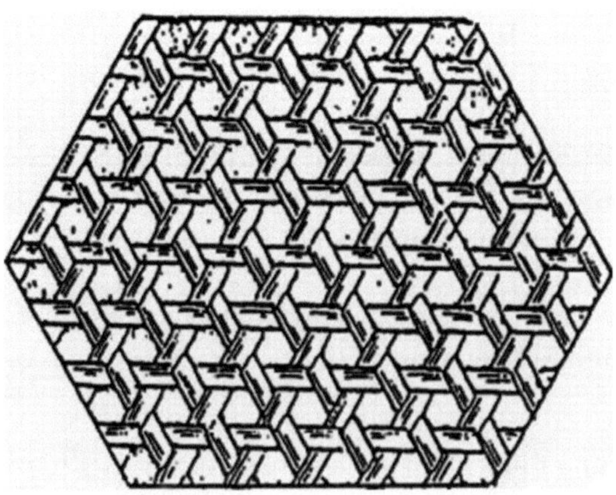

Figure 14. The structure of a triaxial braided cane.

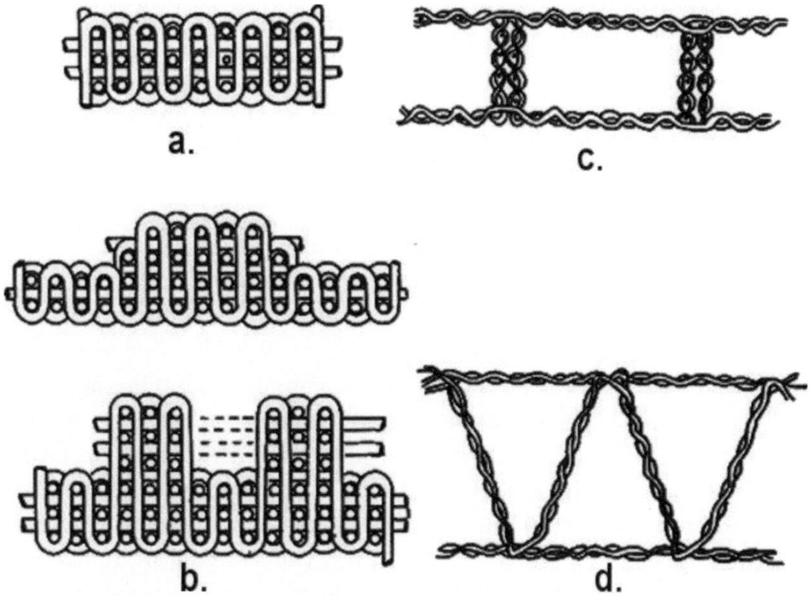

Figure 15. Types of 3D structures used as reinforcing element of composite. a) solid orthogonal panel; b) solid panels with variable thickness; c) panel coreless structures like beams; d) lattice girders.

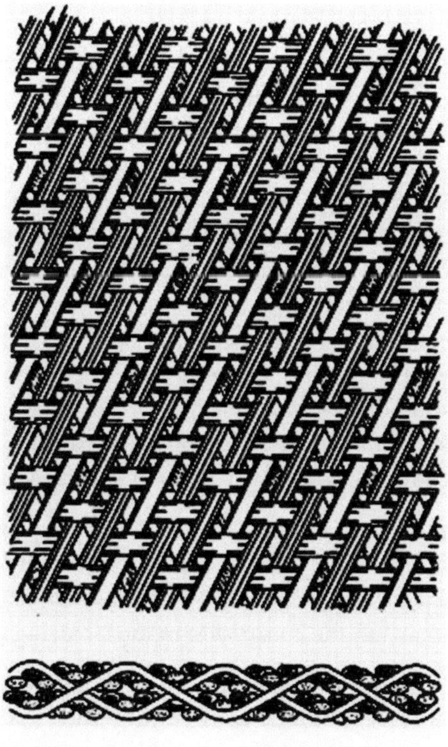

Figure 16. Woven 3D structures, multilayer.

The 3D fabrics, thanks to low draping characteristics, require overlaying multiple layers that allow for a smooth transfer of loads. In the structure of composite material, fabrics used as insertions are placed so that the yarn systems participating in the fabric systems (perpendicular) are to be at a ± 45^0 angle to the composite axis. Thus, textile insertion takes external efforts in a uniform manner in all directions of application, avoiding the different behaviors of the composite during its use. For example, in combination with the unidirectional reinforcing materials are produced the composite materials to achieve tennis rackets, materials with shear strength, and high rigidity [15, 16].

2.3.3. *Biodegradable woven structures for composites*

Plain woven fabric, from bast yarns, can be used to obtain cheap and biodegradable composite materials. Biodegradable composites—currently based on biopolymers with natural fiber reinforcement—are intended to be compostable, after their lifetime use, in order to prevent the growth of permanent environmental pollution. This class of composites is based on some raw materials that are available on our internal market—animal glue (together with some curing and stabilizing agents) and bast fiber fabrics. The samples are firstly preformed, by impregnating the polymeric matrix in the suitable textile reinforcement arrangements, and then consolidated by moderately hot pressing. The evolution of composite tensile properties was studied, in dependence with the parameters (temperature and pressure) of consolidating process, in order to obtain their suitable values for optimizing the composite mechanical response [17].

The biodegradable composite materials were manufactured as rectangular samples of crossply laminates, using a polymeric (proteic) matrix, based on animal glue, initially as an aqueous solution containing some curing and stabilizing agents. For every sample, the reinforcement, in a weight fraction $W_f = 0.49$, was composed by four plies of flax fiber fabrics, alternately disposed with the principal directions (corresponding to the warp and the weft yarns) on the long axis of the composite sample. On the basis of some experimental results, (Mareş et al.), previously obtained by the authors of the present chapter, four different levels were used for the processing temperature, namely 45, 55, 65, and 75°C, successively combined with four pressure level values, 0.15, 0.20, 0.25, and 0.28 [18]. The biodegradable composite materials that are presented herein were intended to be used for components (from the ambient design, for example) that must not have high levels of mechanical strength. In that, it must be said that the composite samples, as resulted from the moderate hot pressing process, are comparable in stiffness with the plywood samples of similar thicknesses.

The composite load-elongation dependence (**Figure 17**) has a pronounced nonlinear aspect, with a down-right oriented convexity, that is typical for woven textile reinforcements, as it could be observed from the load-elongation curve which was obtained for the jute woven, as shown in **Figure 18**, before starting the composite manufacturing process.

As it can be observed on the above presented load-elongation curves, the principal Young's modulus (E_1) of the composite, corresponding to the specimen loading direction in the tensile test, could be considered as increasing with the applied force: the modulus value is relatively low at the beginning of the curve, but it is many times bigger at the last portion of the curve, before its maximum point (F_{max}). One can say that, having in view the values of ultimate ten-

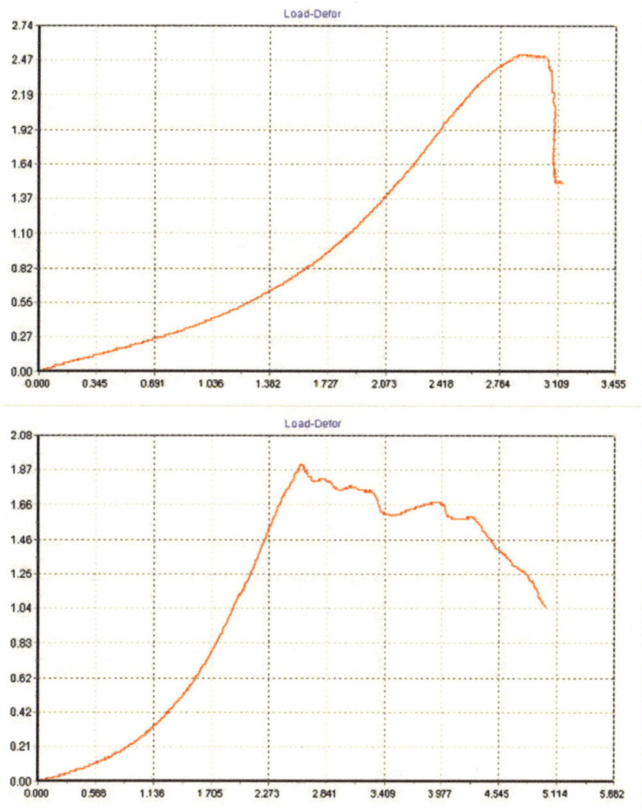

Figure 17. Typical aspects of the load-elongation dependence for the studied composites.

sile strength (24–27 MPa) for the studied composites, it is convenient for these materials to be utilized in samples that have to support, on the principal material direction, normal stresses of approximately 16–22 MPa. Such a mechanical load will lead to a material response corresponding to its maximum stiffness level. In **Figure 19** are briefly presented the evolution of the average tensile strength of the studied biodegradable composites in dependence with the technological parameters (temperature and pressure) of the forming process.

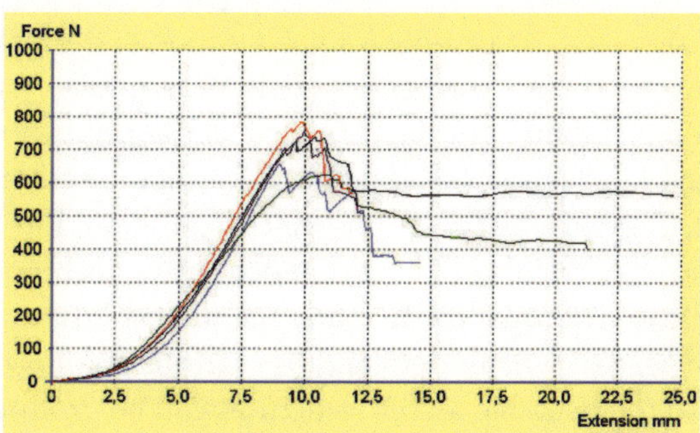

Figure 18. The load-elongation dependence for the jute woven reinforcement.

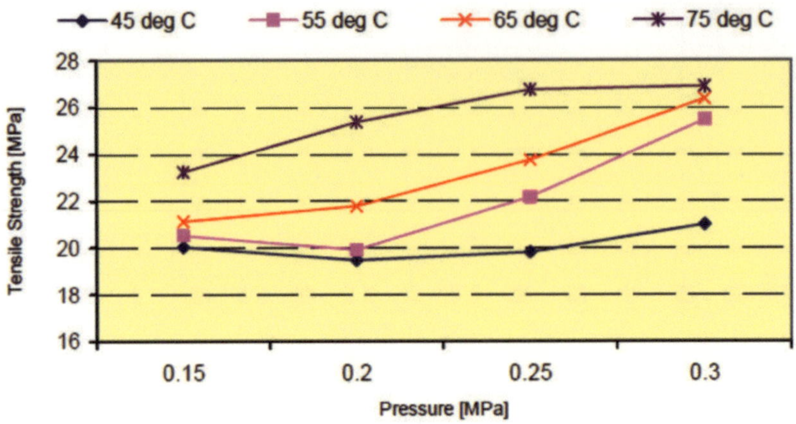

Figure 19. Tensile strength variation as a function of forming pressure, for different values of processing temperature.

Some particular issues could be observed, in the figures from above, for the studied biodegradable composites, regarding their mechanical response:

- as a general tendency, an increase in strength and stiffness could be observed, when temperature and pressure levels are both increasing;
- an interesting effect can be seen, regarding the results that are obtained for the upper levels of temperature—namely those of 65 and 75°C, mainly for the pressure levels overtaking 0.2 MPa; for using these values of the technological parameters, the suitable mechanical properties are corresponding to the temperature of 65°C, instead of the highest level (75°C), as it could be expected;
- on that basis, one can conclude that the best combination of composite mechanical stiffness and strength was obtained for the samples that were pressed at 0.3 MPa and 65°C, that can be retained as the optimum parameter values for composite consolidating process.

2.4. Conclusions

The fabrics presented in this chapter are only a small fraction of the technical fabrics which are produced. Also, these fabrics can have other applications, and they will take into account the fact that the design and development of the technical textile product need basic understanding and application of textile science and technology. Technology advances in the industry are driven by forces outside the pure textile sector, that is, polymer and fiber producers and, in some cases, the machinery producers of fabric manufacturing techniques. There is a growing need for nontextile application know-how in many segments of the industrial textiles market. Textile technologists are needed who understand the various engineering aspects of potential industrial applications so that suitable textile structures can be produced.

Author details

Adrian Buhu* and Liliana Buhu

*Address all correspondence to: adibuhu@yahoo.com

Faculty of Textiles, Leather and Industrial Management, "Gheorghe Asachi" Technical University of Iasi, Iasi, Romania

References

[1] Gries T, Veit D, Wulfhorst B. Textile Technology—An Introduction. 2nd ed. München: Hanser Publications; 2015. p. 441

[2] Beilken D, Hinken JH. Fibre Reinforced Plastic: A Feasibility Study of Microwave Based Non-Destructive Testing [Internet]. 10/2005.Available from: http://www.ndt.net/article/v10n10/hinken15/hinken15.htm [Accessed: 12-04-2016]

[3] Shah N. Carbon Fiber Reinforced Plastic Composites [Internet]. 23 September 2011. Available from: http://engineringcorner.blogspot.ro/2011/09/carbon-fiber-reinforced-plastic.html [Accessed: 12-04-2016]

[4] GermanBelt [Internet]. 2017. Available from: http://www.germanbelt.com/en/products/conveyor-belts/steel-cord-conveyor-belts.html [Accessed: 01-05-2017]

[5] Diandra. The Dirt on Radials VS. Bias-ply Tires [Internet]. July 25, 2014. Available from: http://buildingspeed.org/blog/2014/07/25/the-dirt-on-radials-vs-bias-ply-tires/ [Accessed: 01-8-2017]

[6] Indiamart-Nylon Tyre Cords [Internet]. 2017. Available from: https://www.indiamart.com/vrsimpex/other-products.html [Accessed: 01-8-2017]

[7] BACKED FABRICS [Internet]. 2016. Available from: http://textileschool4u.blogspot.ro/2013/12/backed-fabrics.html [Accessed: 01/12/2017]

[8] Senthil Kumar R. Textiles for Industrial Applications. Boca Raton, FL, USA: CRC Press, Taylor & Francis Group; 2014. p. 370

[9] Morozov VE. Mechanics and analysis of fabric composites and structures. AUTEX Research Journal. 2004; **4**(2): 60-71

[10] Chretien N. Numerical Constitutive Models of Woven and raided textile structural composites [dissertation]. Blacksburg, Virginia, USA: Virginia Polytechnic Institute and State University; 2002. p. 175. Available from: https://scholar.google.ro/scholar?lookup=0&q=N.+Chretien+%E2%80%93+Numerical+constitutive+models+of+woven+and+braided+textile+structural+composites+(2002),+Virginia+Polytechnic+Institute+and+State+University.&hl=ro&as_sdt=0,5

[11] Panigrahi S, Tabil LG, Crerar WJ, Sokansanj S. Application of Saskatchewan Grown Flax. Canadian Society of Agricultural Engineering. CSAE/SCGR, Canada 2002;(Paper No. CSAE 02-302)

[12] Aono, M. Computer-aided geometric design for forming woven cloth composites [thesis]. Rensselaer Polytechnic Institute; 1995. Available from: http://dl.acm.org/citation.cfm?id=220979

[13] Van West BP, Luby S C. Fabric draping simulation in composites manufacturing. Part I: Description and applications. Journal of advanced materials. 1997;**28**(3):29-35

[14] Dow NF, Tranfield G. Preliminary investigations of feasibility of weaving triaxial fabrics (Dow Weave). Textile Research Journal. 1970;**40**(11):986-998. DOI: 10.1177/004051757004001106

[15] Jenkins MJ. Good vibrations—materials swing into action. Materials World. 2000;**8**(6): 11-13

[16] Chou P JC, Ding D, Chen WH. Damping of moisute-absorbed composite rackes. Journal of Reinforced Plastics and Composites. 2000;**19**(11):848-862

[17] Mares M, Racu C, Buhu L, Buhu A. Mechanical properties of some biodegradable polymer matrix composite materials, with natural yarn fabrics reinforcement. Bulletin of the Polytechnic Institute of IASI. 2012;**LVIII,(LXII)**(3):101-108

[18] Mares M. et al. The influence of technological parameters on the mechanical properties of some biodegradable polymer matrix composite materials. Petroleum-Gas University of Ploiesti Bulletin. 2011;**LXIII**(1):141-146

Chapter 3

Multifunctional Foldable Knitted Structures: Fundamentals, Advances and Applications

Alenka Pavko-Čuden and Darja Rant

Additional information is available at the end of the chapter

http://dx.doi.org/10.5772/intechopen.69292

Abstract

Contemporary multifunctional textiles are based on hi-tech functionalization. Knitted structures can be relatively rapidly designed and produced in a variety of textures due to their composition of many interlacing loop elements and their combinations. Foldable weft-knitted structures exist in a wide range of forms from simple rolls, ribs, and pleats to more complex three-dimensional structures. They exhibit new kind of geometry and deformation mechanisms. Some of them exhibit auxetic potential. Foldable knitted structures are multifunctional and widely usable. They can be produced in a variety of structures, qualities, and dimensions: in panels, fully-fashioned, or seamless. Their possible application lies in different fields, such as fashionable and functional clothing, sportswear, medical care, packaging, interior design, sound and shock absorption, etc.

Keywords: knitted structures, textiles, clothing, foldable, collapsible, multifunctional

1. Introduction: new age of multifunctionality

Today, consumers are demanding textiles and clothing with high-performance properties, even in the traditional clothing and home textiles areas. Functional and visual appearance are very significant. Many textile and clothing producers develop products with innovative characteristics that can represent an important added value. Added value can be achieved by complex engineering design and by merging knowledge and skills of experts from various areas including craft, industrial design, materials, production technologies, marketing, psychology, ecology, etc. The properties and the characteristics that were initially developed for products for special use are nowadays often present in functional textiles for everyday use. In contemporary textiles and clothing, modern technologies play an important role. Textiles and clothing are not "just" textiles and clothing anymore. They represent an important hi-tech field even if in the people's minds, they are still often considered as simple traditional everyday objects. With contemporary multifunctional textiles based on hi-tech functionalization, a new textile tradition has been developing, slowly changing the rooted perception of textiles and clothing.

Functionalization of textiles aims to improve native properties as well as to impart new functions in the textile products [1]. For instance, with the selection of raw materials and the design of structural parameters of non-woven, woven and knitted structures, mechanical, permeability, and comfort features of textiles can be optimized and upgraded. Textile finishing can obtain new functional properties such as UV resistance, water repellency, flame retardancy, antibiotic, antistatic, antimicrobial activity, wrinkle recovery, etc. to the fabrics.

Functional properties can be defined as all the effects that are beyond the pure esthetic and decorative functions. As described above, they can be obtained either by:

- the raw material (characteristics of the polymer or additives before fiber forming);
- yarn, fabric or 3D textile construction; or
- textile finishing,

that is by material, mechanical, or chemical functionalization.

Multiple functions are often required, leading to what we can call multifunctional textiles [2]. The term "multifunctional material" is defined to be any material or material-based system which integrally combines two (or possibly more) properties, one of which is normally structural and the other functional. Both active and passive functionality are included [3].

The basic underlying technological need for the development of multifunctional materials is that solutions to particular problems or needs cannot always be found by using a simple combination of materials with different functions, and a technological barrier is reached. Real benefit will often only be found if true multifunctionality can be achieved [3].

The potential to exploit multifunctional materials spreads over a broad range of market sectors and products. Key areas are health care, security, energy, packaging, aerospace and transport, consumer friendly textiles and wearables, defense, etc. [3]. The multifunctionality of materials often occurs at scales from nano through macro and on various temporal and compositional levels [4].

The major barrier to the development of multifunctional materials and systems is, paradoxically, the very thing that gives them their advantage over combinations of single functions—multidisciplinarity. That is, the need to pull together and establish close and sustainable links between often disparate and closed disciplines, including materials scientists, chemists, physicists, engineers, biologists, physicians, and designers. Multifunctional materials have the potential to support the sustainability agenda. For example, multifunctional structures might be designed for re-use or recyclability [3]. After the first life cycle is over, the second one can start by up-cycling.

Multifunctional materials can be both naturally existing and specially engineered [5]. Many of them may draw inspiration from nature, where size and weight are often critical and multifunctionality is a necessity rather than a luxury [3]. Biological materials routinely contain sensing, healing, actuation, and other functions built into the primary structures of an organism [4]. Textiles can take advantage of the very sophisticated and highly efficient mechanisms with which nature protects itself in hazardous environments, based on extraordinary functionalities. Processes used for the functionalization of textile materials need to be increasingly environmental friendly. The use of natural resources, energy, and chemicals needs to be minimized [2].

We have entered the era of hybridization of materials and engineering techniques, for example, highly multifunctional materials, with folded materials and knitted structures among them. Multifunctional objects are as old as mankind, but after the period of excessive consumerism, it seems that we have finally focused on fewer but combined and more efficient multifunctional objects and processes. We are in the era which grants a new meaning to the multifunctionality.

Knitted structures, especially flat knitted structures, can be relatively rapidly designed and produced in a variety of textures due to their composition of many interlacing loop elements and their combinations (front and rear loop, transfer loop, tuck, miss, rack, etc.). Mechanical functionalization involves the design of the structural and geometrical parameters which influence the performance properties of foldable knitted structure. Chemical functionalization upgrades their performance to achieve the planned characteristics. Unlike weaving, flat knitting enables manufacturing of fully fashioned and seamless products. Shaping and real three-dimensional knitting expands the boundaries of their multifunctionality.

2. Foldability as a principle of collapsibility

Collapsibility is an elementary design principle applied to a great many everyday objects [6]. Often, we are not even aware of it. For example, every day, we fold our newspaper or book after reading and fold our clothes to put them in the wardrobe.

Size adjustment to meet functional requirements is a time-honored principle in nature, too. Animals downsize to hide, relax, rest and protect themselves, and upsize to brag, threaten, fly, fight, and court [6]. In modern times, many patents have been applied for collapsible objects with space saving as their primary added value. Minimalization of equipment is very important in some areas, for example, in storage, transport, medicine, aerospace application, etc.

2.1. Genuine or quasi-collapsibles

Collapsibles are smart manmade objects with the capacity to adjust in size to meet a practical need. They are functional doubles with two opposite states, one folded and passive and one (or more) unfolded and active. They grow and shrink, expand and contract, according to functional need. To give collapsibility to an object, its volume must be redistributed in one way or another to occupy less practical space. Practical space is space we want to free up for some other purpose. To qualify as a genuine collapsible, an object must be repeatedly collapsible and expandable. Object designed to fold or unfold only once does not qualify as collapsible. Moreover, two conditions must be met before a manmade collapsible is conceived and created. First, somebody must see advantage in reducing the size of a tool when it is not in use. And second, it must be mechanically possible to reduce the size of that tool [6].

Collapsability *per se* is never the purpose function of a tool. It is always a support function. Collapsibility may never be the most important function of a tool, but it is often the decisive factor when the buyer makes choice [6]. As mentioned above, a genuine collapsible has one-folded passive state and one or more unfolded active states. If the folded state is both active and passive (like a pair of scissors), then the object is not a genuine collapsible. Many active states and no passive state as well as no space saving define a quasi-collapsible [6].

Mollerup [6] defined twelve collapsibility principles. Most describe the action by which an object is collapsed. They include the most frequently applied methods of mechanical size reduction:

- stress,
- folding,
- creasing,
- bellows,
- assembling,
- hinging,
- rolling,
- sliding,
- nesting,
- inflation,
- fanning and
- concertina.

Most of the principles describe the action, while some of them (bellows and concertina) describe the structure. The differences between collapsibility principles are often indistinct. The boundary between collapsing by folding and collapsing by creasing (along pre-folded lines), for instance, is not always clear [6]. In some cases, the collapsibility is a result of two parallel principles.

In textiles and clothing, some principles are more common than the others. The stress and rolling are often applied for reducing the volume, for example, for a sleeping bag storage. Folding technique is used for packing clothes or for adjusting the sails surface. Creasing means folding along preset lines. Pleated window screens are both functional and decorative. Bellows collapsible racks can be used as camping equipment. Rolling is a basic principle of roller blinds operation. Fanning principle is named after a collapsible fan, a fashion accessory. Flap bags are closed by creasing, but their closing principle can be considered hinging as well; in this case, the boundary between the two principles is quite blurred.

From the presented examples, it can be seen that collapsibility can be achieved by many ways, including folding. In recent years, more and more designers of all disciplines have turned to folding to create a wide range of handmade and manufactured objects, both functional and decorative. A little time spent looking through design and style magazines will reveal a significant number of folded products, from apparel to lighting and from architecture to jewellery. Origami is one of the most vibrant buzzwords in contemporary design [7]. It is often used as a synonym for folded structures, in knitting, as well as in other textile techniques [8].

2.2. Self-folding structures

The science and technology associated with origami-inspired engineering are new and developing rapidly. It has evolved from esthetic pursuits to design folding structures across cultures and scales. The underlying principles of origami are very general, which has led to applications ranging from cardboard containers to deployable space structures which can be fabricated, assembled, stored, and morphed only through bending without any cutting and gluing. More recently, researchers have become interested in the use of active materials (i.e., those that convert various forms of energy into mechanical work) to effect the desired folding behavior. When used in a suitable geometry, active materials allow engineers to create self-folding structures. Such structures are capable of performing folding and/or unfolding operations without being kinematically manipulated by external forces or moments. This is advantageous for many applications, including space systems, underwater robotics, small-scale devices, and self-assembling systems [9, 10].

Self-folding in not exactly a new phenomenon. It frequently appears in nature for the efficient fabrication of structures but is seldom used in engineered systems. Recently, self-folding structures were developed, consisting of shape memory composites that are activated with uniform heating in an oven or a heated bath [11].

Self-folding also occurs in textiles. At the fiber level, it is shown as self-curling. In the nature, the curling property of wool (WO) results from its bilateral structure, where ortho and para cortex are arranged in asymmetrical, side-by-side order in the cross-section of the fiber. Wool fibers have, because of this difference, a helical crimped configuration. There are also man-made crimped fibers. There are two groups of spinning methods for producing bi-component fibers with self-crimping ability. In first group, there are methods where special equipment is needed to conjugate two different components together in a side by side order. In the second group of methods, a non-symmetrical character across the cross-section of the filaments is introduced to the filament on the classical spinning devices, without any special additional apparatus. It is also clear that the formation of crimps is a result of the bilateral structure of asymmetrically cooled yarns. The consequence of the bilateral structure is the formation of crimps after drawing [12].

In flat knitting, some links-links structures exhibit self-folding after exiting the take-down zone (**Figure 1**). The folded state represents the relaxed, that is, the passive state.

Origami as an inspiration for hi-tech engineered products has been studied in depth from the practical as well as from theoretical point of view. For instance, Peraza-Hernandez et al. [13] noted that modeling and analysis of origami structures allow for the understanding of

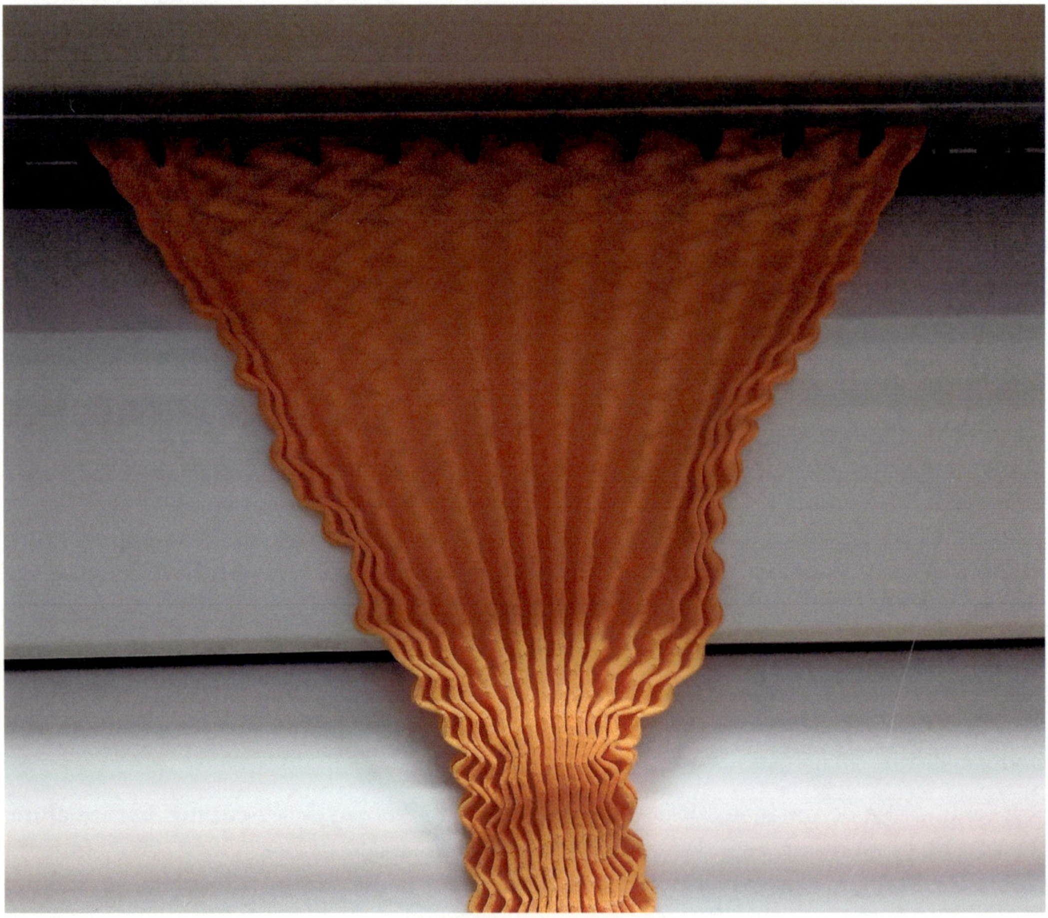

Figure 1. Self-folding of links-links structures after exiting the take-down zone.

their behavior and the development of computational tools for their design. They presented a novel model analogous to that for rigid origami, for origami structures having folds of non-zero surface area that exhibits higher order geometric continuity (termed smooth folds). Modeling of origami structures with smooth folds exhibiting elastic behavior is performed by determining the configuration of the structure that minimizes its total potential energy subject to the derived kinematic constraints. However, a generalized understanding of origami remains uncertain because of the differences between model predictions and experimental confirmations [13].

Smart materials can play a significant role in the realization of self-folding origami-inspired structures. Researchers have demonstrated self-folding behavior in many active material systems with inducing fields that include thermal, chemical, optical, electrical, and magnetic. Several combinations of materials, geometry, and inducing field are feasible, yielding an array of design options [9].

3. Multifunctional foldable textiles

3.1. Textile folding techniques

Fashion designers have long used pleated fabrics for esthetic effect and to introduce disguised fullness to women's clothing [6]. In textiles and clothing, foldable structures are a fundamental element of design. Rolls, folds, ribs pleats, and bubbles make a flat structure three-dimensional. Redistributed volume causes changes in esthetic appearance as well as in functional properties like thermal insulation, sound absorption, compression and support, strength, stiffness, handle, etc.

Origami-inspired folding of textiles can be performed by various techniques. Woven and nonwoven textiles usually exhibit folded look achieved by pressing or finishing. Folded textiles can also be formed by sewing together parallel stripes of fabric alternately on the face and the rare side. On the other hand, knitted products can be designed by integrating folds directly into the knitted structure. Creased or folded knits can involve a wide range of structures from simple ribs and pleats to more complex 3D structures [8]. Knitted pleats are worked into the fabric by varying the tension and knitting tight and loose rows, thus creating a fold line in the fabric. Young London-based Korean designer Hanjoo Kim demonstrates the creativity that can be achieved with the technique of pleating, adding both structure and movement to the fabric [14].

Many traditional textile and clothing objects have been foldable and multifunctional at the same time. In continuation, some examples of folding textiles and garments are presented. For instance, berets are made from felt. They are folded for storage and unfolded when used as a headdress. They are genuine collapsibles. Berets protect from cold and wind. They often represent various institutions as parts of their uniforms. In some periods, a beret was a fashion statement, also a revolutionary symbol.

Surgical masks are made from nonwovens. Usually, they consist of multiple layers; a filter is placed between nonwoven layers to stop bacteria from entering or exiting the mask. Most surgical masks feature pleats or folds allowing the user to expand the mask and cover the lower part of the face. As the surgical masks are made for a single use only, they are unfolded only once in their lifetime, so they are foldable by definition, but they are not considered as genuine collapsibles. Surgical masks protect transmission of body fluids from and to the wearer. They can be used as dust masks. Surgical masks with decorative designs are popular in countries in which they are worn in public to protect ordinary people from infections or allergies. They can also help to conceal a person's identity; pop singer Michael Jackson was often hiding behind a surgical mask when he appeared in public.

A fine example of capitalizing on mass-production technology can be seen in the work of Reiko Suno and Nuno fabrics, one of the most important textile design studios, founded in 1984. Nuno, meaning functional textile, specializes in creating unique fabrics. One of their products is Nuno Circle origami pleated bag, made of polyester that can be recycled [15]. The bag is constructed and sewn, then folded repeatedly at sharp angles and permanently pressed at 200°C in a special pleating process that is patent pending [16].

Japanese artist and designer Issey Miyake is considered revolutionary for his use of materials and its iconoclastic, conceptual approach to fashion. He blends traditional, historical elements of Japanese fashion, such as wrapping and folding, with cutting-edge technological innovation that have revolutionized fabric manufacturing. His designs demonstrate a desire to expand the potential for clothing outside of the purely functional. He is best known for the technique of pleating silk via a heat treatment, first used in his iconic collection "Pleats Please" in 1993 [17]. Even if Miyake is well known for his overarching esthetical-functional concept, in his case, the foldability is often a sophisticated visual effect rather than a (multi) functionality effect.

3.2. Foldable textiles for various applications

Textiles are most appreciated for their softness and pleasant touch; nevertheless, they can be used as substrates for hardening finishes or elements of composites. With appropriate coating, their behavior is similar to that of paper or other stiff foldable materials. Cutting edge technology is often inspired by past inventions. The same would be true for foldable textiles.

3.2.1. Interior and exterior textiles

Interior architecture and contemporary textiles have an odd, somewhat unresolved relationship. Although fabrics hold the potential to structure space, dampen sound, and emanate light, they are usually confined to a secondary role within the interior. Traditionally, interior textiles have been seen as soft furnishing and used as decorative accessories, but this is changing as materials such as glass filaments, carbon fibers, conductive wires, and metal mesh begin to replace architectural substrates. Although the new generation of interior fabrics are regarded as high-tech devices, they can be described as beautiful, too [18].

Many modern interior textiles have creased or pleated structure but in fact, they are not genuine collapsibles. For example, German architect Jürgen Mayer used pleated fabrics to create elliptical colons. The installation was designed for Nya Nordiska's exhibition at a design trade fair in Frankfurt [18]. The purpose of the pleats was purely esthetical, not functional. On the other hand, modern venetian blinds are made of textile composites and are genuine collapsibles [6].

Now, as in the past, a tent is a house for man on the move. Carried on truck, back, or bike, it takes up little practical space. Tent designers must accommodate a number of conflicting demands for easy handling, maximum shelter, and minimum weight, and they take great efforts to find new spatial and technical solutions to meet these demands [6]. Tents are genuine collapsibles.

3.2.2. *Sport textiles*

Backpackers wage a never ending struggle to minimize the weight and volume of their gear. One way of reducing volume is to stow a sleeping bag in a compression sack, which works by stress pressure. Traveler's mattress reduces for storage when the air is pressed out and held out by a stopper. When the stopper is removed, the mattress automatically takes in air and reshapes itself. This principle is stress pressure as well [6]. The flexibility of textiles allows clothes to be folded and stored when not in use or when prepared for traveling. Anoraks, raincoats, and wet-weather jackets can often be folded and stored into a sewn-in pocket.

All sailing boats are equipped with sails which are folded and unfolded to meet the weather and wind changes. Nobody takes as much interest in folding their collapsible tool as parachuters. For them, sloppy folding could be fatal [6].

A balaclava is a form of a cap designed to (partly) cover the face. Usually, only the eyes, mouth and nose, or the front of the face are unprotected. Some versions can be folded into a cap to cover just the top of the head or unfolded to cover the face or being used as a warming collar. Apart from skiing, balaclavas are used today as safety garments for firefighters and race drivers. They are made from a flame-resistant material and can contain a breathing apparatus.

3.2.3. *Transport textiles*

In cars, foldable sunshades made from knitted mesh are used for side windows. Their form is similar to the photographer's collapsible light reflector, invented in 1985 by John Riston. When twisted, the spring coils itself into three smaller rings, making it compact enough to stow away in a small bag [6].

Convertible cars can convert between an open-air mode and an enclosed one. They are equipped by a collapsible textile roof which is hinged to fold. The roofs of baby strollers fold in the same principle.

3.2.4. Contemporary ("technical") fashion

Nowadays, more and more non-textile materials and non-textile technologies are incorporated in textiles and clothing to contribute to their multifunctionality. Some examples of a complex integration of materials and supporting devices can also be found in history. Bustle was a support used to expand the fullness of the back of a woman's dress. It was worn under the skirt at the back, below the waist. Collapsible wire bustles were used in nineteenth century to facilitate seating. They were the early representatives of the "technical fashion."

Hussein Chalayan is regarded as an inventor, philosopher, and architect among fashion designers. He approaches his collections like conceptual artist, frequently interpreting in his designs socially relevant themes such as cultural identity, tradition, and migration [19]. He was one of the first designers to engage with technological systems, and many of his collections have pioneered garments that feature wireless technology, electrical circuitry and embedded connectors [18]. For his 2007 collection entitled "One hundred and eleven," he designed a true collapsible collection of six mechanical transforming dresses, expanding and folding to change the shape and silhouette of the garments with the aid of electronics. The hemlines were raising and lowering, the skirts were expanding and contracting.

3.2.5. Fashion accessories

Fans were once an essential feminine accessory. Foldable fans could be inspired by a peacock tail. They came into use after 1580. They functioned as a temperature controlling device by inducing an airflow over the skin. They could also be used as means for concealing identity [6, 20, 21].

An umbrella is a manmade adaptation to the changes in the weather [6]. It protects from rain, while the parasols were once used as a shield against the sun [20]. The supporting structure of umbrellas is hinged. So too are the legs of a collapsible tripod, though they may additionally expand and collapse by sliding [6]. Some umbrellas can be unfolded manually, while the others can spring open automatically. Umbrellas are genuine collapsibles.

The "chapeau claque" is a foldable hat. It is a genuine collapsible which folds flat for storage and folds out for use [6]. It could be a raw model for future clothes and accessories occupying more and more space in our wardrobes and inducing the need for bigger and bigger living space.

4. Foldable knitted structures

Basic interlacing loop elements of a knitted structure are loop, tuck, and float (**Figure 2**). Loops, tucks, and floats have a very different appearance on the front and back (rear) side of the knitted structure, respectively. Loops are made during a complete knitting process. Tucks are made in a tucking process when the yarn is bent but not intermeshed. Floats are made in a miss process when the knitting needles are out of action. Racking, that is, lateral movement of

http://dx.doi.org/10.5772/intechopen.69292

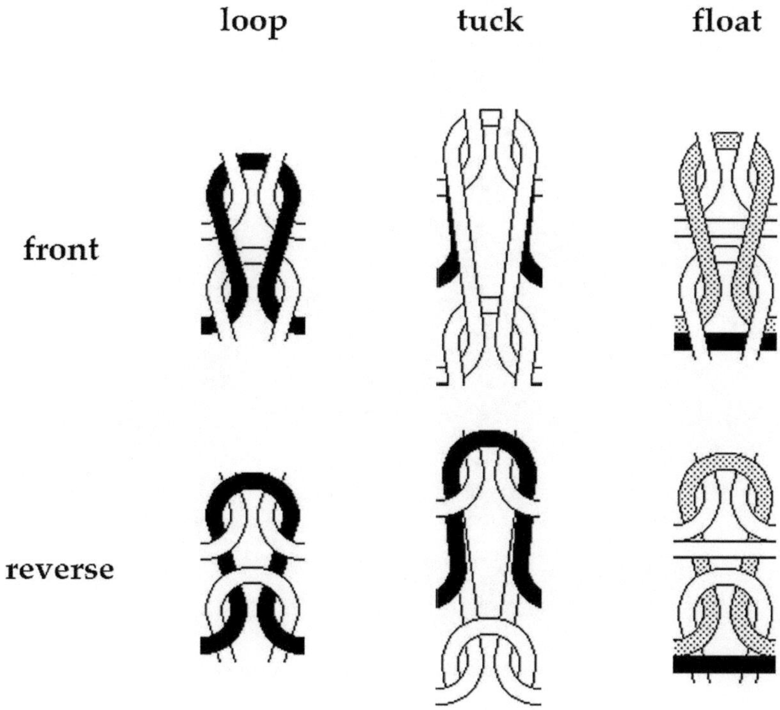

Figure 2. Interlacing loop elements.

the needle-beds results in inclined and/or crossed loops which leads to additional patterning possibilities including cable and aran patterns. Transferring loops from one needle to another in the same or opposing needle bed is used for shaping and for knitting complex-knitted structures.

Single-knitted structures are made on a single needle-bed. The front side of the structure is composed of front interlacing elements, while the back side of the structure is composed of reverse loops, tuck, and floats.

Double-knitted structures are made on two or more needle beds. There are front as well as reverse interlacing elements seen on both sides of the knitted structure.

The simplest double structures are ribs (**Figure 3**) composed of alternating wales of front and reverse interlacing loop elements, respectively. Each wale contains only one type of loop interlacing elements: front or reverse. A special arrangement of front and reverse loop wales results in folding into a pleated structure.

A links-links or a purl structure is made on double-bed knitting machines equipped with double-ended latch needles or on rib machines equipped with transfer needles which enable loop transfer from one needle-bed to the other, combined with needle-bed racking. Links-link structures are composed of front and reverse loop-interlacing elements alternating in both directions, along wales as well as courses.

Figure 3. Folding of ribbed structures with various rib widths made on the same number of knitting needles.

The speciality of knitting technology is the possibility of producing continuous fabrics, knitted panels, whole garment, or seamless products as well as basic and advanced two-dimensional and three-dimensional compositions including composite substrates. Knitting technology supports and upgrades other textile and non-textile technologies. Knitted structures can be combined with other materials to achieve optimal performance [22].

Knitting process allows the production of a vast range of structures. Foldable weft knitted structures exhibit new kind of potential, geometry, and deformation mechanisms. Creased or folded knits can involve a wide range of structures from simple rolls, ribs, pleats, and bubbles (**Figure 4**) to more complex three-dimensional structures. Links-link knitting enables manufacturing of very esthetically intriguing structures which are flat-knitted but crease and fold after relaxation, forming various textures and spatial patterns. Among them, links-link structures with zigzag or other geometrical patterns are particularly promising as they are rather simple to design and produce. Foldable weft-knitted fabrics have potential applications in different fields, such as functional clothing, sportswear, medical care, packaging, sound and shock absorption, etc. [8, 23].

4.1. Rolls

Edge curlings occur in plain-knitted fabrics owing to the unbalanced yarn bending moment existing in the three-dimensional nature of the structure. The curlings occur at the upper and lower edges of a piece of fabric toward the front side and at the left and at the right edges of the fabric toward the back side (**Figure 4**). The yarn in a loop wants to adopt a straight form, but it is prevented from doing so by neighboring loops. Thus, curling can start at the edges

Figure 4. Foldable knitted structures: roll, pleats, bubbles.

as there is no neighboring loop on one side to prevent curling. The edge curlings can create problems during the plain-knitted clothing goods processing. In addition to the problems, there are also positive effects or advantages provided by edge curlings, such as the upper side edge curling being used to form the neck of pullovers [24]. The polo pullover closure can consist of a knitted placket band partially rolled-up to form a welt.

Formation of a curling multiplies the thickness of the knitted edge which can result in adequate breaking strength increase. In the past, this led to specific regulations for the preparation of specially shaped knitted samples for the breaking strength testing.

4.2. Ribs and pleats

Knitting ribs requires two sets of needles operating in between each other so that wales of face stitches and wales of reverse stitches are knitted on each side of the fabric [25]. Ribbed structure has a vertical cord appearance because the face loop wales tend to move over and in front of the reverse loop wales. As the face loops show a reverse loop intermeshing on the other side, ribs have the appearance of the technical face of plain fabric on both sides until stretched to reveal the reverse loop wales in between [25].

Ribs fold by relaxing and unfold by stretching in the direction perpendicular to the fabric formation. Ribbed structure is one of the basic foldable knitted structures which can be used for elastic beginnings and welts for knitwear like socks and pullovers. Due to its elastic recovery potential, ribs can be used for tight as well as compression garments. The folding effect depends on repeat (**Figure 3**), course and wale density of the structure, and material composition.

Knitted pleats (**Figure 4**) can be manufactured on double bed flat knitting machines with special needle arrangements on both beds. As every knitted structure tends to curl or fold toward the reverse side along the length of the fabric, inactive needle in either needle bed creates a wale of single reverse loops within the double structure. The knit folds toward the fabric side with the wale of reverse loops. Various arrangements of inactive needles result in various types of pleats: knife, accordion, or rolled pleats. The width of the vertical ribs influences the folding effect, for example, the extent of width-wise shrinkage.

4.3. Links-link knitted structures

Links-links or purl is the only structure having certain wales containing both face and reverse meshed loops. Although in the past structures of this type were knitted only on flat bed and double cylinder purl machines employing double-ended latch needles, electronically controlled V-bed flat machines with rib loop transfer and racking facilities are now used [25].

The simplest links structure is 1 × 1, which consists of alternate courses of all face and all reverse loops and is produced by the needles knitting in one bed and then transferring over to the other bed to knit the next course. Its lateral stretch is equal to plain, but its length-wise elasticity is almost double. When relaxed, the face loop courses cover the reverse loop courses, making it twice as thick as plain structure [25].

By alternating multiple courses of all face and all reverse loops, horizontal links-link ribs are manufactured. They can fold by relaxing and unfold by stretching in the direction of fabric formation. The height of the horizontal links-link ribs influences the folding effect, for example, the extent of length-wise shrinkage.

There are also other, patterned foldable links-links structures, composed of geometrical elements like, square, diamond, zigzag lines, etc. (**Figure 5**). Each geometrical element is composed of the same type of loops, front, or reverse.

Figure 5. Various foldable links-links knitted structures.

5. Characteristics and properties of links-link foldable structures

5.1. Folding effect

In weft links-links knitting, folding is based on the structural disequilibrium of face and reverse loops which causes the fabric to crease, contract, fold, and form into a three-dimensional structure after the take down and relaxation. Foldable structures shrink in both course and wale directions. Under applied strain in the horizontal or vertical direction, three-dimensional foldable structures smooth into a flat fabric, creases unfold and the structure expands in both directions [8, 26]. In order to establish the influence of yarn composition and structural parameters such as size of the repeating unit cell on the folding effect of links-link knitted fabrics, a set of experiments was designed by Pavko-Čuden et al. The number of the same type of loops in a course direction needed to initiate the structure folding effect was also investigated [8].

Two series of samples with 12 different links-link zigzag structures were knitted. The first series of samples (**Figures 6** and **7**) was produced with varying unit cell sizes both in the course and wale direction (from the smallest 2 × 2 to the biggest 24 × 24 loops in a unit cell). The second series (**Figures 8** and **9**) was produced with varying widths of a zigzag line in a unit cell with a constant number of courses (from the narrower 2 × 24 to the widest 24 × 24 loops in a unit cell). Both series of knitted structures were produced on the knitting machine Shima Seiki SES 122 RT of gage12E. Two yarns of different material compositions were used: WO/PAN and CV/PA.

After relaxation, the dimensions of the samples in horizontal and vertical direction were measured. Considering the repeat sizes (number of loops in each repeat), the width/loop and the height/loop values were calculated to estimate the folding potential of the samples. Smaller value of width/loop or height/loop, respectively, means better folding of the structure.

Zigzag-knitted structures with varying repeating unit cell sizes (1st series of samples) fold in both the course and the wale direction. The folding effect appears in all sizes of a unit cell for structures produced from both yarns, except for the smallest zigzag-knitted structure with a 2 × 2 repeat. The result shows that these structures are more closely folded in the course direction rather than in the wale direction. As the width/loop and height/loop values do not vary substantially for different repeat sizes, it can be assumed that the folding effect of these structures in both directions is good.

The width/loop values of the zigzag structures with varying unit cell sizes increase with decreasing unit cell sizes. It signifies that knitted structures with smaller unit cell sizes are

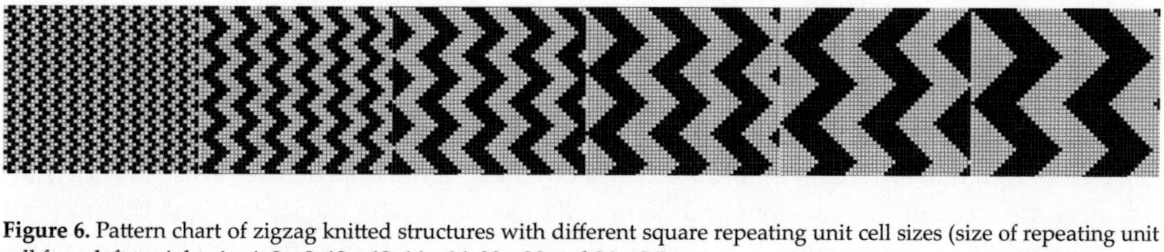

Figure 6. Pattern chart of zigzag knitted structures with different square repeating unit cell sizes (size of repeating unit cell from left to right: 4 × 4, 8 × 8, 12 × 12, 16 × 16, 20 × 20 and 24 × 24).

Figure 7. Zigzag knitted structures with different square repeating unit cell sizes made of WO/PAN yarn (size of repeating unit cell from left to right: 4 × 4, 8 × 8, 16 × 16 and 24 × 24).

Figure 8. Pattern chart of zigzag knitted structures with different widths of zigzag line (size of repeating unit cell from left to right: 4 × 24, 8 × 24, 12 × 24, 16 × 24, 20 × 24 and 24 × 24).

Figure 9. Zigzag knitted structures with different widths of zigzag line and a constant number of courses made of CV/PA yarn (size of repeating unit cell from left to right: 8 × 24, 14 × 24, 16 × 24 and 24 × 24).

less folded in the course direction. Nevertheless, even the knitted structure with the 4 × 4 unit cell size is well folded. The height/loop values of zigzag-knitted structures with varying unit cell sizes are more or less constant and do not change much with the variation in the unit cell size. Knitted structures made of WO/PAN fold better in the wale direction, while in the course direction, the folding is better for the knitted structures made of CV/PA yarn.

Regarding the knitted structures with varying widths of a zigzag line and a constant number of courses (2nd series of samples), the differences in the width/loop values are more substantial; decreasing the width of the zigzag line quickly increases the width/loop values. The 4 × 24 knitted structure is very poorly folded. The height/loop value of structures with varying widths of zigzag lines does not vary considerably.

Structures made of WO/PAN evenly fold from the unit cell sizes 24 × 24 to 16 × 24, while structures made of CV/PA evenly fold from the unit cell size 24 × 24 to 14 × 24. Knitted structures with narrower zigzag lines are poorly folded. Hence, the width of zigzag lines for the structures with a constant number of courses substantially influences the ability to fold. Certain number of the same type of loops in a course direction is needed to provide sufficient folding effect in the relaxation process after knitting. Merely six (for the CV/PA structures), seven (for the WO/PAN structures), or less loops in this kind of arrangement of face and reverse loops do not provide sufficient folding force; thin zigzag lines are too narrow for the fabric to fold.

The same experimental design was used for the preparation of another two series of samples, knitted from the same yarns, but on the coarser machine Stoll CMS 502HP E 2.5.2 [23]. It was concluded that the yarn material composition, the size of the repeating unit cell, and the width of the zigzag line at the constant number of courses in the repeat significantly influence the folding ability of links-links knitted structure.

By experimenting, it was found that the variation of the couliering depth setting of the front and rear cam to different levels leads to even more intriguing foldable structure (**Figure 10**). More open areas contribute to the increased porosity and permeability, while the compact areas contribute to the stiffness and therefore folding potential of the structure. Systematical investigation of links-link weft knitted structures with zigzag patterns in various repeat sizes showed that folds of the viscose/polyamide samples with higher repeats tend to sag. It was also established that additional PA monofilament stiffens the structure and increases the folding tendency (**Figure 11**). The investigation showed that the loop density has a significant impact on the folding tendency of the knitted structure.

5.2. Compression resistance

Compression is one of the important fabric properties, in addition to friction, bending, tension, and shear. Compression may be defined as a decrease in intrinsic thickness with an appropriate increase in pressure. Intrinsic thickness is the thickness of space occupied by a fabric subjected to barely perceptible pressure. The applied compressive force allows the yarn to undergo deformation non-linearly, resulting in the change of fabric thickness [27]. The relationship between the applied force (normal to fabric plane) per unit area and the resulting fabric thickness can be obtained with a simple test. The tested fabric specimen is placed horizontally on a platter and subsequently loaded and unloaded with a presser foot. The fabric thickness, which is the distance between the presser foot and the platter, is recorded as the function of applied pressure. This pressure-thickness relationship describes the compression characteristic of the fabric. The pressure-thickness curve of textile fabrics in lateral compression is highly non-linear [28].

Figure 10. Foldable links-links zigzag structure knitted with different setting of the front and rear cams (left—folded, right—unfolded).

Figure 11. Stiffening folded structure by interlacing additional PA filament (left—no filament, right—incorporated filament).

The compressibility behavior of knitted and woven fabrics depends on a number of factors, that is, fabric tightness, fabric surface irregularity, yarn hairiness, yarn compressibility, fiber material etc. The analyses of the pressure-thickness relationship performed by Alimaa et al. [29] demonstrated a very prominent effect in terms of the knit construction and yarn structure. It was observed that fabric compressibility primarily depended on the fiber material.

The composition properties of knitted fabrics were also essentially due to their knit constructions. Moreover, the loop length determined the compressibility of knitted fabrics to a great extent [29].

In order to evaluate the behavior of links-link weft knitted fabrics with a zigzag structure under compression, the same two series of samples as for the folding effect were examined (Section 5.1). The examined structures could potentially be used as a packaging and mechanical damage protection material. The influence of yarn material composition and the structural parameters of foldable structures, such as repeat size, that is, width/height ratio, on the compression properties of foldable links-link knitted structures were examined in order to evaluate their adequacy for compression-resistant materials. The compression behavior of examined foldable structures was compared to some selected actual compression materials used in packaging, such as bubble foil, textured rubber foam, and woolen felt [30]. Since the width of zigzag lines of the second series of samples distinctively influences the folding ability, only the compression properties of fully folded knitted structures were examined.

First, the compression test was performed on a dynamometer INSTRON 5567 based on the Bluehill® software compression application module. The speed of the movable pressure foot was 0.3 mm/s. The compression load was read when the distance between the movable pressure foot and the fixed flatten reached 1 mm. A circular pressure foot with 9 cm in diameter was used. Ten measurements of the maximum compressive load at the compressed thickness of the knitted structure t_{compr} = 1 mm for each sample were performed.

Then, the thickness of knitted structures was measured in a separate testing procedure where the speed of the movable pressure foot was adjusted to 0.1 mm/s to detect the contact of the movable pressure foot and the fabric surface. When the compression load was detected, the distance between the clamps was read from the compression curve. Five measurements for each sample were performed.

It can be concluded that the fiber and yarn type contribute substantially to the compression behavior of samples as they were all knitted on the same machine and under the same conditions to eliminate the influence of the knitting process.

The maximum compression load of CV/PA foldable knitted structures exceeds the maximum compression load of WO/PAN structures with comparable repeats, although the knitted structures made from WO/PAN yarn are thicker than the comparable knitted structures made from CV/PA yarn. The maximum compression load and fabric thickness decrease with the repeat reduction.

The decrease in compressive stress was not linear with the knitted structure repeat reduction. For the CV/PA knitted structures, the decrease got more distinctive with smaller repeats, whereas for the WO/PAN knitted structures, the compressive stress decreased more in the case of bigger repeats.

The compressive stress decreased similarly for the structures with the repeat widths from 24 loops to 18 loops. The foldable knitted structures with the repeat widths smaller than 18 loops differed substantially; the structures with the square repeat which were all fully folded (first series of samples) exhibited a gradual compressive stress decrease, while for the structures designed with various widths of zigzag ribs (second series of samples), an instant drop of compressive stress was evident. These structures did not fully fold when the rib width was smaller than approximately seven loops (**Figure 12**).

Foldable knitted structures are compressible. To compress the examined foldable knitted structures to the thickness of 1 mm, substantial loads are required.

5.3. Auxetic potential

Auxetic materials are different from most conventional materials in that they exhibit a negative Poisson's ratio (NPR). They expand laterally when stretched and contract laterally when compressed [31]. This counterintuitive behavior gives auxetic materials various beneficial effects, such as enhanced shear stiffness, increased plane strain fracture toughness, increased indentation resistance, and improved energy absorption properties [32, 33]. As the Poisson's ratio is a physical parameter that is independent of the material scales, the auxetic behavior can be achieved at any material level, from molecular to macroscopic [34, 35].

Flat-knitting technology can provide a simple but highly effective way of fabricating auxetic fabrics from conventional yarns. 3D geometry of specially developed links-link knitted

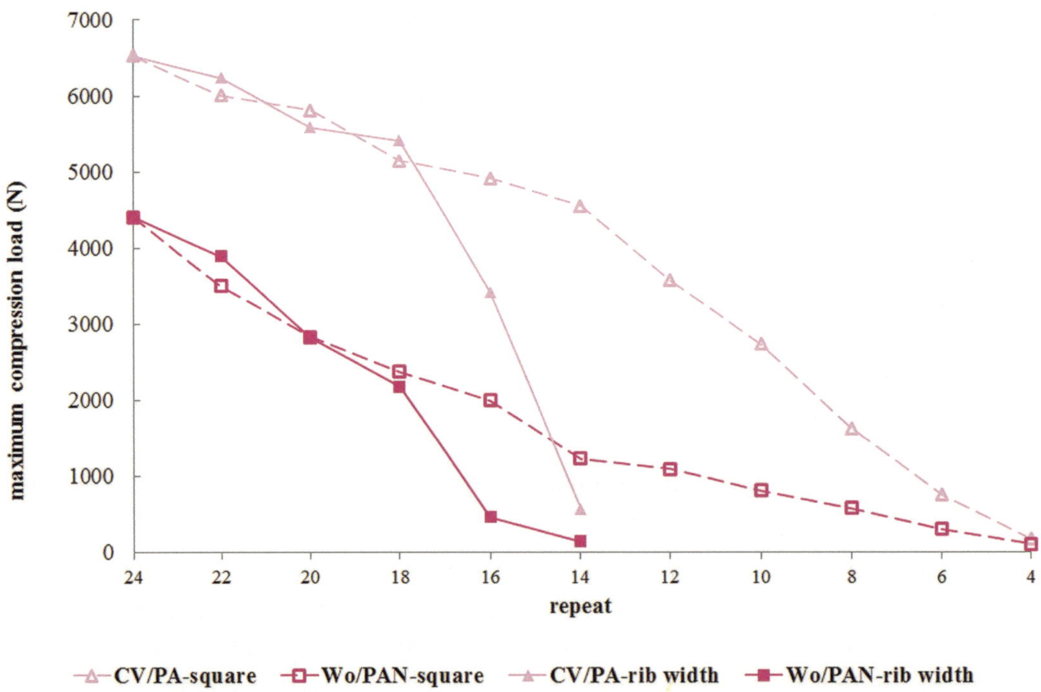

Figure 12. Comparison of compressive stress at maximum compression load for foldable zigzag knitted structures with various repeats: square—same number of courses and wales in repeat; rib width—different widths of zigzag ribs and constant number of courses in repeat [30].

structures enables a new deformation mechanism called "opening of the folded structure." The fabrics that are more closely folded can result in a smaller opening angle and consequently have higher NPR values. A negative Poisson's ratio as low as $v = -0.5$ was reported in the scientific literature for such structures; nevertheless, the examined fabrics exhibited auxetic effect only in one direction [26].

Buhai et al. [36] stated that the structural parameters of the knits are influenced by two factors: yarn fineness and stitch length. They concluded that the auxetic effect is influenced by the stitch length values; as these values are lower, the fabric is tighter and maintains its shape better. The auxetic effect increases as the values of the stitch length decrease.

The purpose of the investigation of the foldable links-link knitted structures by Drol and Pavko-Čuden [37] was to compare auxetic properties of foldable links-link knitted fabrics made of different yarns, on flat knitting machines with different gages, different densities of knitted structure, and different repeats. Foldable zigzag-ribbed structures with auxetic potential were produced from three different conventional yarns. The yarn selection was based on the material composition, which affects the elasticity and stiffness of the yarn and thus the anticipated rigidity, stability and folding of the zigzag rib knit structure with auxetic potential. Due to the expected rigidity, cotton and multifilament viscose yarn were selected, while due to the expected extension and elastic properties of knitted fabrics, a yarn made of a wool and acrylic mix was selected. Different sizes of basic geometric units of knitting, that is, width and height of the zigzag ribs were achieved by knitting on knitting machines with different gage, while different compactness/stiffness of the knitted structures was achieved by knitting in different densities, that is, by setting different couliering depths. The samples were knitted on a flat-knitting machine STOLL CMS 340 TC with gage 8E and flat multigage knitting machine STOLL CMS 340 TC E6.2 that allows knitting with gage 6E (every second needle knitting) and 12E (all needles knitting). Knitted samples were prepared in two densities, respectively, by two positions of the couliering depth for each course density. Auxetic effect of the fabrics was determined on the basis of measurements of the fabrics' dimensions during extension in the knitted courses direction and by Poisson's ratio calculation (**Figure 13**).

It was found that in most cases, the samples exhibit the highest auxetic effect at 60–90% extension. Knitting with 45° inclination of ribs exhibits the best folding tendency. Fabrics produced on knitting machines with finer gage exhibit higher auxetic effect. Material composition and knitting machine gage have a great impact on the Poisson's ratio of foldable links-link knitted fabrics with zigzag ribs.

5.4. Sound insulation

During recent years, the subject of noise has received increasing amount of attention to the scientists, technologists, and public as a whole. For a healthy and a pleasant environment, controlling the sound hazards is an important issue. There is a medical evidence that the human body will take sound as "pollution" if the ambient sound levels exceeds 65 dB. This sound pollution leads to significant health problems including hypertension, dizziness, depression, sleep disturbance, hearing loss, decrease in productivity/learning ability/scholastic performance,

Figure 13. Measuring dimensional changes for evaluating auxetic potential.

increase in stress-related hormones, and most commonly, loss of hearing. Therefore, unwanted and uncontrolled noise should be reduced using noise barriers and noise absorbers. Properly designed textile materials may be considered as noise control elements in a wide range of applications, including wall cladding, acoustic barriers, and acoustic ceilings [38–40]. Sound absorbing materials are commonly used to soften the acoustic environment of a closed volume by reducing the amplitude of the reflected waves. Many natural and manmade raw materials have been used as sound absorbers.

Two methods for measuring acoustical properties of fabric materials have been most common: the impedance tube method and acoustic chamber method. The impedance tube method uses rather small test samples, it is faster and generally reproducible, while for the acoustic chamber method, large reverberation rooms and large test samples are used.

To evaluate the sound absorbing potential of foldable link-links knitted structures, different three-dimensional flat weft knitted fabrics made from various yarns were produced (**Figure 14**), including 100% wool and 50% wool/50% PAN as basic yarns, and 100% polyamide filament which was added to basic yarns as reinforcement thread [41]. The thickness and the mass/unit area of the knitted samples were measured.

For the finishing of knitted fabrics, two hardening agents were used: Tubicoat A 41 (CHT, Bezema, Switzerland) and Beaippret liquid (CHT Bezema Switzerland). After finishing, the handle and esthetic look (change of color, color uniformity, dimensional stability) of the samples were evaluated by a survey. The Tubicoat finishing agent in 40% concentration exhibited the best handle and esthetic appearance. The results of the survey revealed that the knitted

Figure 14. Presentation of foldable links-links structure as used for sound absorption test (left – folded, right – extended).

structure presented in **Figure 13** was the most appropriate for sound absorbing panels. The sound absorption coefficient of the foldable knits was measured by the impedance tube (Kundt tube) method. The results were compared to the sound insulation performance of the commercial woolen felt.

The results showed that the selected foldable structure can be used as sound insulating material as it exhibits good sound absorption properties. Further interesting and attractive foldable structures can be developed for sound insulation with similar thickness, compactness, and mass/unit area. Woolen structure showed the best acoustic properties, followed by wool/PAN structure reinforced with poliyamide filament. 100% wool structure with added polyamide thread and 50% wool/50% PAN structures also exhibited good acoustic properties. Hardening agent significantly reduced the sound absorption coefficient. Incorporating nylon into knitted structure improved the stiffness of the structure; it decreased the sound absorption coefficient in the case of woolen structure and increased the sound absorption coefficient in the case of wool/PAN structure.

5.5. Antibacterial properties

The suitability of foldable seamless knits for the storage of bread and bakery products has been studied by Rant, Pavko-Cuden and Tomsic. It was assessed by testing the antibacterial properties of the selected foldable knitted structures made from various raw materials (Tomsic B 2017, personal communication, January 23).

First, the soil burial test according to SIST EN ISO 11721-1:2001 standard was performed for determining the resistances of foldable knitted fabrics made from various yarns to microbiological deterioration. Apart from a basic single knitted structure, a zigzag links-links knitted structure with 4 × 4 square repeating unit cell was selected. The selected yarns were made of 100% combed cotton, 100% cotton with added polyamide filament, flax/viscose blend (FLAX-CV), natural bamboo/modal blend (BAMB-MOD-N), carded 100% cotton, 100% lyocell (LIO), 100% viscose, wool/viscose blend, lyocell/viscose blend, lyocell/cotton blend (LIO-BW), cashmere/polyamide/viscose (CASH-PA-CV), and 100% wool. The time of exposure was 12 days. Afterward the knitted samples were carefully removed from the soil. The samples were rinsed in water, sterilized by soaking in 70% ethanol at room temperature for 30 min and dried. Initially, the deterioration of samples was visually assessed. After that, the rate of biodegradation of the examined samples was determined by color measurement with a spectrophotometer SPECTRAFLASH 600 PLUS (Datacolor International, USA) using the CIELAB color system. ΔL^* values of the buried and unburied samples were determined and compared.

The results of the study showed that the ΔL^* values of the foldable knitted structures were lower than for the single-knitted structures. The visual assessment of the samples also showed that in most cases, the foldable knitted structures were less deteriorated than the single-knitted structures. The results proved that the foldable links-links knitted structures have better antibacterial properties compared to the single structures.

According to the results, foldable links-links knitted samples from flax/viscose blend (FLAX-CV), natural bamboo/modal blend (BAMB-MOD-N), lyocell/cotton blend (LIO-BW),

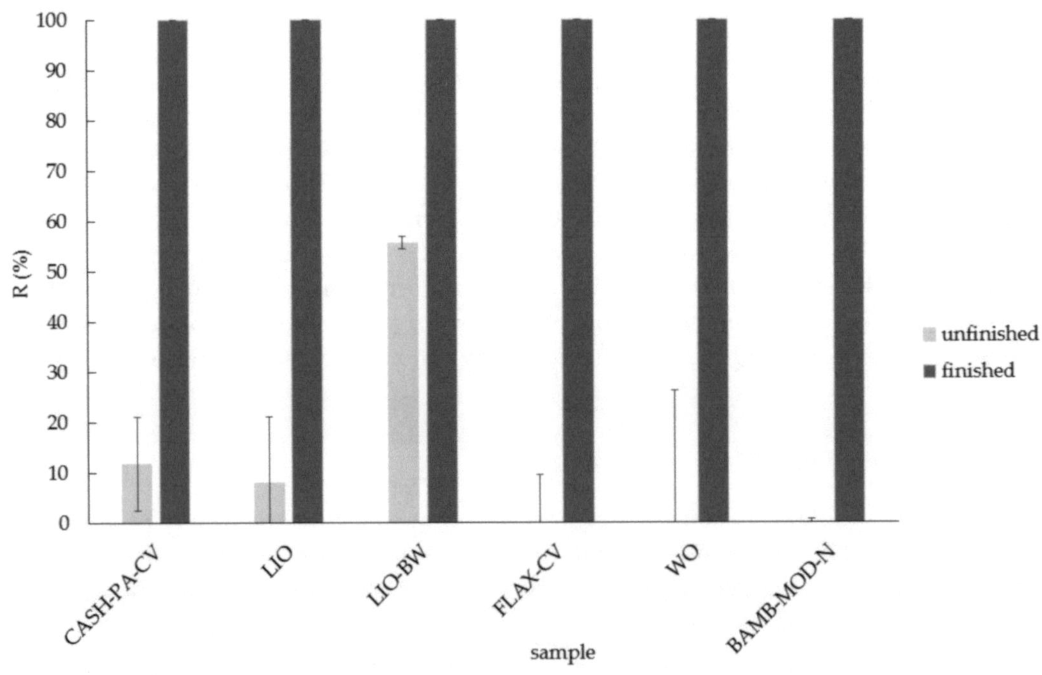

Figure 15. Antibacterial properties of foldable knitted structures made from various yarns before and after treatment with antibacterial finishing agent.

100% lyocell (LIO), cashmere/polyamide/viscose (CASH-PA-CV), and 100% wool (WO) were selected for further investigation.

In order to achieve antibacterial activity, Bioshield Excalibur (Izinova Ltd., Bled) finishing agent was selected, which is chemically alkyl dimethyl (3-trimethoxysilylpropyl) ammonium chloride. Its antibacterial activity is based on bio-barrier formation mechanism. Antibacterial activity of the examined knitted samples was estimated by determination of bacterial reduction according to the ASTM E 2149-01 standard method. Bacterial reduction of the samples was evaluated against Gram-negative bacteria *Escherichia coli* (ATCC 25922).

The results (**Figure 15**) showed that the selected antibacterial finish was fully effective, reflecting in complete growth reduction of the tested bacteria. On the other hand, untreated foldable links-links knitted samples exhibited rather important differences in the reduction of *E. coli* growth. The samples made from lyocell/cotton blend exhibited 55% reduction, the reduction of cashmere/polyamide/viscose blend was 11%, while the reduction of the rest of the samples was less than 10%. Better results for the samples made from natural bamboo/modal yarn were expected due to the original antibacterial properties of the natural bamboo fibers.

6. Potential multifunctional application of foldable knitted structures

Foldable weft-knitted fabrics have a big potential for applications in different fields. They can be used for fashionable knitwear, functional garments, and para-garments (textiles that are put onto the body but are not garments: for example, certain medical textiles) as well as for various non-clothing purposes.

They are compressible. They can be usable in functional knitwear to alleviate pressure on certain parts of the body. As advanced unconventional packaging material, they could protect fragile objects from mechanical damage. They could also replace or complement spacer knitted fabrics in padding, for example in medical cushions, seat covers for the automotive industry, mattresses, etc.

In interior design, they can be used as room partitions when mounted on a frame for thermal insulation and protection from light. They can replace embossed rubber foam which is often used as sound insulating material. For that purpose, they can be designed as modular elements available in various colors and textures to be fixed on walls and ceilings. Due to their featured texture, they can be used as interior decoration. Dimensional changes caused by folding and unfolding make them suitable for lampshades if made from transparent and stiff material or for multifunctional shading-lightening objects if LED elements are incorporated (**Figure 16**).

A combination of innovative mechanical functionalization, that is, selection of optimal raw materials and intriguing knitted structures, and successive high-performance chemical functionalization could result in new, reusable, and recyclable household food packaging, for example for storage of bread and bakery products or fruit.

Figure 16. Multifunctional interior textiles: collapsible sound diffusors/shades/lights (simulation) designed by Andrej Vilar, Slovenia.

The current problem with the wider use of foldable knitted structures lies in their bi-axial extensibility which is substantially reduced when the foldable panels are sewn together. If joined in the extended state, the folding ability is hindered and the esthetic appearance of the sewn together piece is affected. If sewn together in a relaxed state, the extensibility is significantly reduced. Flat knitting allows seamless production, therefore seamless foldable links-links knitted products can be manufactured. For that reason, seamless foldable-knitted structures can be used for casual and functional knitwear as well as for fashion accessories.

7. Future scope

Textiles are transcending their traditional functions and are morphing into uniquely tactile interfaces through which broader sensory stimulus can be perceived. Because fibers, fabrics, and textile techniques are becoming seamlessly integrated with technology, textiles represent an interconnected collective that links many disciplines. Our world seams polarized around sensory extremes: hard and soft, protection and exposure, intransigence, and tactility. As textiles embrace new types of fibers and fulfill new roles, they bridge these polarities better than any other material. Textiles are dramatically transforming the world around us, and as they do so, they also inspire radical new visions for the future [18].

Foldable knitted structures exhibit bi-polar attributes. They can be folded or extended, transparent or opaque, flat or curved. They can be multifunctional. Foldable knits are very complex structures, exhibiting unusual behavior; for example, auxetic behavior was proved for some zigzag foldable links-links structures. Therefore, in-depth research into their characteristics, above all into the impact of material, structural and geometrical parameters, finishing, repeated use, textile care, etc. is anticipated.

3D printing is a rapidly emerging technology which is often claimed to be the base for a new industrial revolution. Integration of 3D printed elements and flat knitted structures into novel textile composites, including foldable composites is another area of the future research.

8. Conclusion

Foldable knitted structures are multifunctional and widely usable. They can be produced in a variety of structures, qualities and dimensions: in panels, fully-fashioned or seamless. They exhibit a supreme esthetics and have a big potential for the use in multiple areas. Some of the foldable knitted structures exhibit auxetic properties which have lately become a subject of extensive research. Foldable knitted structures, links-links knits among them, can be considered a promising development line of sustainable hi-tech knitting technology and design, especially if combined with other technologies. The development of sustainable, re-usable and up-cyclable, and genuine self-folding knitted collapsibles should be encouraged.

Author details

Alenka Pavko-Čuden* and Darja Rant

*Address all correspondence to: alenka.cuden@ntf.uni-lj.si

Department of Textiles, Graphic Arts Design, Faculty of Natural Sciences Engineering, University of Ljubljana, Ljubljana, Slovenia

References

[1] SciTechnol. Multifunctional textiles. Available from: https://www.scitechnol.com/scholarly/multifunctional-materials-journals-articles-ppts-list.php [Accessed 17th February 2017]

[2] Roy PK, Malik T, Sinha TK. Technical textiles. Functionalisation of textiles—Future perspectives. Available from: http://www.technicaltextile.net/articles/clothing-textiles/detail.aspx?article_id=3174&pageno=1 [Accessed 10th February 2017]

[3] Materials UK. Multifunctional materials. Reports 2006. Available from: http://www.matuk.co.uk/docs/Multifunct.pdf [Accessed 11th February 2017]

[4] Nemat-Nasser S, Nemat-Nasser SC, Plaisted T, Starr A, Amirkhizi AV. Multifunctional materials in biomimetics. In: Bar-Cohen Y, editor. Biologically Inspired Technologies. Boca Ranton: CRC Press; 2005. pp. 309-341.

[5] The Pennsylvania State University. EME 807: Technologies for sustainability. Multifunctional materials and their impact on sustainability systems. Available from: https://www.e-education.psu.edu/eme807/node/698 [Accessed 12th February 2017]

[6] Mollerup P. Collapsibles: A Design Album of Space Saving Objects. London: Thames & Hudson; 2001

[7] Jackson P. Folding Techniques for Designers. From Sheet to Form. London: Laurence King Publishing; 2011

[8] Pavko-Čuden A, Rant D, Vilar A. Origami knits. Knitting International. 2014;**120**(2):28-31

[9] Peraza-Hernandez EA, Hartl DJ, Malak RJ Jr, Lagoudas DC. Origami-inspired active structures: A synthesis and review. Smart Materials and Structures. 2014;**23**(9):094001. Available from: DOI: 10.1088/0964-1726/23/9/094001 [Accessed 13th February 2017]

[10] An N, Li M, Zhou J. Predicting origami-inspired programmable self-folding of hydrogel trilayers. Letter. Smart Materials and Structures. 2016;**25**(11LT02):8 pp. Available from: DOI: 10.1088/0964-1726/25/11/11LT02 [Accessed 14th February 2017]

[11] Tolley MT, Felton SM, Miyashita S, An B, Rus D, Wood RJ. Self-folding origami: Shape memory composites activated by uniform heating. Smart Materials and Structures. 2014;**23**(9):094006. Available from: DOI: 10.1088/0964-1726/23/9/094006 [Accessed 18th February 2017]

[12] Demšar A, Sluga F. Crimped polypropylene yarns. Kovine Zlitine Tehnologije. 1999;**33**(6): 523-526

[13] Peraza-Hernandez EA, Hartl DJ, Aklemanb E, Lagoudasa DC. Modeling and analysis of origami structures with smooth folds. Computer-Aided Design. 2016;**89**:93-106

[14] Brown C. Knitwear Design. London: Laurence King Publishing; 2013

[15] Black S.. Eco-chic. The Fashion Paradox. London: Black Dog Publishing; 2011

[16] Nuno. Nuno bag. Available from: https://www.nunoonline.com/products/nuno-bag-crystal-organdy-origami-sky-blue [Accessed 17th February 2017]

[17] Smith MO, Kubler A. Art/Fashion in the 21st Century. London: Thames & Hudson; 2013

[18] Quinn B. Textile Futures. Fashion, Design and Technology. Oxford, New York: Berg; 2010

[19] Schmidt P, Stattman N. Unfolded. Paper in Design, Art, Architecture and Industry. Basel, Boston, Berlin: Birkhäuser; 2009

[20] Ginsburg M, Hart A, Mendes VD. In: Rothstein, editor. 400 Years of Fashion. London: V&A Publishing; 1984

[21] Cumming V, Cunnington CV, Cunnington PE. The Dictionary of Fashion History. Oxford, New York: Berg; 1976

[22] Rant D, Pavko-Čuden A. Foldable weft knitted structures with auxetic potential. In: Textiles of the Future: ISKA 2013: Proceedings of the International Symposium in Knitting and Apparel, ISKA 2013, 21-22 June 2013, Iaşi, Romania. Iaşi: Faculty of Textiles—Leather and Industrial Management; 2013. pp. 45-49

[23] Blaga M, Ciobanu AR, Pavko-Čuden A, Rant D. Production of foldable weft knitted structures with auxetic potential on electronic flat knitting machines. Melliand International. 2013;**19**(4):220-223

[24] Kurbak Aand Ekmen O. Basic studies for modeling complex weft knitted fabric structures. Part I: A geometrical model for widthwise curlings of plain knitted fabrics. Textile Research Journal. 2008;**78**(3):198-208. Available from: DOI: 10.1177/0040517507082352 [Accessed 18th February 2017]

[25] Spencer DJ. Knitting Technology. A Comprehensive Handbook and Practical Guide. Cambridge: Woodhead Publishing; 2001

[26] Liu Y, Hu H, Lam JKC, Liu S. Negative Poisson's ratio weft-knitted fabrics. Textile Research Journal. 2010;**80**(9):856-863

[27] Murthyguru. Novel approach to study compression properties in textiles. Autex Research Journal. 2005;**5**(4):176-193

[28] Huang W, Ghosh TK. Online measurement of fabric mechanical properties: Compressional behavior. In: IEEE Annual Textile, Fiber and Film Industry Technical Conference: Proceedings, 4-6 May 1999, Atlanta, USA. Available from: DOI: 10.1109/TEXCON.1999.766185 [Accessed 10th February 2017]

[29] Alimaa D, Matsuo T, Nakajima M, Takahashi M, Yamada EY. Pressure-thickness relation and compression mechanism of plain and rib knitted fabrics. Journal of Textile Machinery Society of Japan (English edition). 2000;**46**(1):7-10

[30] Rant D, Ciobanu R, Blaga M, Pavko-Čuden A. Compression of foldable links-links knitted structures. Texktil ve Konfeksiyon. 2014;**24**(4):349-355

[31] Evans KE, Nkansah MA, Hutchinson IJ, Rogers SC. Molecular network design. Nature. 1991;**353**(6340):124. Available from: DOI: 10.1038/353124a0 [Accessed 5th May 2015]

[32] Evans KE, Alderson A. Auxetic materials: Functional materials and structures from lateral thinking!. Advanced Materials. 2000;**12**(9):617-628

[33] Yang W, Li ZM, Shi W, Xie BH, Yang MB. Review on auxetic materials. Journal of Materials Science. 2004;**39**(10):3269-3279

[34] Grima JN, Farrugia PS, Gatt R, Attard D. On the auxetic properties of rotating rhombi and parallelograms: A preliminary investigation. Physica Status Solidi (b). 2008;**245**(3):521-529

[35] Liu Y, Hu H. A review on auxetic structures and polymeric materials. Scientific Research and Essays. 2010;**5**(10):1052-1063

[36] Buhai C, Blaga M, Ciobanu R, Budulan C. Development of auxetic weft knitted fabrics. In: 46th International Congress IFKT: Proceedings, 6-8 September 2012, Sinaia, Romania. Iaşi: Faculty of Textiles—Leather and Industrial Management; 2013. pp. 758-763

[37] Drol P, Pavko-Čuden A. Auxetic properties of zigzag links-links knitted structures with various material composition. In: Simončič B, Tomšič B, editors. Napredne tekstilije za družbo in gospodarstvo: Proceedings, Zbornik izvlečkov, 45. simpozij o novostih v tekstilstvu, Ljubljana, 3. June 2014, Ljubljana, Slovenia. Ljubljana: Naravoslovnotehniška fakulteta, Oddelek za tekstilstvo; **2014**. p. 30

[38] Shoshani Y, Yakubov Y. Numerical assessment of maximal absorption coefficients for nonwoven fiber webs. Applied Acoustics. 2000;**59**(1):77-87

[39] Hong O. Hearing loss among operating engineers in American construction industry. International Archives of Occupational and Environmental Health. 2005;**78**(7):565-574

[40] Mikolajczyk H, Cieslewicz A. Auditory threshold and the degree of its temporary and permanent shifts in the textile industry workers. Medycyna Pracy. 1982;**33**(1-3):57-64.

[41] Zakrajšek Š, Vilar A, Tomšič B, Simončič B, Sadar A, Pavko-Čuden A. 3D foldable knitted structures for sound absorption. In: CORTEP 2016: Proceedings: CORTEP 2016, 16th Romanian Textiles and Leather Conference, 27-29 October 2016, Iaşi, Romania. Iaşi: Performantica; 2016. pp. 216-221

Section 3

Textiles for Composite Applications

Chapter 4

Textile Reinforced Structural Composites for Advanced Applications

Nesrin Sahbaz Karaduman, Yekta Karaduman,
Huseyin Ozdemir and Gokce Ozdemir

Additional information is available at the end of the chapter

http://dx.doi.org/10.5772/intechopen.68245

Abstract

Textile-reinforced composites are increasingly used in various industries such as aerospace, construction, automotive, medicine, and sports due to their distinctive advantages over traditional materials such as metals and ceramics. Fiber-reinforced composite materials are lightweight, stiff, and strong. They have good fatigue and impact resistance. Their directional and overall properties can be tailored to fulfill specific needs of different end uses by changing constituent material types and fabrication parameters such as fiber volume fraction and fiber architecture. A variety of fiber architectures can be obtained by using two- (2D) and three-dimensional (3D) fabric production techniques such as weaving, knitting, braiding, stitching, and nonwoven methods. Each fiber architecture/textile form results in a specific configuration of mechanical and performance properties of the resulting composites and determines the end-use possibilities and product range. This chapter highlights the constituent materials, fabric formation techniques, production methods, as well as application areas of textile-reinforced composites. Fiber and matrix materials used for the production of composite materials are outlined. Various textile production methods used for the formation of textile preforms are explained. Composite fabrication methods are introduced. Engineering properties of textile composites are reviewed with regard to specific application areas. The latest developments and future challenges for textile-reinforced composites are presented.

Keywords: textile-reinforced composites, fiber architectures, textile production methods, advanced applications, engineering properties

1. Introduction

A composite material can be defined as a combination of a reinforcement material and a matrix. The properties of a composite are superior to the properties of the individual components. Reinforcement is the main load-bearing component and is responsible for the strength and stiffness of the composite material. Reinforcement forms include fibers, particles, and flakes. Matrix, on the other hand, keeps the reinforcement in a given orientation and protects it from chemical and physical damage. It is also responsible for the homogeneous distribution of an applied load between the reinforcement elements. Composite materials are generally employed when traditional materials such as metals, ceramics, and polymers do not satisfy the specific requirements of a certain application. One of the main advantages of composite materials is that they can be designed to obtain a wide range of properties by altering the type and ratios of constituent materials, their orientations, process parameters, and so on. Composites also have high mechanical properties with a low weight which makes them ideal materials for automotive and aerospace applications. Other advantages of composites include high fatigue resistance, toughness, thermal conductivity, and corrosion resistance. The main disadvantage of composites is the high processing costs which limit their wide-scale usage.

Fiber reinforcements basically cover short and continuous fibers and textile fabrics. Textile-reinforced composites consist of a textile form as the reinforcement phase and usually a polymer for the matrix phase. 2D or 3D woven fabrics, knitted fabrics, stitched fabrics, braids, nonwovens, and multiaxial fabrics can be used as textile materials. Each of these textile forms has its own fiber architecture and combination of properties such as strength, stiffness, flexibility, and toughness which are translated to composite performance to a certain extent. Different textile architectures offer an enormous potential for designing the composite properties. The first textile structure to be used in composite reinforcement was 2D biaxial fabric to produce carbon-carbon composites for aerospace applications. However, multilayered 2D fabric structures suffer from poor interlaminar properties and damage tolerance due to lack of through-the-thickness fibers (z-fibers). 3D textile fabrics with through-the-thickness fibers have improved interlaminar strength and damage tolerance. Therefore, 3D textile composites have attracted great interest in the aerospace industry since the 1960s in order to produce structural parts that can withstand multidirectional mechanical and thermal stresses [1]. Advantages of 3D textile-reinforced composites are their high toughness, damage tolerance, structural integrity and handleability of the reinforcing material, and suitability for net-shape manufacturing. Today, composites reinforced with 2D and 3D fabrics are in common use in various industries including aerospace, construction, automotive, sports, and medicine.

This chapter reviews the fabrication, properties, and application areas of textile-reinforced composites. Fiber and matrix types used for composite production are presented. Various textile forms and their production methods are outlined. Properties and performance of textile composites are reviewed with regard to specific application areas. The future possibilities and challenges for textile-reinforced composites are discussed.

2. Materials and structures used

2.1. Fibers

2.1.1. Carbon fibers

Carbon fibers are one of the oldest and most common classes of high-performance fibers used in composite production. The most important carbon fiber types with respect to carbon source are polyacrylonitrile (PAN)-based and pitch-based carbon fibers. Other types include vapor-grown fibers and carbon nanotubes. Graphite fiber refers to a specific member of carbon fibers whose atomic structure is similar to that of carbon; both consist of sheets of carbon atoms arranged in a regular hexagonal pattern (graphene sheets). The only difference is that in graphite, adjacent aromatic sheets overlap with one carbon atom at the center of each hexagon. **Table 1** lists the mechanical properties of selected carbon fibers [2].

PAN-based carbon fibers account for most of the carbon fibers (≅90%) in commercial use. The first step in PAN-based carbon fiber production is the preparation of a suitable precursor which is critical for the quality of the resulting fibers. For this purpose, acrylonitrile monomers are mixed with plasticized acrylic comonomers and a catalyst, such as itaconic acid or methacrylic acid in a reactor. Free radicals are formed within the molecular structure of acrylonitrile by continuous stirring of the ingredients which leads to polymerization. The obtained acrylonitrile in powder form is then dissolved in either organic or aqueous solvents to prepare the spin "dope." The next step is the spinning of PAN precursor fibers. Either wet spinning or dry spinning techniques can be employed. Wet spinning is the preferred method for the production of high-performance fibers. In the wet spinning process, the spin dope is forced through a spinneret into a coagulating bath, and then stretching is applied to form the fibers. In dry spinning technique, the spin dope enters a hot gas chamber after passing through the spinneret. The obtained PAN precursor fibers are then oiled, dried, and wound

Type	Manufacturer	Product name	Tensile strength (GPa)	Young's modulus (GPa)	Strain to failure (%)
PAN	Toray	*T300*	3.53	230	1.5
		T1000	7.06	294	2.0
		M55J	3.92	540	0.7
	Hercules	*IM7*	5.30	276	1.8
GP-Pitch	Kureha	*KCF200*	0.85	42	2.1
HP-Pitch	BP Amoco	*Thornel P25*	1.40	140	1.0
		Thornel P75	2.00	500	0.4
		Thornel P120	2.20	820	0.2

Table 1. Mechanical properties of selected carbon fibers [2].

onto bobbins. These fibers are fed through a series of ovens at a temperature of 200–300°C for the oxidation step which combines oxygen molecules from the air with the PAN fibers and causes the cross-linking of polymer chains. The oxidized PAN fibers contain approximately 50–65% carbon molecules together with hydrogen, nitrogen, and oxygen. After oxidation, PAN fibers are subjected to carbonization in an oxygen-free atmosphere by going through a series of specially designed furnaces with progressively increasing temperatures (from 700–800 to 1200–1500°C). Fibers must be kept under tension throughout the production process. The resulting carbon fibers contain more than 90% of carbon. The main difference between carbon and graphite is that the former is carbonized at about 1315°C and contains 93–95% of carbon, whereas the latter is graphitized at 1900–2480°C and contains more than 99% of carbon. PAN-based carbon fibers are the material of choice to obtain high-strength composites.

Pitch-based carbon fibers fall into two categories such as the low-strength general-purpose fibers and high-performance fibers. Low-strength fibers are produced from isotropic pitch which is obtained from high-boiling fractions of crude oil. Two different spinning techniques such as centrifugal spinning and melt blowing can be utilized to produce low-strength carbon fibers. In centrifugal spinning, molten pitch is forced through small holes in a rotating bowl. The pitch stream is converted into fibers by centrifugal forces. The fibers are obtained in the form of tows or mats [2]. In melt blowing, on the other hand, molten stream of pitch is extruded into a high-velocity gas stream which converts the pitch into fiber form [3]. High-performance fibers are produced from mesophase pitch by melt spinning process followed by stabilization and carbonization. Mesophase pitch, a liquid crystalline material, is synthesized from pure aromatic hydrocarbons such as naphthalene and methylnaphthalene. Mesophase pitch has high purity and aromaticity which leads to high orientation in the final material [4]. Liquid crystalline materials readily orient during fiber formation and create a high degree of molecular orientation which leads to fibers with high moduli and thermal conductivity [5].

2.1.2. Aramid fibers

The development of aramid fibers (aromatic polyamides) dates back to as early as the 1960s when DuPont released its meta-aramid fiber with product name Nomex® in 1967. Later, in 1971, the same company developed and commercialized a para-aramid fiber, Kevlar®, with high strength and modulus [6]. Then, in the late 1980s, another para-aramid fiber, Twaron® was developed (Twaron® is a registered product of Teijin). Aromatic copolyamide fiber Technora® which is a more flexible fiber having high tenacity appeared on the market in 1987 [7].

Aromatic polyamides are synthesized using aromatic diamines and diacids or diacid chlorides [8]. There are two different well-established methods for the polymerization of aramid fibers such as low-temperature (<50°C) polycondensation [9] and direct polycondensation in solution using phosphites in the presence of metal salts [10]. Low-temperature polycondensation is generally preferred for the production of meta- and para-aramid fibers because of its high efficacy. Aramid fibers can be spun by using wet spinning or dry-jet wet spinning techniques. The wet spinning technique is used to produce meta-aramid fibers. It results in a semicrystalline fiber with partially oriented molecular chains. The dry-jet wet spinning technique is more effective in producing highly oriented high-strength/high-modulus fibers and usually preferred to produce para-aramid fibers. In this technique, the anisotropic polymer dope is forced through spinnerets at approximately 100°C through a narrow air gap into cold water (0–4°C). The spinnerets and air gap ensure the alignment of liquid crystal domains and result in highly crystalline and oriented aramid fibers. The attenuated filaments are washed, neutralized, and dried to obtain high-strength and high-modulus aramid fibers. The as-spun fibers are subjected to heat treatment under tension to further improve their tenacity and modulus [11].

Meta-aramids such as Nomex® are highly resistant to temperature, chemical degradation, and abrasion. They are generally used for heat-resistant workwear and firefighter clothing. Para-aramids such as Kevlar®, on the other hand, have high tensile strength, modulus, and toughness. Applications of para-aramid fibers include composite reinforcement, ballistic protection, wire and cable, protective gloves, and so on.

2.1.3. Glass fibers

Modern glass fibers that we know today were discovered in the early 1930s when Dale Kleist of Owens-Illinois Glass Company accidentally produced glass fibers during his attempts to seal architectural glass blocks together by melting and spraying glass. This crucial breakthrough paved the way for the mass production of insulation-quality glass fibers. In 1938, Owens-Illinois joined with Corning Glass Company to form Owens-Corning Fiberglass Corporation which is still one of the leading glass fiber manufacturers today. During the same years, they patented the product "Fiberglas" which is the origin of generic reference to fiberglass. Later developments led to the mass production of continuous glass filaments for composite reinforcement and other advanced applications [12].

The main ingredient of all glass fibers is silica (SiO_2). Other ingredients such as Al_2O_3 (aluminum oxide), CaO (calcium oxide), and MgO (magnesium oxide) are incorporated for additional functionality and process viability. For example, B_2O_3 (boron oxide) is added to increase the margin between the melting and crystallization temperatures of E-glass in order to avoid nozzle clogging during the fiber formation step. Another example is S-glass which contains a higher percentage of SiO_2 for an enhanced tensile strength. The production of E-glass fiber takes place in five basic steps such as mixing, melting, fiber formation, coating, and drying/packaging. In the first step, the starting materials are thoroughly mixed with the aid of an automated blender. Extreme care must be taken when weighing and adding the ingredients since the slightest deviation could affect the properties of the resulting fibers. Then, the mixture is melted at a temperature of about 1400°C in a sectional refractory furnace. The next step is the fiber formation which involves extrusion and attenuation. In extrusion, the molten glass from furnace is delivered through a bushing with very fine orifices. Molten glass streams coming out of the bushing are cooled by water jets and attenuated by mechanically drawing with the aid of a high-speed winder in order to obtain the fine glass filaments with a diameter between 4 and 34 μm. Then, size is applied which typically contains lubricants for abrasion protection and adhesion promoters or coupling agents to increase the fiber/matrix adhesion in composite materials. Glass fibers can be produced in various forms such as rovings, chopped strand mats, and milled fibers [13]. There are various types of glass fibers such as E-glass, i.e., general-purpose glass for low electrical conductivity, S-glass for high strength,

R-glass for high strength and acid corrosion resistance, Te-glass for high strength at elevated temperatures, D-glass for low dielectric constant, ECR-glass for high corrosion resistance, and ultrapure silica fibers, hollow fibers, and trilobal fibers for more advanced applications [14]. **Table 2** lists the specific properties of glass and other fibers in comparison to conventional materials [15].

2.1.4. Ultrahigh-molecular-weight polyethylene (UHMWPE) fibers

Polyethylene (PE) is produced through polymerization of ethylene monomers via either free radical polymerization or ionic polymerization. Free radical polymerization results in low-density polyethylene (LDPE) which has a branched structure and low mechanical properties. Ionic polymerization of PE leads to the formation of linear chains with little or no branching with a high level of crystallinity [16]. This structure is referred to as high-density polyethylene (HDPE) and is used for the manufacture of high-performance PE fibers. Another important characteristic of PE fibers is the molecular weight. PE with low molecular weight is of no interest in fiber production since it has low mechanical properties and melting point. As the molecular weight increases, mechanical properties and thermal stability increase as a consequence of enhanced molecular entanglement and intermolecular interactions. PEs with molecular weight in the range of 10^4–10^5 Da are used to produce commercial products like injection-molded plastics, beverage container films, and melt-spun PE fibers used in high-tenacity ropes. When the molecular weight reaches approximately 2–6 million Da, PE is referred to as ultrahigh-molecular-weight polyethylene (UHMWPE). This structure is composed of very long chains of PE with a very high level of orientation and crystallinity. UHMWPE has

	Young's modulus (GPa)	**Tensile strength (GPa)**	**Density (g/cm³)**	**Specific modulus (Mm)**	**Specific strength (km)**	**Failure strain (%)**	**Fiber diameter (μm)**
E-Glass	72	1.5–3.0	2.55	2.8–4.8	58–117	1.8–3.2	10–20
S-Glass	87	3.5	2.5	3.5	140	4.0	12
S2-Glass	86	4.0	2.49	3.5	161	5.4	10
Carbon	220–350	2.3–3.7	1.8–2.0	12–18	130–190	0.7–1.7	7
Aramid	60–180	2.65–3.45	1.44–1.47	4.0–12.2	180–235	4–1.9	12
PBT	250	2.4	1.5	17.0	160	1.0	20
PE	60–120	1–3	1.0	6–12	100–300	–	–
Steel	210	0.34–2.1	7.8	2.7	4.3–27	–	–
Aluminum	70	0.14–0.62	2.7	2.6	5–22	–	–
Bulk glass	60	0.05–0.07	2.6	2.3	1.9–2.7	0.08–0.12	–
Epoxy	2–3.5	0.05–0.09	1.2	0.16–0.29	4–7.5	1.5–6	–
HDPE	1.3	0.027	0.96	0.135	2.8	–	–

Table 2. Specific properties of glass and other fibers in comparison to conventional materials [15].

high strength-to-weight ratio as well as high resistance to abrasion, chemicals, and fatigue. Commercial UHMWPE fibers include DSM's Dyneema and Honeywell's Spectra fibers.

UHMWPE fibers are manufactured in a gel spinning process which enables draw ratios in excess of 50× [16]. The PE gel precursor is formed by swelling UHMWPE in solvent at high temperatures [17]. The gel state reduces the level of polymer entanglement, thus enhancing the drawability of the polymer. It has been found that 5–8 wt% polymer in solution results in optimal drawability [18]. The gel is extruded through a quenching bath which is followed by heating and drawing. The optimum drawing temperatures are above 110°C at which the alpha relaxation occurs promoting the unraveling of folded chains and facilitating the orientation of the crystals [16]. UHMWPE fibers are used in a range of applications such as body armor and composite helmets, climbing ropes, cordages, sails, etc.

2.1.5. Ceramic fibers

Ceramic fibers in continuous form are commercially available in two different classes such as oxide fibers and nonoxide fibers. Oxide fibers are based on the alumina-silica (Al_2O_3-SiO_2) system and on α-alumina (α-Al_2O3), whereas the nonoxide fibers are based on β-phase silicon carbide (SiC) [19]. Two important ceramic fibers in commercial production are silicon carbide (SiC) and aluminum oxide or alumina (Al_2O_3); other types are available on a much smaller scale. Ceramic fibers are polycrystalline in structure and known for their outstanding temperature resistance and strength retention at high temperatures. For example, SiC retains its strength above 650°C, and Al_2O_3 up to about 1370°C. Ceramic fibers are mainly used as reinforcement in metal and ceramic matrix composites. Ceramic fibers can be manufactured by using either of the two techniques such as chemical vapor deposition (CVD) and spinning. In CVD, ceramic is vapor-deposited on a heated substrate usually made of tungsten or carbon filaments. For example, silane and hydrogen gases are reacted in a tubular glass reactor at about 1300°C to produce β-SiC vapors which is then deposited on a tungsten or carbon substrate filament. Spinning technique involves the preparation of a precursor polymer which is then converted into a precursor filament by using conventional spinning techniques such as melt, dry, or wet spinning or the sol-gel method. This precursor filament is finally converted cohesively to the desired ceramic structure via controlled pyrolysis [13].

2.2. Matrices

2.2.1. Thermosetting matrices

Thermosetting polymers have long been used as matrix materials for textile composites, and they are far more common compared with thermoplastics. Processing with thermosetting matrices takes place in two basic steps such as the application of the liquid resin to the textile material and curing the resin in the presence of heat and pressure. The resin hardens gradually upon the application of heat owing to cross-linking of its molecules. The processing temperature typically ranges from room temperature to about 200°C [20]. The curing process is irreversible unlike thermoplastics. The most important thermosetting matrices include epoxy, unsaturated polyester (PES), vinylester, and phenolics. Advantages of thermosets include low processing temperatures, low viscosity/high wettability of the resins, high mechanical properties, high-temperature resistance, and good strength retention at high temperatures.

2.2.2. Thermoplastic matrices

Thermoplastic matrices melt when they are heated and harden upon cooling. Thermoplastic composites are produced in a heat-form-cool cycle, and thus they can potentially be recycled at the end of life. The process is reversible unlike thermosets. Thermoplastics such as polypropylene (PP), polyethylene (PE), polyamide (PA), polyester (PET), polystyrene (PS), and polyether ether ketone (PEEK) are increasingly used in reinforced composites. Among the numerous advantages of thermoplastics over thermosets are lower material and manufacturing costs, short processing times, unlimited shelf-life, reprocessability/recyclability, fewer safety risks, thermoformability, ductility, high toughness, and environmental tolerance [20].

2.3. Textile reinforcement forms

2.3.1. Two-dimensional (2D) fabrics

2.3.1.1. 2D woven fabrics

A woven fabric consists of two or more sets of yarns interlaced together to form a continuous 2D surface. The most common 2D woven fabric is the biaxial (orthogonal) fabric which is composed of longitudinal (warp) yarns and transverse (weft or filling) yarns interlaced at right angles [21]. **Figure 1** shows the schematic view of a loom used for the fabrication of woven fabrics [21]. Basically, four distinct motions are required to complete the fabric production cycle such as (1) shedding, (2) filling insertion, (3) beat-up, and (4) warp and fabric control. Shedding involves the movement of harnesses which contain warp yarns in order to open a path for weft insertion. The vertical movement of harnesses is achieved by using separate cams. In addition to the cam system, there are other mechanisms for shedding control. For example, the dobby head mechanism uses a maximum of 24 harnesses, thus enabling the control of interlacing 24 different groups of warp yarns. Jacquard looms, on the other hand, enable the control of every individual yarn across the fabric width and hence provide greatly enhanced design possibilities. The number of harness frames depends on the weave type. For example, two harness frames are used for plain-weave fabrics. Other weave types used in composites such as twill, basket, and satin require more harness frames. **Figure 2** shows the basic weave types used for composite reinforcement [22]. For the filling insertion, filling yarn is fed through the shed opening across the fabric width. There are different alternatives for carrying the filling yarn such as projectiles, rapiers, air jets, and water jets. The final operation is beat-up which incorporates the filling yarn into the fabric structure. A wire grate called a reed is used for this purpose. Biaxial woven fabric has a good dimensional stability and balanced properties in the fabric plane. Another advantage of this fabric type is the ease of handling and low fabrication cost. Disadvantages include poor in-plane shear resistance, lack of through-the-thickness reinforcement, and poor fiber-to-fabric tensile strength translation

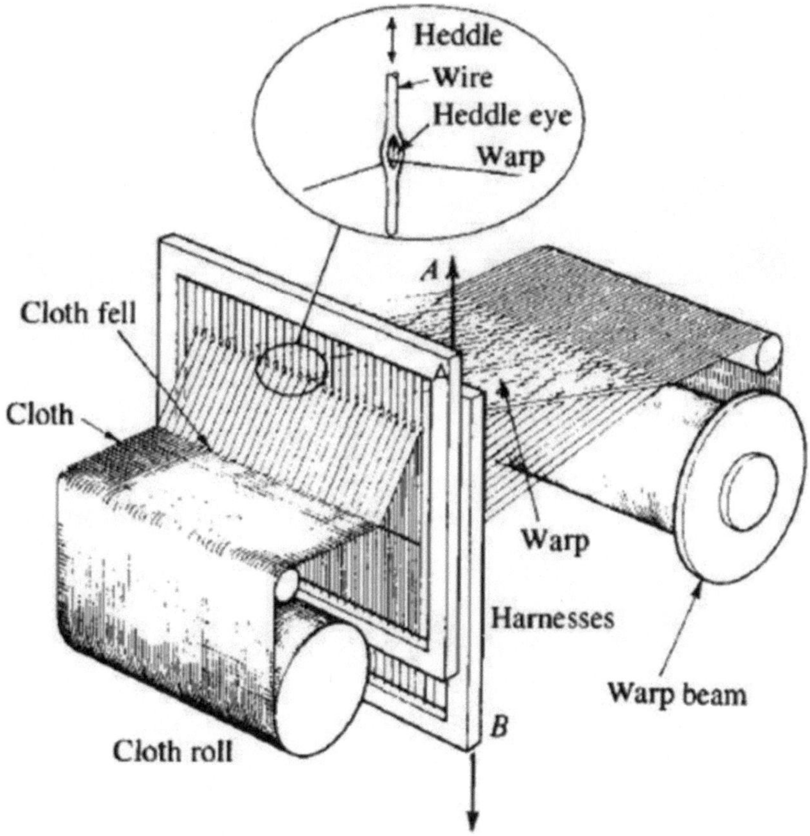

Figure 1. Schematic view of a weaving loom [21].

efficiency due to yarn crimp. Another type of 2D woven fabrics is triaxial fabric which consists of three yarn sets interlaced together at 60° angles. Triaxial fabrics are more isotropic and possess higher in-plane shear rigidity compared with the orthogonal woven fabrics [23].

2.3.1.2. 2D knitted fabrics

A knitted fabric is constituted from interlacing loops in transverse or longitudinal direction. There are two basic types of 2D knitted fabrics such as weft knitted and warp knitted fabrics. In weft knitting, the loops are formed in widthwise direction by a single yarn. In warp knitting, loops are formed in lengthwise (**Figure 3**) [24]. There are four primary base weft knitted structures such as plain, rib, interlock, and purl from which all the other weft knitted structures are derived. Horizontal rows of knitted loops are termed "courses," whereas the vertical columns of intermeshed loops are termed "wales." The length of the yarn in a loop is referred to as "stitch length." The total number of loops in a given fabric area is termed "stitch density" which is a very important fabric parameter for composite applications. Stitch density is the product of the number of courses in one inch and the number of wales in one inch. The most

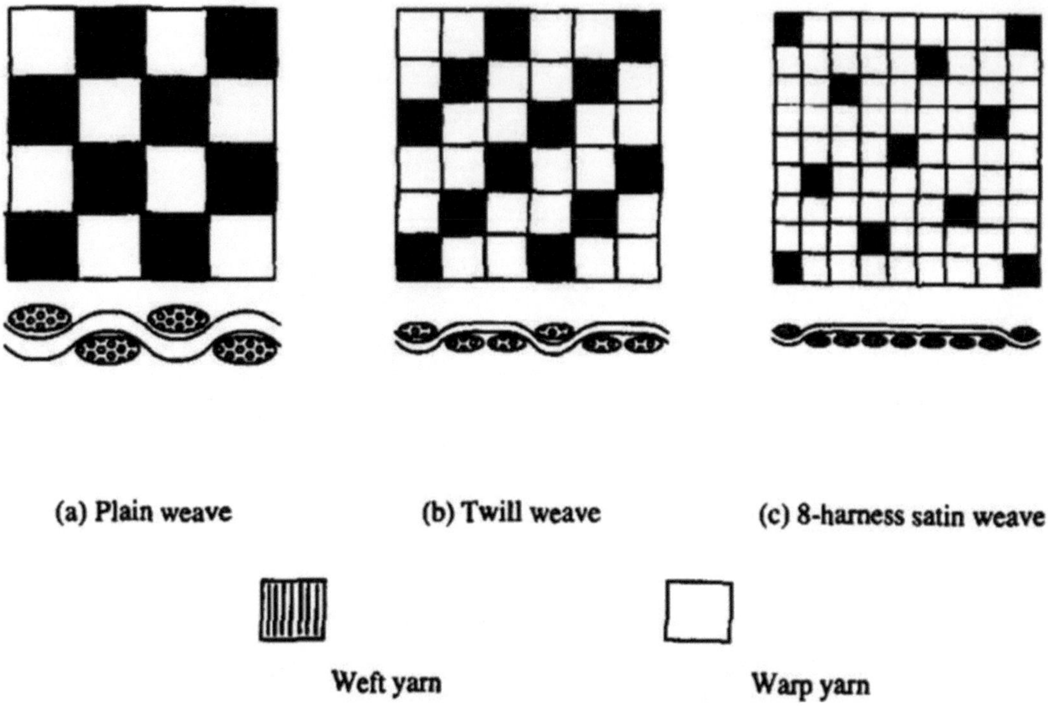

Figure 2. Basic weave types: (a) plain weave, (b) twill weave, and (c) eight-harness satin weave [22].

important elements of the knitting machine are the yarn supply unit which consists of the yarn package or beam accommodation, tensioning devices, yarn feed control, and yarn feed carriers; knitting elements; the fabric takeaway mechanism, and quality control systems such as stop motions and fault detectors [24].

Weft and warp knitted fabrics can readily extend in all directions and are thus suitable for the manufacture of complex-shaped composite parts and deep-draw molding applications. Knitted fabrics have a great extensibility under tension which gives them a strong energy

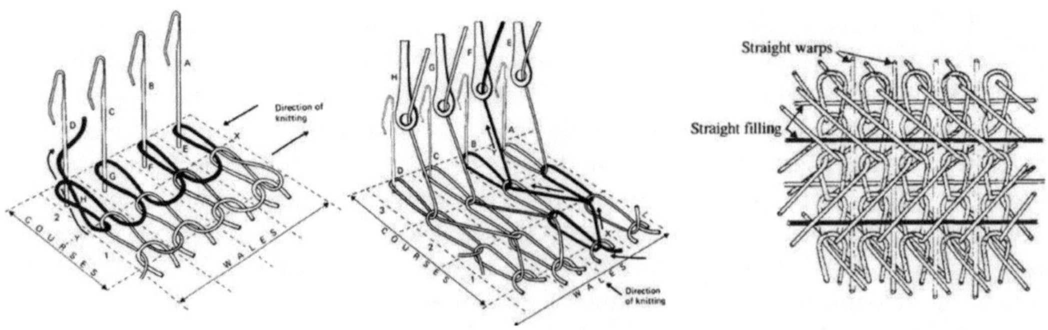

Figure 3. Weft knitting (left), warp knitting (middle) [24], and laid-in fabric [25].

absorption and dissipation capability when used in composite materials. Inlaid knitted fabrics offer dimensional stability and high mechanical properties of the resulting composites. These fabrics consist of a base weft or warp knitted structure together with non-knitted yarns which were incorporated (laid in) into the structure during the same knitting cycle (**Figure 3**) [25]. The base fabric structure holds in position the laid-in yarns. An inlay yarn may be 6–8 times coarser than the optimum yarn count for that particular machine type [24]. It is possible to introduce the laid-in yarns in weft knitting or warp knitting by using traditional knitting machines or by employing special equipment. Especially, weft-inserted warp knits with laid-in warp yarns offer high yarn-to-fabric translation efficiencies and greater in-plane shear resistance compared with their woven equivalents [23].

2.3.1.3. 2D braided fabrics

A braided fabric is produced by intertwining yarns about each other. 2D braiding technique can be used to produce complex-shaped tubular preforms. The process is best described by Maypole process in which the yarns are braided over a mandrel by yam carriers moving in a rotational fashion around the mandrel. The braiding machine consists of three main components: (1) yarn carriers, (2) interlinking mechanism, and (3) take-up mechanism (**Figure 4**) [26]. The braiding fiber tows are placed onto spools that are then loaded onto the carriers. The carriers are connected to the braiding machine through "horndogs" and "horngears," which propel the carriers into their rotational path. Different braid patterns can be obtained by modifying the horngears. Axial yarns can be introduced in the braids for enhanced stiffness by using hollow horngears [26].

There are three basic braided structures such as diamond, regular, and Hercules braid. The regular braid is the most common form in production. In diamond braid, any given yarn passes over and under one opposing yarn, and this pattern repeats itself. Following the same notation, regular braids can be designated as 2/2, whereas Hercules braids are 3/3. Triaxial braided fabrics can be formed by incorporating longitudinal (triaxial) yarns within the tubular biaxial fabric structure. The most important parameters of a braided fabric are braid angle, θ, (which can vary between 10 and 80°), cover factor, and the volume percent of longitudinal yarns in the structure. Braided fabrics are generally used for complex-shaped round composite parts due to their high extensibility in fabric state and also in tubular form [27]. 2D braiding is well established in composite industry in terms of knowledge and expertise unlike 3D braiding which is still in its infancy.

2.3.1.4. 2D nonwoven fabrics

A nonwoven fabric is basically a web structure made of fibers which is consolidated by employing various techniques. The production of nonwovens involves two basic steps such as the web formation and consolidation. For web formation various techniques can be employed:

- The drylaid system (carding or airlaying)
- The wetlaid system
- The polymer-based system (spunbonding, melt blowing, flashspun)

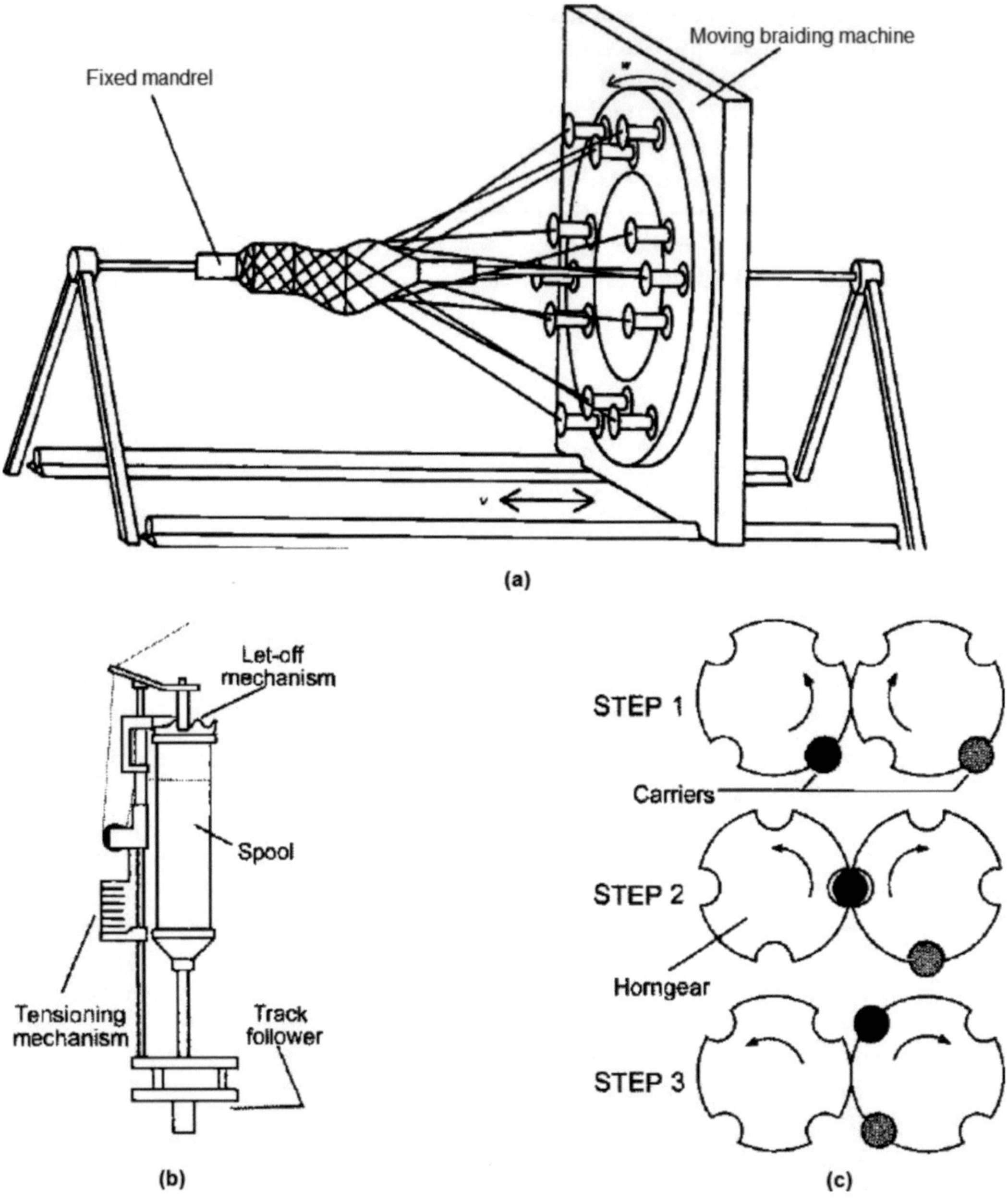

Figure 4. (a) Schematic diagram of a typical braiding machine, (b) yarn carrier, and (c) horngear movement [26].

Drylaid system has its origins in the textile processing. In this method, the fibers are carded or aerodynamically formed as loose webs with no practical strength. Then, these webs are bonded by various techniques such as needle punching, thermobonding, chemical bonding, hydroentanglement, etc. In the wetlaid system, first the fibers are dispersed in water, and the fiber/water suspension is transferred on a moving screen which can filtrate the water for web formation. Then, the web is dried and bonded. In spunbonding process, synthetic filaments are extruded from polymer onto a conveyor in the form of a randomly oriented web. In melt blowing, on the other hand, molten thermoplastic resin coming out of an extruder die tip is blown onto a conveyor by using air to produce self-bonded webs.

After the formation, the webs are bonded by using mechanical, chemical, or thermal means. The extent of bonding determines the strength, flexibility, porosity, and density of the resulting nonwovens. Mechanical bonding covers the needle punching, stitchbonding, and hydroentanglement techniques. In the needle punching process, the fiber web is entangled by the action of reciprocating barbed needles. In stitchbonding, fiber webs are bonded using stitching yarns. Hydroentanglement employs high-velocity water jets and turbulent water flow to entangle the fiber web. In chemical bonding, adhesive binders are applied to webs by spraying, impregnating, foaming, or printing techniques, or the fiber surfaces are partially solvated by using suitable chemicals, and the fibers are bonded through these surfaces (a process called "solvent bonding"). Thermal bonding applies heat and pressure to soften and then fuse fibers together to form the final fabric [28]. Thermal energy can be transferred by calendering, welding, or using hot air.

2.3.2. *Three-dimensional (3D) fabrics*

3D fabrics offer through-the-thickness reinforcement and thus improve the interlaminar properties and damage tolerance of the resulting composites [29]. They can be also formed to near-net shape with complex architectures, thus eliminating the labor-intensive layering processes to create a composite part [30].

2.3.2.1. *3D woven fabrics*

3D woven fabrics are produced using multiple warp layers. The movement of each group of warp yarns is governed by separate harnesses so that some are formed into layers, while others weave these layers together. The most common classes of 3D weaves are angle-interlock, orthogonal, and fully interlaced weaves. Angle-interlock fabrics fall into two main categories depending on the number of layers that the warp weavers travel such as through-the-thickness angle interlock and layer-to-layer angle interlock. In through-the-thickness fabric, warp weavers pass through the entire thickness of the preform, while in layer-to-layer structure, they bind only two filling layers. Orthogonal interlock weaves, on the other hand, are characterized by warp weavers oriented from orthogonal to other in-plane directions and run through the thickness of the preform. The x- and y-yarns of the angle-interlock and orthogonal structures are not interlaced, whereas the x-yarns are interlaced with both y- and z-yarns in the fully interlaced structure [31]. **Figure 5** shows the schematic views of orthogonal, angle-interlock, and fully interlaced 3D woven structures [32]. 3D weaving is capable of producing a wide range of architectures [33]. The main limitations of 3D woven fabrics include the lack of in-plane bias reinforcement and long preparation and processing times.

2.3.2.2. *3D braided fabrics*

3D braiding technology is an extension of the well-established 2D braiding technology. 3D braids can be produced in either horngear or Cartesian machines. The main difference between

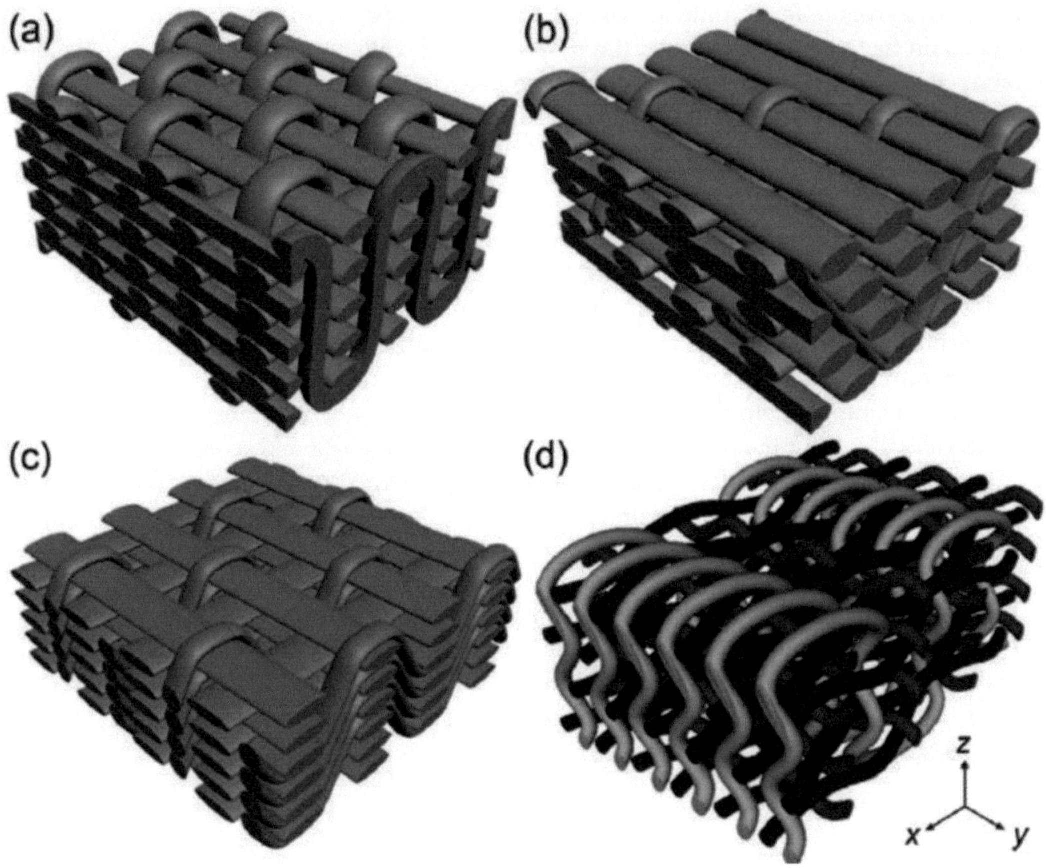

Figure 5. 3D woven structures: (a) orthogonal, (b) through-the-thickness angle interlock, (c) layer-to-layer angle interlock, and (d) fully interlaced [32].

these two types of machines originates from their yarn carrier displacement methods. Horngear-type machines have shorter braiding times, whereas the Cartesian machines are more compact and offer more design possibility for braid architecture. The first successful production of 3D braids dates back to the 1960s by using a technique referred to as four-step braiding (i.e., row and column). There are three major techniques of 3D braiding such as solid braiding, four-step braiding, and two-step braiding [34].

Solid braiding uses multiple interlocked braided layers. Complex 3D preforms can be produced by selective interlocking of braided layers which can be achieved through controlled movement of carrier yarns during braiding [35]. Some common braided structures that can be produced by solid braiding include two diagonal, three diagonal, four diagonal, spiral, and round braided structures.

Four-step braiding (also known as row and column) uses a flat bed which contains rows and columns of yarn carriers for forming the desired preform shape [36]. The number of steps refers to the number of movements required for the yarn carriers to return to their original positions [37]. Additional carriers can be added outside of the array depending on the shape of the preform. As the name suggests, there are four different sequences of row and column motion which collaboratively intertwine the yarns and produce the braided preform. Various braid patterns can be obtained by changing the motion of rows/columns and take-up [38]. In the first step, alternate columns move a predetermined relative distance. In the next step, alternate rows shift. The third and fourth steps rearrange the rows and columns to return the device to its original configuration [37]. **Figure 6** shows the steps carried out in four-step braiding [38].

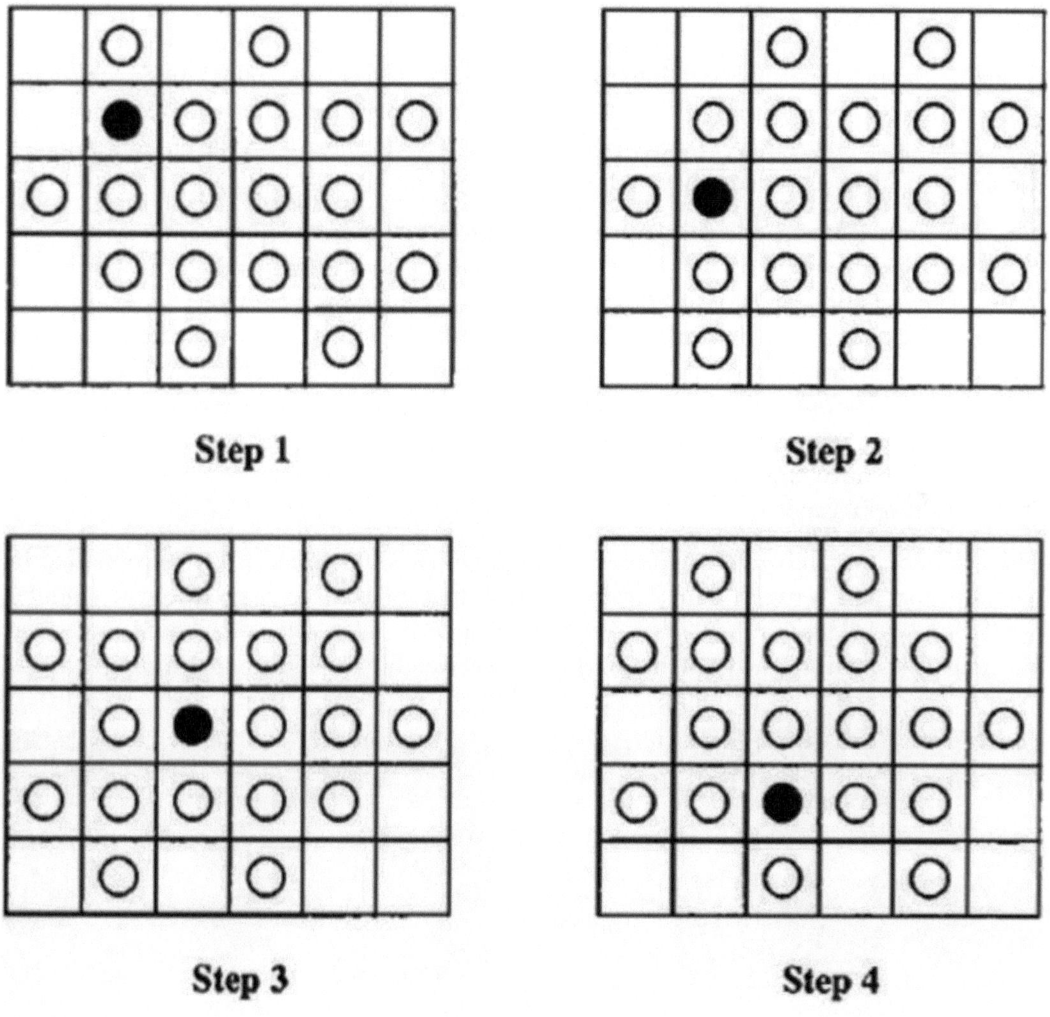

Figure 6. Schematic of the four-step braiding process [38].

Two-step braiding also uses a flat bed configuration, but unlike the four-step braiding, the two-step process utilizes a large number of yarns fixed in the axial direction and a smaller number of braiding yarns. Braiding carriers are positioned around the perimeter of the axial carriers. The final architecture of the braided preform is determined by the arrangement of axial carriers. The process is conducted in two steps in which the braiding carriers move completely though the structure between the axial carriers. This method is capable of producing a large number of shapes including circular and hollow fabrics. Complex-shaped parts can be produced by both four-step and two-step braiding processes [39]. It is also possible to braid holes into the preform structure which eliminates the drilling process. Sectional parts can be produced by adjusting the braiding process parameters which greatly reduce the problems related to section joints.

2.3.2.3. 3D knitted fabrics

3D knitted fabrics can be divided into three main categories: (1) integrally knitted fabric preforms produced by computer-controlled weft knitting machines, (2) multiaxial warp knit (MWK) fabrics, and (3) spacer fabrics which consist of two knitted fabric layers connected by through-the-thickness yarns between them.

Integrally knitted fabrics enable complex 3D shapes to be produced, eliminating labor-intensive sewing processes. Near-net-shape structures can be produced by using this fabric preforms. The first application of this type of fabric composites was a composite antenna made by Courtaulds Advanced Materials in the 1980s. Another example of near-net-shape structure was knitted by what is called the Pressure Foot® process [40]. This structure was used for carbon-carbon aircraft brakes in a collapsed form. The main advantage of integrally knitted preforms is their conformability which enables the production of very complex shapes. However, they have limited mechanical performance which restricts their wide-scale usage in composite applications.

Spacer fabrics are sandwich structures that consist of two knitted fabric layers connected by pile yarns that pass between them (**Figure 7**) [41]. A spacer fabric can be produced on a double needle bar Raschel machine. The center of the fabric may be filled with solid, liquid, or gaseous materials. The main advantage of this fabric is its low weight, flexibility, high bending stiffness,

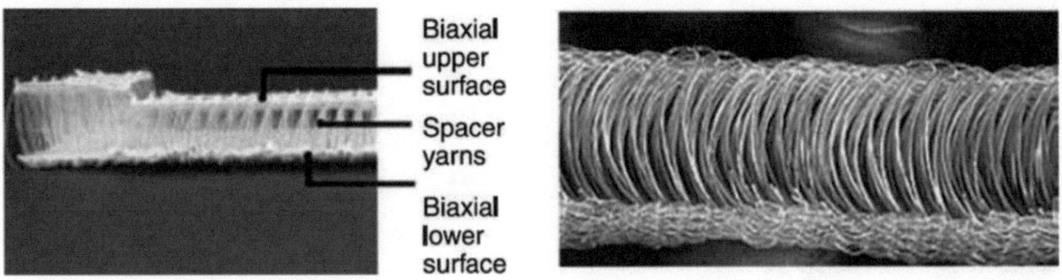

Figure 7. 3D spacer warp knitted fabrics [41].

and good acoustic and thermal insulating properties. The spacing can be up to 60 mm and widths up to 4400 mm [24]. Spacer fabrics are used in civil engineering as thin sheet-cement components, such as wall panels, claddings, and exterior siding [42, 43].

In MWK fabrics, several yarn layers each oriented in warp (0°), weft (90°), and bias (±θ) directions are stacked one on top of the other and bonded by a through-the-thickness loop system to provide structural integrity (**Figure 8**) [44]. A chain or tricot stitch can be used for this purpose. MWK fabrics are designed to bear various loads in all in-plane directions as well as in through-the-thickness direction. These fabrics are also referred to as non-crimp fabrics since the constituent yarns lie straight in the fabric, making no interlacements or intertwine-type interactions. This non-crimp nature ensures the effective translation of yarn-to-fabric strength properties [45].

2.3.2.4. 3D stitched fabrics

Multi-stitched 3D fabric preforms can be obtained by simply stitching 2D fabrics of any kind in the thickness direction [46]. The main reason for using stitching in this way is to impart through-the-thickness reinforcement to 2D multilayered fabric constructions [47]. Stitching operation can be conducted manually or by using suitable stitching machines. The stitching can be applied in longitudinal (0°), transverse (90°), and bias (±θ°) in-plane directions. The most important processing parameters are the stitch density (stitch/unit length), the type and size of the stitching yarn, and the stitch type used. Lockstitch, modified lockstitch, and chain stitch are generally used [26]. Modified lockstitch is generally preferred because in this stitch type, the crossover knot between the bobbin and needle threads is positioned at either laminate surface, thus minimizing in-plane fiber distortion [26]. Aramid yarns are generally used for stitching although other yarns such as glass, carbon, and nylon have also been used. Stitching substantially increases the interlaminar delamination resistance of composite laminates under mode I and mode II loadings. They also exhibit higher postimpact

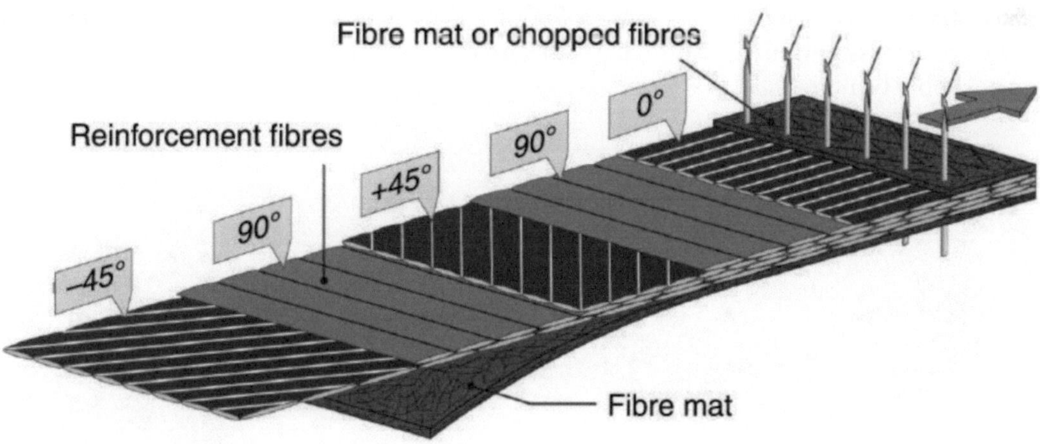

Figure 8. Multiaxial warp knit system [44].

residual mechanical properties compared with their unstitched counterparts [48]. Stitching also improves shear lap joint strength under both static and cyclic loadings due to reduced peel stresses. Therefore, stiffeners stitched onto a panel are more resistant to disbanding with reported improvements in load-carrying capability of up to 15% [26]. The main limitation with stitching, however, is the low in-plane properties and fatigue performance of the resulting structure because of the stitching holes which induce resin cracking, fiber breakages, and fiber misalignment [49].

2.3.2.5. *3D auxetic fabrics*

An auxetic material has negative Poisson's ratio which means that they laterally expand upon stretching and laterally contract when compressed in longitudinal direction. Auxetic textile structures include fibers, yarns, and fabrics [50, 51]. 2D auxetic fabrics can be produced with auxetic yarns and fibers by using traditional textile technologies like weaving and knitting [52, 53]. Auxetic materials have several advantages over conventional materials such as high shear modulus and fracture toughness. They have synclastic property which means that they take dome shape when fixed on a curved surface rather than saddle shape in the case of conventional materials. Auxetic materials also have better impact and ballistic resistance due to the fact that in the event of an impact, the material flows toward the impact point unlike conventional materials which flow away from it [54]. These properties make auxetic composites quite attractive for automotive, aerospace, and civil engineering applications. Nonwoven and stitching technologies can be combined to produce 3D auxetic fabric structure [55]. The fabric consists of three yarn systems such as warp, weft, and stitch yarns. There is no interlacement between warp and weft yarns; they are held in position by out-of-plane stitch yarns. There is half-yarn spacing between the warp yarn sequences on alternating weft layers. When the fabric is compressed in the through-the-thickness direction, the warp yarns remain straight, whereas the weft yarns get crimped so that the structure remains unchanged in warp direction while shrinking in weft direction, thus exhibiting auxetic property (**Figure 9**) [55]. It was stated that the auxetic effect is mainly produced by spaces between the warp yarns. In order to ensure more strong auxetic effect, the weft yarns used should be flexible, whereas the warp yarns should be more rigid.

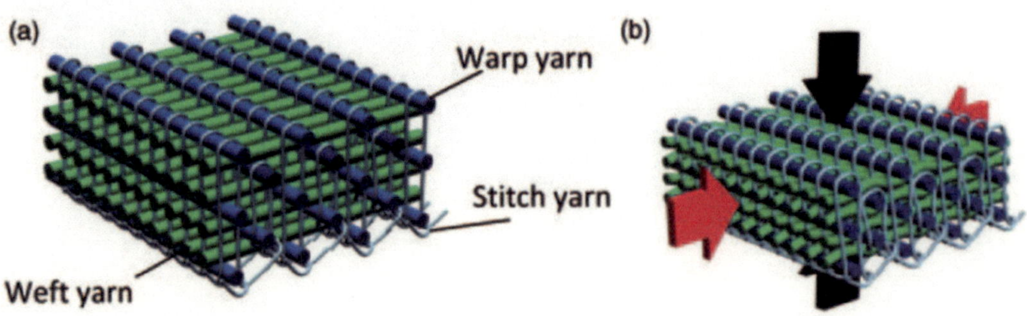

Figure 9. 3D auxetic textile structure: (a) initially and (b) under compression [55].

3. Composite manufacturing techniques

The type of matrix material, i.e., thermosetting or thermoplastic, is the main factor that determines the manufacturing technique used for the production of composites. Other parameters include the matrix material used, reinforcement form, fiber volume fraction, dimensions of the part to be produced, and complexity of the part shape. In the thermosetting resin-based methods, the matrix material is generally used in liquid resin form, whereas thermoplastic-based composite processing requires the melting of the polymer material. Thermosetting composites can be manufactured using a range of methods such as hand lay-up, resin transfer molding (RTM), autoclave molding, compression molding, filament winding, resin infusion, and pultrusion. For thermoplastic matrix-based composites, injection molding and thermoforming are the most commonly used techniques.

3.1. Hand lay-up

Hand lay-up is the simplest and most common manual technique for composite manufacture. The technique does not require much experience to apply, and the equipment cost is lower than that of other more advanced methods. Wooden, plastic, or metal molds in the shape of the composite part to be made are used. The first step in composite production is the application of surface-release agents on the surface of the mold to ensure an easy removal of the produced part. After the application of the release agent, a gel coat can be applied to the mold surface for high-quality part surface. Then, the reinforcement fabric or mat is placed inside the mold, and the resin-hardener mixture is applied on the surface of this reinforcement with the aid of a brush. Then, the entrapped air is removed, and the fabric surface is evened using a roller. This procedure is repeated after placing each layer of the reinforcement. The number of layers used depends on the desired part thickness. After laying-up is finished, the part is left for curing process. Finally, the cured part is removed from the mold. Hand lay-up is a labor-intensive and time-consuming technique with relatively poor quality results.

3.2. Resin transfer molding (RTM)

Resin transfer molding (RTM) produces composite parts with a good finish on both surfaces. The mold consists of male and female parts. The reinforcement fabrics are stacked one on top of the other and placed inside the female mold part. Then, the male part of the mold is closed on the fabric. The resin-hardener mixture is then pumped into the mold cavity. The mixture is pumped until the mold is completely filled after which the resin is allowed to cure. After curing, the mold is opened and the product is removed. RTM is less labor intensive and faster compared with hand lay-up. An advanced form of RTM is referred to as vacuum-assisted resin transfer molding (VARTM) which can apply vacuum to remove the entrapped air bubbles and improve the wetting of the fabric.

3.3. Autoclave molding

Autoclave is a high-temperature pressure vessel in which the produced composite parts are cured to obtain a high-quality product. In this technique, a composite part produced with hand lay-up can be used. The part is covered with a vacuum bag, and vacuum is applied to remove the air bubbles. With high temperature and pressure, the curing of the composite parts takes place in a slow manner. This technique yields to high-quality products with good surface finish.

3.4. Compression molding

Compression molding can be used for the manufacture of thermoplastic composites as well as thermosets. In this technique, a compression molding press capable of applying high pressures and temperatures is used. Pre-impregnated fibers, fabrics, and/or mats as well as sheet molding compounds (SMCs) and bulk molding compounds (BMCs) can be used as the starting material. The curing process typically starts upon the application of heat and pressure. Metallic dies are used due to high-pressure and high-temperature conditions. The curing reaction generally takes a few minutes. After curing, the die is opened and the part is removed. The product has a good surface finish on both sides.

3.5. Filament winding

Filament winding is generally used to produce cylindrical parts such as long composite pipes for liquid transportation. Fiber volume fraction ratios of up to about 70% can be attained in this method. Fiber rovings are first impregnated with resin and then wound on a rotating mandrel. The winding angle which depends upon the speeds of the mandrel and fiber feeder determines the directional and overall properties of the composite part. Various combinations of properties can be attained by changing the fiber volume ratio and winding angle depending on the end-use requirements.

3.6. Vacuum infusion (VI)

In this process, a large mold made of wood or plastic is generally used. The fabric is placed inside the mold cavity and covered with a polyethylene bag. The bag is sealed to the mold to prevent the entrance of air from the sides. A vacuum pump evacuates the air inside the bag creating atmospheric pressure on the fabric and at the same time sucks the resin from a container and right into the mold cavity. The surface of the part experiences a pressure of 1 atm. After the fabric is sufficiently wetted out by the resin, the vacuum is stopped, and the part is allowed for curing in room temperature. The process is suitable for the manufacture of very large but not complex composite parts such as boat hulls and automobile hoods.

3.7. Pultrusion

This method is used to manufacture long composite profiles of constant cross section such as beams used in roof structures, bridges, ladders, frameworks, and so on. In pultrusion process, the fibers are pulled from a creel through a resin bath where they are impregnated with a thermosetting resin. Then, they are pulled through a heated die for curing. The finished profiles are cut to length with the aid of a saw at the end of the process line [56]. Fabrics may also be introduced into the die to provide directional reinforcement. Pultrusion is a very fast and economic composite production method. High fiber volume fractions can be attained.

3.8. Injection molding

Injection molding is one of the most common techniques for the production of thermoplastic-based composites. Short fibers or particulates are used as the reinforcing phase. Complex-shaped parts can be produced at high production rates. There are several factors which limit the mechanical properties of the resulting composites such as short fiber lengths, low fiber volume fraction ratio, and random fiber positioning. The molding compound in pellet form which is composed of short fibers and thermoplastic matrix is heated in the injection chamber of the extruder. The material is melted by the heat and shearing action of the screw. The melted material is then injected into a die under high pressure. After a short cooling cycle, usually 20–120 s, the part is removed from the mold.

3.9. Reinforced reaction injection molding (RRIM) and structural reaction injection molding (SRIM)

Reaction injection molding (RIM) is a process for producing unreinforced thermoset parts. Polyurethanes are the most commonly used resin type, whereas nylons, acrylics, polyesters, and epoxies can also be used. For the production of polyurethanes-based composites, a two-component resin made up of isocyanate and polyol is used. These two components are mixed in a dynamic mix head under high pressures and injected into the mold cavity. Reinforced reaction injection molding (RRIM) is similar to RIM except that short fibers are incorporated to one of the resin components. The fibers must be extremely short (e.g., 0.03 inches) to allow for easy flowing of the resin. In structural reaction injection molding (SRIM), a preform is placed in the die before injection [57].

3.10. Thermoforming

In the thermoforming process, fiber-reinforced thermoplastic sheet is preheated using contact heating panels or rods, ovens, or IR heaters above the glass transition temperature of the thermoplastic matrix. The preheated sheet is then compressed within a preheated mold to form the final composite part. The final part is trimmed from the sheet.

4. Fiber-matrix interface in composite materials

Fiber-matrix interface has a profound effect on the mechanical properties and performance of composite materials. The main tasks of fiber-matrix interface are to transfer the loads from matrix to stronger fibers and distribute the loads across the main body of the material. Fiber-matrix bonding can take place through chemical bonding such as ionic, covalent, and metallic bonds; through intermolecular interactions such as ion-dipole forces, van der Waals forces, and hydrogen bonds; through coulombic interactions if the surfaces of the fibers and resin have a net charge of opposite signs; and through mechanical locking. There are various methods for improving fiber-matrix interface such as surface coating, plasma treatment, and chemical modification which are described briefly in this section.

4.1. Surface coating

This method involves coating of the fiber surface for a better fiber-matrix bonding. The most common coating techniques include electrodeposition, chemical vapor deposition (CVD), metallorganic deposition, and vacuum deposition. Electrodeposition is used to deposit metals onto carbon fibers. In CVD, the ingredients in a gaseous mixture react with one another and with the fiber surface to produce a solid film having the desired properties. Metallorganic deposition uses a liquid precursor that contains metallorganic species dissolved in an organic solvent. This precursor is applied onto the fiber surface for the desired coating. Vacuum deposition includes sputtering, physical vapor deposition, e-beam evaporation, plasma-assisted CVD, and ion-plating techniques [58].

4.2. Plasma treatment

In plasma treatment, plasma of different gases is used to alter the surface of the fibers. In this technique, an ionized region including excited species such as ions and radicals is formed around the fiber surface, thereby increasing its reactivity with polymer matrices and leading to an improved fiber-matrix bonding.

4.3. Chemical modification

Chemical modification techniques aim to change the chemical structure of the fibers and/or matrix in order to improve the fiber-matrix compatibility and adhesion.

4.3.1. Silane treatment

Organosilanes are generally used as coupling agents to improve the strength and durability of glass fiber-reinforced composite materials, and they are still the largest group of coupling agents used in composite industry today. A silane that contains at least one carbon-silicon bond (Si–C) structure is referred to as an organosilane. The organosilane molecule can be represented by the formula $R\text{–}(CH_2)_n\text{–}Si(OR')_3$ where n = 0–3; R is a non-hydrolyzable functional organic group that is reactive toward various groups such as amino, epoxy, vinyl, methacrylate, sulfur; and OR′ is a hydrolyzable group like an alkoxy group that can react with hydroxyl groups present in inorganic or organic substrates. Organosilanes can act as bridges between fibers and polymer matrices and can substantially improve adhesion between them.

4.3.2. Alkali treatment

In this method, natural fibers are treated with a dilute solution of sodium hydroxide (NaOH) with varying concentration, temperature, and duration. Alkali treatment has been shown to remove the impurities from the fiber surface, thus creating a rougher surface morphology. Consequently, the number of available sites for matrix penetration is increased which results in an improved mechanical interlocking between fibers and matrix.

4.3.3. *Esterification-based treatments*

In esterification technique, hydroxyl groups, —OH, in natural fibers are reacted with the carboxyl groups, —COOH, which is the functional group found in carboxylic acids. As a result, the —OH groups, which give the fibers their hydrophilic character, are eliminated, and the fibers gain a more hydrophobic character and hence become more compatible with most polymer resins. Esterification techniques include acetylation, benzylation, propionylation, and treatment with stearates.

4.3.4. *Graft copolymerization*

In this method, the cellulose material is treated with an aqueous solution of selected ions and then exposed to high-energy radiation. This process results in the cleavage of cellulose macromolecules and formation of radical groups. Then, the cellulose material is treated with a suitable polymer that is compatible with polymer matrix. The most important grafting method is the treatment of natural fibers with maleic anhydride-grafted polypropylene (MAPP) copolymers which results in the formation of covalent bonds across fiber-matrix interface.

4.3.5. *Treatment with isocyanates*

Polymethylene polyphenyl isocyanates (PMPPIC) can make strong covalent bonds with —OH groups of cellulose through their —N=C=O functional groups. The isocyanate treatment is very effective and can be used to modify both fibers and the polymer matrix.

5. Testing and modeling methods for textile-reinforced composites

It is important to have a good understanding of the stiffness and strength properties of textile-reinforced composites for successful design and application of these materials. Basically, there are two different approaches for determining the physical and mechanical properties of composite materials: (1) experimental measurement using suitable testing standards and (2) theoretical prediction from the properties of the constituent materials (fiber and matrix).

5.1. Testing methods

In this section, the most common testing methods for the evaluation of the physical and mechanical properties of textile composites are listed. A detailed discussion of the test procedures is beyond the scope of this chapter, and the interested reader is referred to the related standards and very detailed texts on the experimental characterization of composite materials [59, 60].

5.1.1. *Testing for physical properties*

Physical characterization of composites includes the determination of density (ASTM specification D792); fiber volume ratio (ASTM D2584 for burn-out method and ASTM D3171 for the acid digestion method and also optical methods based on image analysis); void content (ASTM D2734 and image analysis method); coefficients of thermal expansion (interferometric, dilatometric, optical noninterferometric, or strain gage methods); coefficients of moisture expansion; and heat conduction coefficients (for detailed information on the last two tests; see Ref. [60]).

5.1.2. Testing for mechanical properties

Mechanical characterization includes the determination of in-plane tensile (ASTM D3039), flexural (ASTM D790), compressive (ASTM D3410M), and shear (Iosipescu-type beam test with V-notches described in ASTM D5379M or the standard two-rail or three-rail shear tests described in D4255M) properties; determination of interlaminar properties (i.e., tensile, compressive, shear, toughness); material behavior under special conditions of loading (e.g., multiaxial, fatigue, creep, impact, and high-rate loading); stress and failure analysis of composite structures with structural details and nondestructive testing [60].

5.2. Modeling methods

Modeling studies basically cover three aspects such as the determination of physical properties, stiffness properties, and strength properties of the material. Approaches range from simple procedures based on micromechanics and classical laminate theory (CLT) to more sophisticated ones developed specifically for textile-reinforced composites. There are various methods to predict the behavior of the composites from the properties of its constituent materials. The most commonly used methods are the basic rule of mixtures (ROM), Halpin-Tsai method which is a refinement of the rule of mixtures, and a more sophisticated method referred to as Hashin's composite cylinder assemblage model.

5.2.1. Modeling for physical properties

ROM approach can be used to determine various physical properties of the composites using the properties of the constituent materials. Fiber (V_f) and matrix volume fraction (V_m) can be calculated using the following relations [61]:

$$V_f = \frac{v_f}{v_c} \quad (1)$$

$$V_m = \frac{v_m}{v_c} \quad (2)$$

where $\upsilon_{f,m,c}$ = volume of fibers, matrix, and composite, respectively.

The weight fraction of fibers (W_f) and matrix (W_m) is given by

$$W_f = \frac{w_f}{w_c} \quad (3)$$

http://dx.doi.org/10.5772/intechopen.68245

$$W_m = \frac{w_m}{w_c} \tag{4}$$

where $w_{f,m,c}$ = mass of fibers, matrix, and composite, respectively.

Density of the composite can be calculated using ROM:

$$\rho_c = \rho_f V_f + \rho_m V_m \tag{5}$$

where $\rho_{f,m,c}$ = density of fibers, matrix, and composite, respectively.

The volume fraction of the voids (V_v) is given by

$$V_v = \frac{\rho_{ct} - \rho_{ce}}{\rho_{ct}} \tag{6}$$

where ρ_{ce} and ρ_{ct} are the experimental and theoretical density of a composite.

5.2.2. Modeling for mechanical properties

5.2.2.1. Simple micromechanical approaches

The ROM approach is particularly useful for obtaining the longitudinal modulus and major Poisson's ratio of a unidirectional composite, whereas it gives less accurate results for transverse and shear moduli. Longitudinal modulus (E_1), major Poisson's ratio (ν_{12}), transverse modulus (E_2), and in-plane shear modulus (G_{12}) can be obtained using the following relations:

$$E_1 = E_f V_f + E_m V_m \tag{7}$$

where E_f and E_m are the elastic moduli of the fiber and matrix, respectively; v_f and v_m are the Poisson's ratio for the fibers and matrix, respectively; and G_f and G_m are the shear moduli for fibers and matrix, respectively.

$$\nu_{12} = \nu_f V_f + \nu_m V_m \tag{8}$$

$$\frac{1}{E_2} = \frac{1}{E_f} V_f + \frac{1}{E_m} V_m \tag{9}$$

$$\frac{1}{G_{12}} = \frac{1}{G_f} V_f + \frac{1}{G_m} V_m \tag{10}$$

The Halpin-Tsai equations are a refinement of ROM. In fact, the longitudinal modulus and major Poisson's ratio are the same as those calculated by ROM approach. However, for the transverse and shear moduli, the following relation is used:

$$\frac{M}{M_m} = \frac{1+\xi\eta V_f}{1-\eta V_f} \tag{11}$$

$$\eta = \frac{\frac{M_f}{M_m}-1}{\frac{M_f}{M_m}+\xi} \tag{12}$$

where $M = E_2, G_{12}$, and G_{23} for composite; M_f = fiber property E_f or G_f ; M_m = matrix property E_m or G_m ; and ξ = a constant defining the way the load is shared between fiber and matrix.

Poisson's ratio in the transverse plane (ν_{23}) may be calculated using the following relationship:

$$\nu_{23} \cong 1 - \frac{E_2}{G_{23}} \tag{13}$$

Hashin's composite cylinder assemblage (CCA) model takes into account the fiber anisotropy. The expressions are rather lengthy, and the interested reader is referred to the work of Hashin [62].

5.2.2.2. *Sophisticated models for textile-reinforced composites*

The simple models presented above are for unidirectional lamina where the fibers all lie in the same direction. Textile-reinforced composites generally have complex fiber architectures, and more sophisticated analysis methods have to be used for their successful modeling. We can divide all models into two main categories such as analytical/semi-analytical and numerical models based on finite element analysis (FEA).

The simplest textile structure is non-crimp textile fabrics. Traditional angle-ply laminate procedure can be used directly since the fibers remain reasonably straight. However, woven, braided, and knitted textiles have more complex structures, and their composites require more sophisticated modeling treatments. There are various models developed for 2D and 3D woven fabric composites. Modeling methods for 2D woven composites include 1D methods such as mosaic model [63], fiber undulation model [64], and bridging model [65] developed by Ishikawa and Chou in the early 1980s. 2D models were developed by Naik and coworkers [66, 67] in 1992 by refining the fiber undulation model. The model included the fiber undulation and continuity in both warp and weft directions and the real cross-sectional geometry of the yams. These 1D and 2D models can predict the in-plane elastic properties only. 3D models taking into account the out-of-plane elastic properties include a 3D representative volume element (RVE) proposed by Hahn and Pandey [68] and a fabric geometry model (FGM) which is originally developed by Ko and Chou [69] and used by Vandeurzen and coworkers [70, 71] for woven fabric composites. Models for 3D woven composites include orientation averaging models [72], mixed iso-stress and iso-strain models [73, 74], and finite element applications. Models for braided fabric composites include a FGM model proposed by Pastore and Gowayed [75], fiber inclination model [76], and finite element models. There has been limited attention to the modeling of knitted fabric composites due to their limited usage in structural applications. There are a number of models proposed to model knitted composites [77, 78].

5.2.2.3. Prediction of strength and failure

In general approximation, failure can occur in yarns, matrix, and fiber-matrix interface in a textile composite subject to a general type of loading. Failure strength prediction of textile-reinforced composites can be carried out by determining failure criteria for yarns, matrix, and interface using various methods. Maximum stress and maximum strain criteria and the Tsai-Wu failure criterion may be used to predict the yarn strength. Matrix failure can be predicted by standard failure criteria for a homogeneous and isotropic material. Failure that occurs along the interface may be predicted using failure criteria for predicting interlaminar delamination in composite laminates [38].

6. Applications of textile-reinforced composites

6.1. Aerospace applications

The aerospace industry is among the largest consumers of advanced composites due to the low weight and high mechanical properties of these materials such as static strength, fatigue performance, fracture toughness, damage tolerance, high impact resistance, and resistance to high temperatures. In addition, composite components can be produced with fewer joints and rivets, leading to lower susceptibility to the structural fatigue. Weight reductions of up to 40% could be achieved when aluminum is replaced by composite materials. A reduction of 1 kg in weight corresponds to $50–500 fuel saving depending upon the type of aircraft and fuel prices. Traditionally, 2D woven fabric prepregs are the primary choice, but the trend has then shifted toward the use of unidirectional prepregs due to strength and stiffness considerations. However, the high production cost and low out-of-plane properties of 2D laminated structures are the two main driving forces for the implementation of 3D textiles in aircraft and aerospace applications. Multilayer composite laminates made up of 2D unidirectional prepregs or textile fabrics lack the through-the-thickness reinforcement resulting in low interlaminar strength and fracture resistance. In addition, laminated 2D composites are made thicker than needed to withstand residual stresses after impact which in turn increases the weight and production cost of the components. 3D fiber architectures in which a portion of the fibers are oriented in through-the-thickness or z-direction offer a way to improve the interlaminar strength and fracture resistance. 3D textile composites are also low cost due to elimination of labor-intensive laminating process. Among the 3D textile preforms, multiaxial non-crimp carbon fiber fabric is the material of choice since it provides mechanical properties similar to those of unidirectional prepregs with the added advantage of through-the-thickness reinforcement.

Today, modern aircrafts use composite materials up to 50% by their weight [79]. The composites are used in commercial and military aircrafts for several nonstructural and structural parts such as spoilers, elevators, horizontal stabilizer boxes, radomes, rocket motor castings, tail sections (Airbus), rudder (Boeing 767), fuselage (V22), antenna dishes, engine nacelles, landing gear doors, engine cowls, wings (Prototype ATF, V22), cargo liners, and so on [80]. Carbon fibers dominate the aerospace sector due to their high strength/stiffness and low density.

Other fibers like aramid and glass also have a limited usage, but low strength and high density of glass fiber limit its usage in aircraft structures, whereas aramid fibers have the moisture absorption problem. Regarding the matrix systems, epoxy resin is the most preferred matrix material due to its high mechanical properties and durability. Space applications of advanced composites include launch systems and self-contained space modules such as satellites, space crafts, and space stations. Composites allowed 30% weight savings when used in expendable launch vehicles (ELV) for Atlas, Delta, and Titan rocket launch systems. Composites produced with less labor-intensive techniques such as filament winding and automated tape winding proved more economical than costly joint welding of metals. Composite motor cases from carbon fiber/epoxy with high strength and stiffness provided substantial weight reductions. Nozzles such as exit cone and throat elements are composed of carbon fiber/phenolic or carbon-carbon composites to withstand hot exhaust gases of burning propellant [80]. Hubble Space Telescope contains a carbon fiber/epoxy metering truss with near-zero coefficient of thermal expansion (CTE) which provides dimensional stability required for precise alignment of the optics aboard the telescope. These trusses are also designed to cope with various compressive loads. Composites are also used in space shuttles. Payload bay doors made of carbon fiber/epoxy (T-300) composite parts provided weight savings of over 400 kg [80].

One of the earliest applications of 3D textile composites involved the use of 3D stitching which is described in detail in previous sections. 3D stitching technique was used in a NASA-sponsored demonstrator program to manufacture low-cost and damage-tolerant composite wings. For this purpose, a 28-m-long sewing machine was developed by Boeing. This machine was able to stitch carbon fabric layers of over 25 mm in thickness at a rate of over 3000 stitches a minute. In addition to stitching the skin preform, the blade stiffener flanges made up of tubular braided fabrics are also stitched to the skin. Resin infusion technique was used for final composite production. The resulting panel was reported to be 25% lighter and 20% cheaper than an equivalent aluminum part [26].

One example of 2D layered woven structures was manufactured by Kawasaki Heavy Industries, Japan, by using twill carbon fiber fabric. The prepreg developed was used in various applications including the Embraer ERJ 170 inboard flaps, the Embraer ERJ190 outboard flaps and wing stubs, and the Boeing 737-300 winglets [81]. 3D woven preforms greatly reduce the production time by eliminating the lay-up of individual fabric layers. It is also possible to manufacture net-shape preforms that can be easily handled during the process. 3D woven fabrics can also be manufactured in complex shapes eliminating the assembling costs. The main limitation of 3D woven fabrics is that standard looms cannot introduce in-plane yarns at angles other than 0 and 90°. Therefore, the resulting structures possess very low shear and torsion properties, thus limiting their usage in most aircraft structures where high shear strength is required. 3D woven composites were first developed in the 1970s as a replacement for expensive high-temperature metal alloys in aircraft brakes. Carbon fibers were used for the production. Hollow cylindrical preforms were produced on a modified weaving loom. Carbon-carbon composites reinforced with these fabrics displayed high specific mechanical properties as well as outstanding heat resistance [82]. These composites also have low thermal expansion, good thermal conductivity, high heat capacity, and wear resistance. It was reported that carbon-carbon composite brakes exhibit nearly four times the braking power of the forerunner steel brakes [80]. 3D woven composites are used in H-joint connectors for joining honeycomb sandwich wing panels on the Beech Starship. It was reported that the connector was crucial for the low-cost production of the wing. It also improved the stress transfer at the joint, thereby reducing the peeling stress [83]. 3D woven composites are also used by Lockheed Martin for the air inlet duct in the F35 military fighter jet. In this case, the stiffeners are integrally woven with the duct shell, reducing the need for secondary fastening. It was reported that 95% of the fasteners through the duct are eliminated, thus enhancing aerodynamic and signature performance, minimizing the risk of fasteners being ingested by the engine, and simplifying manufacturing assembly [26]. Other 3D woven composites generally include demonstration structures such as turbine engine thrust reversers, rotors, rotor blades, insulation, structural reinforcement and heat exchangers, rocket motors, nozzles and fasteners, engine mounts, T-section elements for primary fuselage frame structures, and multiblade stiffened panels [84]. 3D carbon-woven/ceramic sandwich composites are used in the combustion chamber of prototype scramjet engines [85].

2D braided composites generally display similar stiffness and strength when compared to those of equivalent 2D layered woven composites. When the damage resistance and tolerance are taken into account, 2D braided composites have proved superior to both unidirectional and 2D woven composites. This is due largely to partly 3D nature of 2D braided composites which help limit the damage growth. 2D braided composites are used in various components such as aircraft propellers, rocket launchers, aircraft fuselage frames, and helicopter rotor blade spars. The use of 3D braiding, on the other hand, has been limited due to the fact that most industrial braiding machines are only capable of producing braid preforms with small cross sections. Very large and expensive machines are necessary to manufacture preforms which are large enough for aircraft components. 3D braided fabrics were first developed around the 1960s to manufacture carbon-carbon composites as a replacement for high-temperature metal alloys in rocket motor components. With composite components, weight savings of 30–50% were attained [38]. Composite parts with holes, bends, and bifurcations can be produced with 3D braiding which allow production of junctions and complex shapes for aircraft structures [26]. Several demonstrator components have been produced using 3D braiding process such as T-section panels, I-beams, bifurcated beams, airframe spars, F-section fuselage frames, fuselage barrels, and rocket engine nozzles [26]. In general, the in-plane properties of 3D braided composites are lower compared with those of the equivalent 2D stacked laminates owing to the yarn crimp introduced during braiding process. Their out-of-plane properties and impact resistance, however, have been shown to be superior when compared to those of conventional laminates. A&P Technology, which is a leading company for braiding technology for the aerospace, manufactured various parts for aerospace applications. One example is a Vectron sock of 2 m in diameter and 3 m in length manufactured using an 800-carrier braiding machine [86]. The developed braided sock was used for a prototype airlock developed for NASA. A&P Technology also produced braid-reinforced wing flaps for Bombardier using RTM [87]. The same company produced the Honeywell jet engine stator vanes using a layer of aramid braid followed by an overlay of carbon fiber braid (**Figure 10a**) [88] and fan case of GEnx jet engine which is a fuel efficient, quiet, and low-emission jet engine developed for the Boeing 787 aircraft and the Boeing 747-8 (**Figure 10b**) [89].

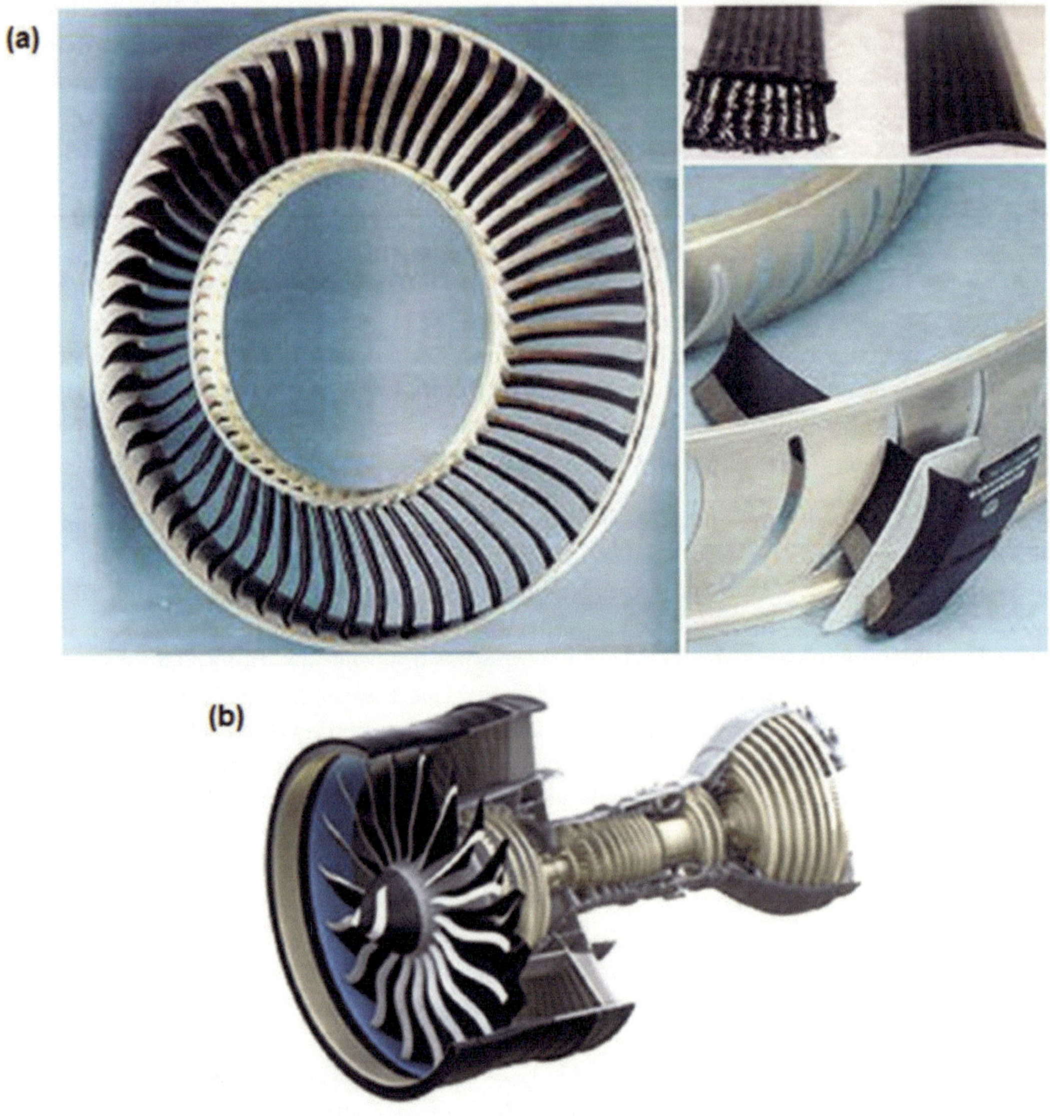

Figure 10. Braided composite components manufactured by A&P Technology: (a) jet engine stator vanes [88] and (b) fan case of GEnx turbine [89].

Multiaxial warp knit (MWK) fabrics (also known as non-crimp fabrics) attracted a special attention in aerospace industry, and today it is the most common form of 3D textile structures used in aerospace applications. The main advantage of this technique, as previously mentioned, is the ability to introduce in-plane yarns without crimping which helps achieve a better yarn-to-fabric property translation efficiency. Moreover, the through-the-thickness yarns provide improved out-of-plane properties. There is a continuing effort within the aerospace industry to use MWK fabrics, and a number of demonstrator components such as wing stringers and wing panels have been produced using MWK fabric composites. One example of MWK composites is the Airbus A380 rear pressure bulkhead (RPB) shown in **Figure 11** [90].

Figure 11. Rear pressure bulkhead of the Airbus A380 [90].

This structure is produced using multiaxial carbon fiber fabrics supplied by Saertex. The part is manufactured using resin film infusion followed by autoclave process. Another successful application is the Airbus A400M cargo door which was manufactured using multiaxial carbon fiber fabrics with additional monoaxial fabrics for directional reinforcements and skin lay-up. Vacuum-assisted infusion was used for the production [91]. One of the problematic issues with MWK fabric is the stitching process which causes misalignments, fiber disruption, and so-called fish eyes in the structure which lowers the mechanical properties. These problems can be largely eliminated by using finer polyester yarns and improving pinpointing of the stitching yarn path to avoid misplacements and yarn damage.

Novel auxetic composite structures were developed for a better sound insulation in aircraft fuselages [92]. The noises generated by the engine are generally transmitted to the cabin through the stiffeners. A novel auxetic composite consisting of an auxetic core, a damping layer, and a constraining layer was developed for stiffeners. It was stated that this new type of material could save the weight as well as enable better sound insulation when compared to conventional stiffeners with thick metallic layers.

6.2. Civil engineering applications

Textile-reinforced composites offer numerous advantages for civil engineering applications such as low weight, off-site mass production of complex-shaped components, added functionality and directional design of mechanical properties, cost reductions related to easy transportation and reduction in construction time, and also desirable aesthetic properties.

One of the earliest applications of composites is the reinforcement of concrete. Steel has traditionally been used to reinforce concrete for many years. Although steel-reinforced concrete has high strength and stiffness, the major limitation of this material is its high weight and susceptibility of steel to corrosion and subsequent deterioration. Moreover, rigid steel bars limit the possibilities of shaping concrete products. Textile materials can be used instead of steel to reinforce concrete. The outstanding durability of composites greatly extends the service life of the structure and reduces the inspection and maintenance costs, making them cost-effective [93]. Some of the advantages of composites over traditional construction materials are high strength-to-weight ratio, ability to be manufactured in complex shapes, ability to tailor their mechanical properties according to specific needs, their noncorrosive nature, low thermal conductivity, and outstanding fatigue performance. One example is the use of GFRP and CFRP reinforcement in the Wotton Bridge deck [93]. Another example of bridge construction is the pedestrian bridge in Oschatz, Germany. This bridge consisted of four layers of woven alkali-resistant (AR) glass fiber fabrics along with steel tendons. The bridge spans 8.60 m and consists of U-shaped segments (**Figure 12**) [94]. Besides bridges, very thin plates of textile-reinforced concrete are used for façades for reduced material and transport costs and an elegant appearance. Sandwich constructions consisting of two AR glass fiber concrete panels enclosing polyurethane foam can also be used for an improved sound and thermal insulation. An example of façade application is the betoShell© developed by Hering Bau (**Figure 13**) [95]. Another example is the award-winning restoration of a train station platform by Hering Bau in Walleshausen, Germany. For this purpose, two layers of carbon fiber textiles were embedded into the concrete. The concrete panels with a dimension of 2.5 × 1.35 m were put on a frost-proof concrete layer and fixed with two stainless steel mandrels [96].

In addition to concrete reinforcement, composites are also used for shear and flexural strengthening and the confinement of the existing concrete structural members through a process called externally bonded reinforcement (EBR). In this process, woven or nonwoven fabrics are applied

Figure 12. First segmental TRC Bridge in Oschatz, Germany [94].

Figure 13. betoShell© façade application developed by Hering Bau [95].

on the surface of the structure through a technique similar to hand lay-up. Alternatively, composite strips consisting of unidirectional fibers embedded in an epoxy matrix can be directly bonded on the surface of the structure by using an epoxy adhesive (**Figure 14**) [93].

Other applications include the I-beams made with a 3D woven composite which is used in the roof of a ski chair-lift building as a replacement of heavy steel beams. 3D woven composite beams demonstrated lower cost and enhanced performance compared with steel and conventional composite beams [97].

6.3. Automotive applications

In recent years, the depletion of petroleum resources and strict environmental regulations forced the scientists and engineers in automotive industry to search for new ways of reducing the fuel consumption and also CO_2 emission. Therefore, the automotive industry is in a constant struggle to reduce the weight of the vehicles in order to decrease the fuel consumption without sacrificing passenger safety and comfort. Lightweight and strong textile composites offer a solution to this problem. Replacing cast iron, steel, and aluminum components with composites can decrease component weight by 10–75%. It is estimated that a 10% reduction in vehicle weight corresponds to a 6–8% reduction in fuel consumption [99]. Composites are gradually replacing steel for several body parts such as bumpers, hatch doors, cabin components, spare wheel containers, and so on. Other applications of composites include underbody panels, instrument panel carriers, battery trays, front-end modules, sunroof beams, door modules, seat structures, convertible header bows, door plate carriers, roof and trunk lids, exterior claddings, skid plates, running boards, step assists, front-end carriers, bumper beams, lift gates, spare wheel tubs, and other structural parts [100]. For example, Turner et al. investigated carbon fiber composites for automotive body panel applications. They produced a full-scale front wing-fender component using two different composite manufacturing processes such as a semi-impregnated system and a novel directed fiber preforming RTM process. The semi-preg system consisted of a surface film (thin woven glass either side of an epoxy film) and a bulk ply (woven 3000 filament, 300 g/m^2 carbon fiber on one side, 400 g/m^2 woven glass on the other side, and an epoxy glass microsphere core) The resulting composite was compared with an existing stamped steel component for mechanical properties, weight

Figure 14. Strengthening of existing concrete structural members using unidirectional carbon and aramid sheets [98].

saving, and cost by employing a technical cost modeling procedure. It was demonstrated that carbon fiber composites can provide 40–50% weight saving compared with steel panels for an equivalent bending stiffness and they have greatly improved dent resistance. The steel part was shown to be more cost-effective at volumes above around 9000 parts/annum [101]. Advanced composite materials are also considered for high-performance racing cars since weight reduction is crucially important, whereas cost issues are not a primary concern in this area. Feraboli et al. investigated the performance of carbon fiber composites used in body panels and integrated chassis components for the Murcièlago Roadster [102]. The laminate is made up of three plies oriented in the 0/90 direction, such as a 0.2 mm thick, 2 × 2 twill at the surface, followed by a 0.4 mm thick five-harness satin, and another 2 × 2 twill that is 0.7 mm thick. It was concluded that carbon fiber composites offer a considerable amount of weight saving as well as design flexibility [102].

3D textile-reinforced composites are also used in various applications such as floors and floor beams in trains and fast ferries; flat load trays in trucks; crash members in cars, buses, and trucks; shipping containers; and other container transport applications. 3D composites reinforced with spacer fabrics made of two parallel skins of 2D glass fabrics connected by glass yarns are used in car and truck spoilers/fairings and luggage floors of automobiles [84]. Another example of 3D textile composites is the oil pan produced using 3D woven interlock fabrics. Three different 3D woven structures were used for this purpose such as layer-to-layer, orthogonal, and through-the-thickness woven fabrics [99].

In the US, several government/private sector projects made use of composite materials in vehicles. These include Ford Motor's "all composite car," the Automotive Composites Consortium (ACC), the Partnership for the New Generation of Vehicles (PNGV) program [103], and the Sunrise™ electric vehicle (EV) program. Among these projects, Sunrise™ electric vehicle (EV) program in 1996 made the most extensive use of textile composites in a well-integrated program. This program is co-funded by the National Institute of Standards and Technology (NIST) together with a consortium of eight industry partners. Nearly all types of textile preforms such as woven, knitted, braided, and nonwoven fabrics were used for reinforcement (**Figure 15**) [104].

Other automotive applications of composites include car noses and monocoques of Formula One car bodies from carbon textile structures, aprons and spoilers of sport cars, braided carbon composites for car bumpers (e.g., BMW M6, Lotus Elise), the crash beams of the McLaren SLR which consist of foam cores overbraided with carbon rovings, and the car roof of the BMW M6 which consists of carbon-woven fabrics [105].

6.4. Applications in medicine

Composite materials are extensively used in medicine owing to their numerous advantages. Among the most important applications of composite materials in medicine is bone repair. Bone fractures can be treated by two main methods such as external fixation and internal fixation. The purpose of external fixation is to hold the bone fragments in alignment by using various materials such as splints, casts, braces, and external fixator systems. Plaster of Paris (calcium sulfate) reinforced with cotton-woven fabrics is still the most commonly used casting

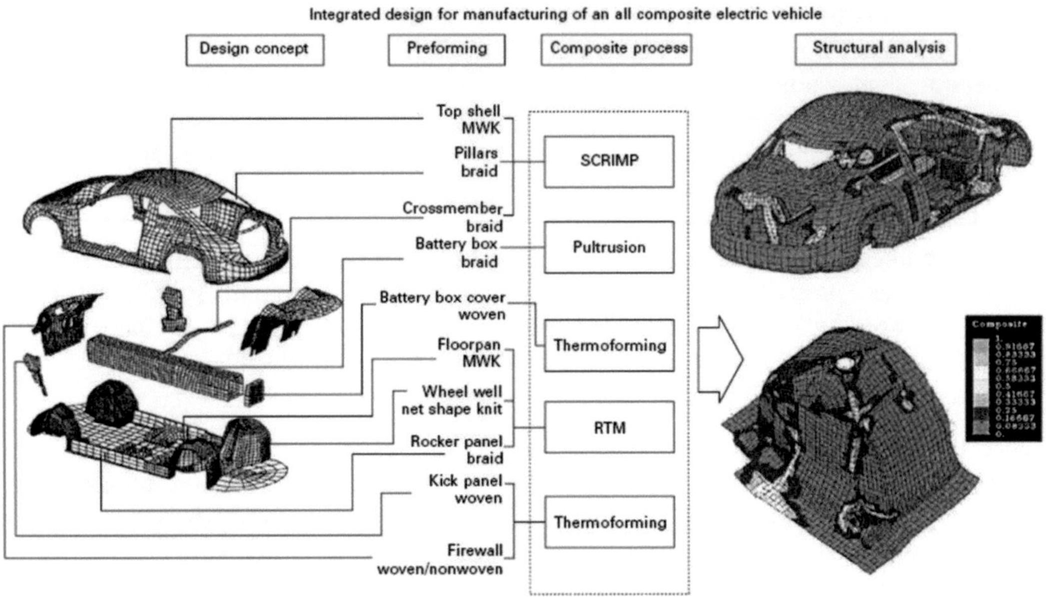

Figure 15. Composite parts breakdown of Sunrise™ BIW according to textile preforming processes and composite manufacturing processes [104].

material, but it has a high failure rate, high density, and low specific strength/stiffness and is sensitive to water. Recently, casts made of glass or polyester fiber fabrics and water-activated polyurethanes are being used to overcome these problems. An ideal splinting material must be lightweight, stiff, strong, and also comfortable. It must be also fit to the complex contours of the limbs. The fabric must be open structured to allow impregnation with a large quantity of plaster. Leno-woven structures as well as warp and weft knitted fabrics can be used [106]. In the internal fixation method, the bone fragments are held together by using wires, pins, screws, plates, and intramedullary nails. Bone plates made of stainless steel and Ti alloy are conventionally used. The stiffness of these materials is way higher than that of the bone, and this stiffness mismatch causes much of the load to be carried by the plate. This causes increased bone porosity (also known as bone atrophy), and the bone becomes less dense and weak. Less rigid plates that are more bone-like in mechanical properties can be used to overcome this problem. Various composite bone plates were developed for the usage of such as CF/epoxy, GF/epoxy, CF/PMMA, CF/PP, CF/PE, CF/nylon, and CF/PEEK [107]. **Figure 16** shows composite bone plates made of various textile composites [107].

Joint replacements such as total hip replacement (THR) and total knee replacement (TKR) also employ composite materials. Conventional THRs use stainless steel, Co-Cr, and Ti alloys for the femoral shaft and neck and Co-Cr alloy or ceramics such as alumina and zirconia materials for the head or ball [107]. CF/PEEK and CF/epoxy systems and UHMWPE-based composites were used for THR and TKR with successful results (**Figure 17**) [108]. Other applications of composites include bone cement, bone grafts, dental post, dental implant and bridge, bracket and archwire, ureter prosthesis, catheters, tendons and ligaments, prosthetic limbs, and medical equipment such as walking support frames and sticks.

http://dx.doi.org/10.5772/intechopen.68245

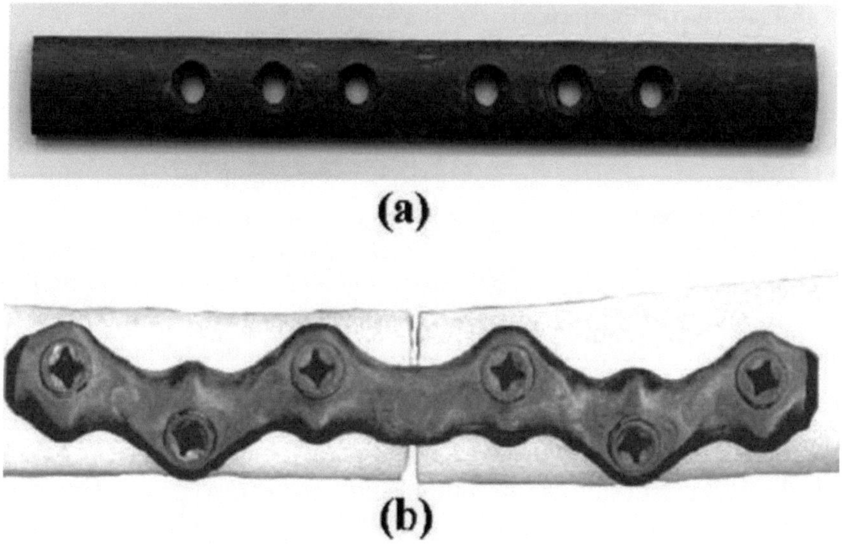

Figure 16. (a) A typical composite bone plate made using braided carbon fiber fabric and epoxy matrix and (b) composite bone plate made of carbon/PEEK material system [107].

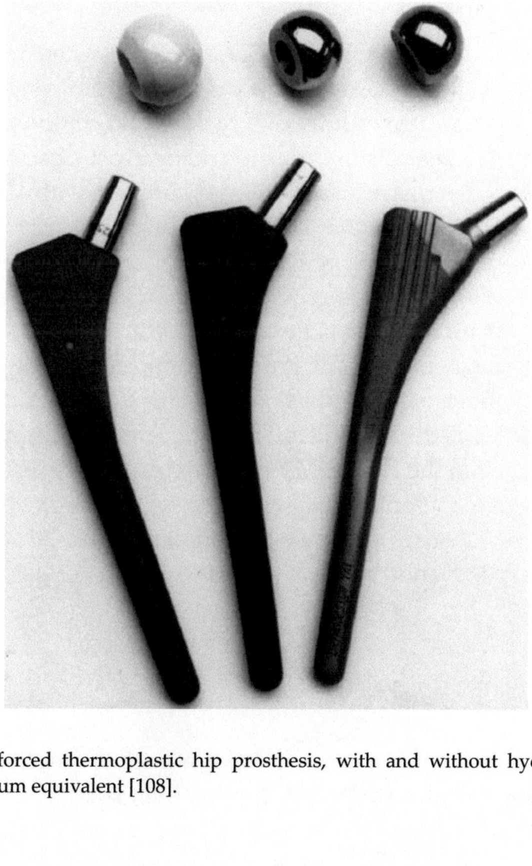

Figure 17. Carbon fiber-reinforced thermoplastic hip prosthesis, with and without hydroxyapatite coating, shown alongside conventional titanium equivalent [108].

6.5. Sports and leisure applications

Composite materials allow for the directional design of mechanical properties which is a great advantage in various sports equipments. Lighter and stronger golf clubs made of graphite fiber-reinforced epoxy are in common use. Carbon fiber composites are used for the production of poles used in pole vaulting. The pole must be lightweight, flexible yet stiff, and torsion resistant. In order to meet these demands, a layered design is used [109]. The outer layer is made up of unidirectional carbon fiber/epoxy system, whereas the middle and inner layers are made up of glass fiber web/epoxy and wound glass fibers/epoxy, respectively. Carbon/epoxy composite layer provides maximum stiffness, while glass fiber composites ensure high torsion resistance [109]. Other applications of carbon fiber composites include bicycle frames, front forks, and seat posts. Carbon fiber composites are lightweight and provide good bending stiffness and fatigue resistance. In addition, glass fiber-reinforced nylon wheels are used as good shock absorbers. Carbon fiber composites are also used to produce tennis racket frames for high strength and stiffness. In addition, the handles can be made by wrapping multiple composite layers around a soft core such as PUR for the purpose of vibration damping [110]. Other applications include kayaks made of carbon and kevlar fibers with epoxy resin, skis and snowboards made of a soft core, and composite layers of various constructions [111].

7. Future scope

Future investigations will focus on novel fiber and matrix systems, fiber architectures and processing techniques, as well as improving the existing technologies. The successful implementation of 3D textile composites requires more work on the production economies and engineering design of these materials in order to create a cost-effective and sustainable production technology. Fabric geometry model (FGM) coupled with CAD-CAM technologies and analyzing systems such as the finite element method (FEM) can create a useful framework which can help integrate fabric design and processing parameters into structural analysis and ultimately efficiently design the required properties for a specific end use. Another interesting topic for the future research is the design of fiber-matrix interface in composites. The quality of the fiber-matrix interface is crucially important for high-standard composite properties. Increasing the fiber-matrix adhesion by using suitable coupling agents and interface modification techniques is the all-time hot topic in composite industry, and it is apparent that this trend will continue in the future. More research is also necessary on the dynamics of matrix infiltration into textile materials during composite processing as well as on the characteristics of curing process. The use of nanofibers and nanotextile structures for the reinforcement of composite materials is another area which is full of potential for future applications.

8. Summary

Composites reinforced with textile materials have attracted great attention as lightweight and strong materials for various advanced applications such as aerospace, construction, automotive, medicine, and sports. Traditionally, multilayered structures consisting of unidirectional prepregs have been used for the production of advanced composite materials. However, these material systems lack through-the-thickness fibers and consequently exhibit poor out-of-plane properties such as interlaminar strength and damage tolerance. In addition, the layering process is labor intensive and time consuming increasing the overall cost. Textile materials in the form of short and continuous fibers as well as 2D and 3D fabric preforms offer enormous design possibilities through the proper manipulation of fiber architecture. 2D fabrics can be produced at a much lower cost compared with unidirectional prepregs and are easy to handle during composite manufacturing. However, they also require the time-consuming layering procedure similar to unidirectional laminates, and they also have poor out-of-plane properties due to lack of z-fibers. 3D woven, braided, stitched, and non-crimp fabrics possess through-the-thickness reinforcement and thus have good out-of-plane properties. In addition, the costly layering step is eliminated with these structures. They are suitable for net-shape manufacturing and inclusion of a wide range of structural details such as holes, bifurcations, and junctions which significantly reduce the production time and cost as well as eliminating the weaknesses in the resulting structures originating from these modifications. The major limitation of 3D fabric manufacture is the high production costs due to poorly developed machine technologies and economies of scale issues. These limitations can be eliminated with the implementation of more efficient production technologies and effective engineering design of 3D textile materials. New developments in fiber and matrix systems, interface modification methods, as well as in nanotechnology will increase the use of textile composites in various applications. It is expected that textile materials will continue to play an important role for the reinforcement of advanced composite materials in the future.

Author details

Nesrin Sahbaz Karaduman[1]*, Yekta Karaduman[1], Huseyin Ozdemir[2] and Gokce Ozdemir[3]

*Address all correspondence to: nesrin38@gmail.com

1 Department of Textile, Clothing, Footwear and Leather, Akdagmadeni Vocational High School, Bozok University, Akdagmadeni, Yozgat, Turkey

2 Department of Textile, Clothing, Footwear and Leather, Vocational High School of Technical Sciences, Gaziantep University, Gaziantep, Turkey

3 Department of Fashion and Textile Design, Faculty of Fine Arts, Gaziantep University, Gaziantep, Turkey

References

[1] Ko FK. 3-D textile reinforcements in composite materials. In: Miravete A, editor. 3-D Textile Reinforcements in Composite Materials. Cambridge: Woodhead Publishing Limited; 1999. pp. 9-42

[2] Lavin JG. Carbon fibers. In: Hearle JWS, editor. High-Performance Fibers. Cambridge: Woodhead Publishing Limited; 2001. pp. 156-190

[3] Shambaugh RL. A macroscopic view of the melt-blowing process for producing microfibers. Industrial & Engineering Chemistry Research. 1988;**27**:2363-2372. DOI: 10.1021/ie00084a021

[4] Mochida I, Korai Y, Ku CH, Watanabe F, Sakai Y. Chemistry of synthesis, structure, preparation and application of aromatic-derived mesophase pitch. Carbon. 2000;**38**:305-328. DOI: 10.1016/S0008-6223(99)00176-1

[5] Edie DD. The effect of processing on the structure and properties of carbon fibers. Carbon. 1998;**36**:345-362. DOI: 10.1016/S0008-6223(97)00185-1

[6] Kwolek SL, Morgan PW, Sorenson WR, inventors. Process of making wholly aromatic polyamides. 1962. US Patent 3063966 A

[7] Kwolek SL, inventor. Optically anisotropic aromatic polyamide dopes. 1972. US Patent 3671542 A

[8] Rebouillat S. Aramids. In: Hearle JWS, editor. High-performance Fibres. Cambridge: Woodhead Publishing Limited; 2001. pp. 23-61

[9] Morgan PW, Kwolek SL. Interfacial polycondensation 2. Fundamentals of polymer formation at liquid interfaces. Journal of Polymer Science. 1959;**40**:299-327

[10] Morgan PW, Kwolek SL. Interfacial polycondensation XII. Variables affecting stirred polycondensation reactions. Journal of Polymer Science. 1962;**62**:33-58. DOI: 10.1002/pol.1962.1206217304

[11] Jassal M, Ghosh S. Aramid fibres—An overview. Indian Journal of Fibre & Textile Research. 2002;**27**:290-306

[12] History of the Owens-Corning Fiberglas Corporation. [Internet]. 2016. Available from: http://media.owenscorning.com/history [Accessed: 08 January 2016]

[13] Kumar S, Gupta VB. Manufactured fibres for high performance, industrial and non-conventional applications. In: Gupta VB, Kothari VK, editors. Manufactured Fibre Technology. London: Chapman & Hall; 1997. pp. 514-559

[14] Wallenberger FT. Structural silicate and silica glass fibers. In: Wallenberger FT, editor. Advanced Inorganic Fibers Processes, Structures, Properties, Applications. Dordrecht: Kluwer Academic Publishers; 1999. pp. 129-168

[15] Jones FR. Glass fibres. In: Hearle JWS, editor. High-Performance Fibres. Cambridge: Woodhead Publishing Limited; 2001. pp. 191-238

[16] Deitzel JM, McDaniel P, Gillespie JW. High performance polyethylene fibers. In: Bhat G, editor. Structure and Properties of High-Performance Fibers. Cambridge: Woodhead Publishing Limited; 2016

[17] Barham PJ, Keller A. High-strength polyethylene fibres from solution and gel spinning. Journal of Materials Science. 1985;**20**:2281-2302. DOI:10.1007/BF00556059

[18] Kavesh S, Prevorsek DC, inventors. High tenacity, high modulus polyethylene and polypropylene fibers and intermediates therefore. 1982. US Patent 4413110

[19] Wilson DM. Ceramic fibers. In: Miracle DB, Donaldson SL, editors. ASM Handbook Volume 21: Composites. Ohio: ASM International; 2001

[20] Chung DDL. Composite Materials: Science and Applications. London: Springer; 2010. 371 p

[21] Lord PR, Mohamed MH. Weaving: Conversion of Yarn to Fabric. Durham: Merrow Publishing Company; 1982. 368 p

[22] Tan P, Tong L, Steven GP. Modelling for predicting the mechanical properties of textile composites-A review. Composites Part A: Applied Science and Manufacturing. 1997;**28**:903-922. DOI: 10.1016/S1359-835X(97)00069-9

[23] Chou TW. Microstructural Design of Fiber Composites. New York: Cambridge University Press; 1992. 592 p

[24] Spencer DJ. Knitting Technology: A Comprehensive Handbook and Practical Guide. Cambridge: Woodhead Publishing Limited; 2001. 386 p

[25] Wray GR, Vitols R. Advances in stitch-bonding, warp- and weft-knitting systems, and automated knitwear manufacture. In: Happey F, editor. Contemporary Textile Engineering. London: Academic Press; 1982. pp. 375-409

[26] Baker A, Dutton S, Kelly D. Composite Materials for Aircraft Structures. Reston, Virginia: American Institute of Aeronautics and Astronautics, Inc.; 2004. 599 p

[27] Pastore C, Gowayed Y. Structure and mechanics of 2D and 3D textile composites. In: Schwartz P, editor. Structure and Mechanics of Textile Fibre Assemblies. Cambridge: Woodhead Publishing Limited; 2008. pp. 141-189

[28] Wilson A. The formation of dry, wet, spunlaid and other types of nonwovens. In: Chapman RA, editor. Applications of Nonwovens in Technical Textiles. Cambridge: Woodhead Publishing Limited; 2010. pp. 3-17

[29] Bluck RM, inventor. High speed bias weaving and braiding. 1969. US Patent 3426804

[30] Florentine RA, inventor. Apparatus for weaving a three dimensional article. 1982. US Patent 4312261

[31] Bilisik K, Sahbaz Karaduman N, Bilisik NE, Bilisik HE. Three dimensional fully interlaced woven preforms for composites. Textile Research Journal. 2013;**83**:2060-2084. DOI: 10.1177/0040517513487791

[32] Stig F, Hallström S. Spatial modelling of 3D-woven textiles. Composite Structures 2012;**94**: 1495-1502. DOI: 10.1016/j.compstruct.2011.12.003

[33] Bilisik K, Sahbaz Karaduman N, Bilisik NE, Bilisik HE. Three dimensional (3D) circular various weave patterns in woven preform structures. Textile Research Journal. 2014;**84**:638-654. DOI: 10.1177/0040517513499437

[34] Bilisik K. Three-dimensional braiding for composites: A review. Textile Research Journal. 2013;**83**:1414-1436. DOI: 10.1177/0040517512450766

[35] Tada M, Uozumi T, Nakai A, Hamada H. Structure and machine braiding procedure of coupled square braids with various cross sections. Composites Part A: Applied Science and Manufacturing. 2001;**32**:1485-1489. DOI: 10.1016/S1359-835X(01)00047-1

[36] Bilisik K, Sahbaz N. Structure-unit cell base approach on three dimensional (3D) representative braided preforms from 4-step braiding: Experimental determination of effect of structure-process parameters on predetermined yarn path. Textile Research Journal. 2012;**82**:220-241. DOI: 10.1177/0040517511404597

[37] Cox BN, Flanagan G. Handbook of Analytical Methods for Textile Composites. Washington: NASA Contractor Report 4750; March 1997

[38] Tong L, Mouritz AP, Bannister MK. 3D Fibre Reinforced Polymer Composites. Oxford: Elsevier; 2002. 254 p

[39] Ko FK. Three-dimensional fabrics for composites. In: Chou TW, Ko F, editors. Composite Materials Series Vol. 3: Textile Structural Composites. Amsterdam: Elsevier Science Publishers; 1989. pp. 129-171

[40] Williams DJ. New knitting methods offer continuous structures. Advance Composite Engineering. 1978; Summer: 12-13

[41] Fangueiro R, Soutinho F. Textile structures. In: Fangueiro R, editor. Fibrous and Composite Materials for Civil Engineering Applications. Cambridge: Woodhead Publishing Limited; 2011. pp. 62-91

[42] Roye A, Gries T, Peled A. Spacer fabrics for thin walled concrete elements. In: 6th RILEM Symposium on Fiber-Reinforced Concrete (FRC); 20-22 September 2004; Varenna, Italy. pp. 1505-1514

[43] Roye A, Gries T. 3-D textiles for advanced cement based matrix reinforcement. Journal of Industrial Textiles. 2007;**37**:163-173. DOI: 10.1177/1528083707078136

[44] Schnabel A., Gries T. Production of non-crimp fabrics for composites. In: Lomov SV, editor. Non-crimp Fabric Composites. Cambridge: Woodhead Publishing Limited; 2011. pp. 3-41

[45] Hu J. 3-D Fibrous Assemblies: Properties, Applications and Modelling of Three-Dimensional Textile Structures. Cambridge: Woodhead Publishing Limited; 2008. 280 p

[46] Bilisik AK, Turhan Y. Multidirectional stitched layered aramid woven fabric structures and their experimental characterization of ballistic performance. Textile Research Journal. 2009;**79**:1331-1343. DOI: 10.1177/0040517509104275

[47] Bilisik K, Yolacan G. Short beam strength properties of multistitched biaxial woven E-glass/polyester nano composites. Journal of Industrial Textiles. 2015;**45**:199-221. DOI: 10.1177/1528083714528017

[48] Dow MB, Smith DL. Damage tolerant composite materials produced by stitching carbon fabrics. In: Proceedings of the 21st International SAMPE Technical Conference; 25-28 September 1989; Atlantic City, NJ. pp. 595-605

[49] Bilisik K, Karaduman NS, Bilisik NE. 3D fabrics for technical textile applications. In: Jeon HY, editor. Non-woven Fabrics. Rijeka: InTech; 2016. pp. 81-141. DOI: 10.5772/61224

[50] Wang Z, Hu H. Auxetic materials and their potential applications in textiles. Textile Research Journal. 2014;**84**:1600-1611. DOI: 10.1177/0040517512449051

[51] Sloan MR, Wright JR, Evans KE. The helical auxetic yarn—A novel structure for composites and textiles; geometry, manufacture and mechanical properties. Mechanics of Materials. 2011;**43**:476-486. DOI: 10.1016/j.mechmat.2011.05.003

[52] Ugbolue SC, Kim YK, Warner SB, Fan Q, Yang CL, Kyzymchuk O, Feng Y. The formation and performance of auxetic textiles. Part I: theoretical and technical considerations. The Journal of The Textile Institute. 2010;**101**:660-667. DOI: 10.1080/00405000902733790

[53] Steffens F, Rana S, Fangueiro R. Development of novel auxetic textile structures using high performance fibres. Materials and Design. 2016;**106**:81-89. DOI: 10.1016/j.matdes.2016.05.063

[54] Wang Z, Zulifqar A, Hu H. Auxetic composites in aerospace engineering. In: Rana S, Fangueiro R, editors. Advanced Composite Materials for Aerospace Engineering. Cambridge: Woodhead Publishing Limited; 2016. pp. 213-240

[55] Ge Z, Hu H. Innovative three-dimensional fabric structure with negative Poisson's ratio for composite reinforcement. Textile Research Journal. 2013;**83**:543-550. DOI: 10.1177/0040517512454185

[56] Groover ME. Fundamentals of Modern Manufacturing- Materials, Processes, and Systems. New Jersey: Prentice-Hall; 1996. 1128 p.

[57] Campbell FC. Manufacturing Processes For Advanced Composites. Oxford: Elsevier; 2004. 532 p

[58] Islam MS, Roy AK. Surface preparation of fibres for composite applications. In: Rana S, Fangueiro R, editors. Fibrous and Textile Materials for Composite Applications. Singapore: Springer; 2016. pp. 301-315

[59] Adams DF, Carlsson LA, Pipes RB. Experimental Characterization of Advanced Composite Materials. Boca Raton, FL: CRC Press; 2003. 256 p

[60] Daniel IM, lshai O. Engineering Mechanics of Composite Materials. New York: Oxford University Press; 2006

[61] Kaw AK. Mechanics of Composite Materials. Boca Raton, FL: CRC Press; 2006

[62] Hashin Z. Theory of Fibre Reinforced Materials. CR-1974. Washington: NASA; 1972

[63] Ishikawa T, Chou TW. Elastic behavior of woven hybrid composites. Journal of Composite Materials. 1982;**16**:2-19. DOI: 10.1177/002199838201600101

[64] Ishikawa T, Chou TW. One-dimensional micromechanical analysis of woven fabric composites. AIAA Journal. 1983;**21**:1714-1721. DOI: 10.2514/3.8314

[65] Ishikawa T, Chou TW. Nonlinear behavior of woven fabric composites. Journal of Composite Materials. 1983;**17**:399-413. DOI: 10.1177/002199838301700502

[66] Naik NK, Shembekar PS. Elastic behavior of woven fabric composites: I—Lamina analysis. Journal of Composite Materials. 1992;**26**:2196-2225. DOI: 10.1177/002199839202601502

[67] Shembekar PS, Naik NK. Elastic behavior of woven fabric composites: II—Laminate analysis. Journal of Composite Materials. 1992;**26**:2226-2246. DOI: 10.1177/002199839202601503

[68] Hahn HT, Pandey R. A micromechanics model for thermoelastic properties of plain weave fabric composites. Journal of Engineering Materials and Technology. 1994;**116**:517–523. DOI: 10.1115/1.2904322

[69] Ko FK, Chou TW. Composite Materials Series 3—Textile Structural Composites. Amsterdam: Elsevier Science; 1989

[70] Vandeurzen P, Ivens J, Verpoest I. A three-dimensional micromechanical analysis of woven-fabric composites: I. Geometric analysis. Composites Science and Technology. 1996;**56**:1303-1315. DOI: 10.1016/S0266-3538(96)00092-9

[71] Vandeurzen P, Ivens J, Verpoest I. A three-dimensional micromechanical analysis of woven-fabric composites: II. Elastic analysis. Composites Science and Technology. 1996;**56**:1317-1327. DOI: 10.1016/S0266-3538(96)00091-7

[72] Cox BN, Dadkhah MS. The macroscopic elasticity of 3D woven composites. Journal of Composite Materials. 1995;**29**:785-819. DOI: 10.1177/002199839502900606

[73] Tan P, Tong L, Steven GP. Models for predicting thermomechanical properties of three-dimensional orthogonal woven composites. Journal of Reinforced Plastics and Composites. 1999;**18**:151-185. DOI: 10.1177/073168449901800204

[74] Tan P, Tong L, Steven GP. Micromechanics models for mechanical and thermomechanical properties of 3D through-the-thickness angle interlock woven composites. Composites Part A: Applied Science and Manufacturing. 1999;**30**:637-648. DOI: 10.1016/S1359-835X(98)00176-6

[75] Pastore C, Gowayed Y. A self-consistent fabric geometry model: Modification and application of a fabric geometry model to predict the elastic properties of textile composites. Journal of Composites, Technology and Research. 1994;**16**:32-36. DOI: 10.1520/CTR10392J

[76] Yang JM, Ma CL, Chou TW. Fiber inclination model of three-dimensional textile structural composites. Journal of Composite Materials. 1986;**20**:472-484. DOI: 10.1177/002199838602000505

[77] Rudd CD, Owen MJ, Middleton V. Mechanical properties of weft knit glass fibre/polyester laminates. Composites Science and Technology. 1990;**39**:261-277. DOI: 10.1016/0266-3538(90)90045-7

[78] Ramakrishna S, Hull D. Tensile behaviour of knitted carbon-fibre-fabric/epoxy laminates—part II: Prediction of tensile properties. Composites Science and Technology. 1994;**50**:249-258. DOI: 10.1016/0266-3538(94)90146-5

[79] Brown G. The use of composites in aircraft construction. [Internet]. 2014. Available from: https://vandaair.com/2014/04/14/the-use-of-composites-in-aircraft-construction/ [Accessed: 29 January 2017]

[80] Pilato LA, Michno MJ. Advanced Composite Materials. Heidelberg: Springer; 1994. 208 p

[81] Liu Y, Araujo M, Hu H. Advanced fibrous architectures for composites in aerospace engineering. In: Rana S, Fangueiro R, editors. Advanced Composite Materials for Aerospace Engineering: Processing, Properties and Applications. Oxford: Elsevier; 2016. pp. 17-58

[82] Mullen CK, Roy PJ. Fabrication and properties description of Avco 3D carbon–carbon cylindrical composites. In: National SAMPE Symposium; 11-13 April 1972; Los Angeles, California. pp. III-A-Two-l–8

[83] Wong R. Sandwich construction in the Starship. In: Proceedings of the 37th International SAMPE Symposium; 9-12 March 1992; Anaheim, California. pp. 186-197

[84] Mouritz AP, Bannister MK, Falzon PJ, Leong KH. Review of applications for advanced three-dimensional fibre textile composites. Composites Part A: Applied Science and Manufacturing. 1999;**30**:1445-1461. DOI: 10.1016/S1359-835X(99)00034-2

[85] Kandero SW. France, Russia to join in scramjet flight tests. Aviation Week and Space Technology. 2001 60-62

[86] Potluri P, Nawaz S. Developments in braided fabrics. In: Gong RH, editor. Specialist Yarn and Fabric Structures. Cambridge: Woodhead Publishing Limited; 2011. pp. 333-353

[87] Technology A. Bombardier Wing Flap. [Internet]. 2017. Available from: http://www.braider.com/Case-Studies/Bombardier-Wing-Flap.aspx [Accessed: 29 January 2017]

[88] Technology A. Honeywell Jet Engine Stator Vane. [Internet]. 2017. Available from: http://www.braider.com/Case-Studies/Jet-Engine-Stator-Vane.aspx [Accessed: 29 January 2017]

[89] Technology A. GEnx Engine. [Internet]. 2017. Available from: http://www.braider.com/Case-Studies/GEnx-Engine.aspx [Accessed: 29 January 2017]

[90] Lowe J. Aerospace applications. In: Long AC, editor. Design and Manufacture of Textile Composites. Cambridge: Woodhead Publishing Limited; 2005. pp. 405-423

[91] Middendorf P, Metzner C. Aerospace applications of non-crimp fabric composites. In: Lomov SV, editor. Non-crimp Fabric Composites. Cambridge: Woodhead Publishing Limited; 2011. pp. 441-448

[92] Liguore SL, Montgomery JM, inventors. Systems and Methods for Reducing Noise in Aircraft Fuselages and Other Structures. 2012. US Patent 8297555

[93] Motavalli M, Czaderski C, Schumacher A, Gsell D. Fibre reinforced polymer composite materials for building and construction. In: Pohl G, editor. Textiles, Polymers and Composites for Buildings. Cambridge: Woodhead Publishing Limited; 2010. pp. 69-128

[94] Ehlig D, Schladith F, Frenzel M, Curbach M. Textilbeton—ausgefuhrte projekte im überblick. Beton-und Stahlbetonbau. 2012;**107**:777-785. DOI: 10.1002/best.201200034

[95] Hering. BetoShell: Textile-reinforced concrete as curtain facade. [Internet]. 2017. Available from: https://www.heringinternational.com/en/products-services/architectural-concrete/material-concepts/textile-reinforced-concrete/ [Accessed: 29 January 2017]

[96] Raupach M, Cruz CM. Textile-reinforced concrete: Selected case studies. In: Triantafillou T, editor. Textile Fibre Composites in Civil Engineering. Duxford: Elsevier; 2016. pp. 275-299

[97] Müller J, Zulliger A, Dorn M. Economic production of composite beams with 3D fabric tapes. Textile Month. 1994; September: 9-13

[98] S&P. FRP strengthening systems. [Internet]. 2017. Available from: http://www.sp-reinforcement.eu/en-CH/products [Accessed: 29 January 2017]

[99] Dufour C, Pineau P, Wang P, Soulat D, Boussu F. Three-dimensional textiles in the automotive industry. In: Chen X, editor. Advances in 3D Textiles. Cambridge: Woodhead Publishing Limited; 2015. pp. 265-291. DOI: 10.1016/B978-1-78242-214-3.00010-3

[100] Vaidya U. Composites for Automotive, Truck and Mass Transit-Materials, Design, Manufacturing. Lancaster, PA: DEStech Publications, Inc.; 2011. 419 p

[101] Turner TA, Harper LT, Warrior NA, Rudd CD. Low-cost carbon-fibre-based automotive body panel systems: A performance and manufacturing cost comparison. Proceedings of the Institution of Mechanical Engineers, Part D: Journal of Automobile Engineering. 2008;**222**:53-63. DOI: 10.1243/09544070JAUTO406

[102] Feraboli P, Masini A, Bonfatti A. Advanced composites for the body and chassis of a production high performance car. International Journal of Vehicle Design. 2007;**44**:233-246. DOI: 10.1504/IJVD.2007.013641

[103] Chapman GB. The need and requirements driving the development of polymer composites for automotive applications. In: Proceedings of ICCM-11; 1997; Gold Coast, Australia

[104] Ko FK. Textile composites for automotive structural components. In: Shishoo R, editor. Textile Advances in the Automotive Industry. Cambridge: Woodhead Publishing Limited; 2008. pp. 320-343

[105] Gries T, Stueve J, Grundmann T, Veit D. Textile structures for load-bearing applications in automobiles. In: Shishoo R, editor. Textile Advances in the Automotive Industry. Cambridge: Woodhead Publishing Limited; 2008. pp. 301-319

[106] Ellis JG. Textile reinforced composites in medicine. In: Long AC, editor. Design and Manufacture of Textile Composites. Cambridge: Woodhead Publishing Limited; 2005. pp. 436-443

[107] Ramakrishna S, Huang ZM, Kumar GV, Batchelor AW, Mayer J. An Introduction to Biocomposites. London: Imperial College Press; 2004. 223 p

[108] Evans SL, Gregson PJ. Composite technology in load-bearing orthopaedic implants. Biomaterials. 1998;**19**:1329-1342. DOI: 10.1016/S0142-9612(97)00217-2

[109] McCormick M. Soaring to new heights: The evolution of pole vaulting and pole materials. Illumin. 2010;**12**:1-5

[110] Froes FH. Is the use of advanced materials in sports equipment unethical? Journal of the Minerals, Metals, and Materials Society. 1997;**49**:15-19. DOI: 10.1007/BF02915473

[111] Van de Velde K. Textile composites in sports products. In: Long AC, editor. Design and Manufacture of Textile Composites. Cambridge: Woodhead Publishing Limited; 2005. pp. 444-457

Chapter 5

Hybrid Yarn Composites for Construction

Mutlu Kurban, Osman Babaarslan and
İsmail Hakkı Çağatay

Additional information is available at the end of the chapter

http://dx.doi.org/10.5772/intechopen.69034

Abstract

Textile-reinforced concrete (TRC) is a new innovative construction material that leads to light-weight and cost-effective construction. TRC consists of a finely grained cement-based matrix and high-performance, continuous multifilament yarns made of alkali-resistant glass, carbon, or polymer. Using these fibers provides superior mechanical properties and corrosion resistance in comparison with ferroconcrete. The application of epoxy resin coating to the textile materials improves the utilization of mechanical performance and handling properties as well. In recent years, researchers have studied alternative methods because coating process is very detailed and epoxy resin is of high cost. The experimental part of this chapter focuses on the experimental investigation carried out on high-strength concrete reinforced with hybrid yarns. Braiding technology was used to manufacture hybrid yarn from alkali-resistant glass fiber (ARG) and polypropylene (PP) filament. Next step, thermoplastic part of braided yarn was melted on press heating. Finally, TRC was produced from ARG, coated ARG, carbon fiber, coated carbon fiber, and heated hybrid yarns. Although the contribution of the heated hybrid yarn is limited, it is expected that the desired results will be obtained by changes in braiding yarn production and yarn composition ratios.

Keywords: technical textiles, hybrid yarn, composites, buildtech, textile-reinforced concrete

1. Introduction to textile composites for construction

The principal necessities of human are nutrition, clothing, and shelter. The clothing comes after nutrition and has got many selection factors such as social, economic, environmental, and physiological. The main material of clothing is textile. Textiles always have played an important role in protecting from different environmental conditions and making feel comfortable. Nowadays, the textile industry has to accelerate research on innovative and attractive products in many fields from agriculture to space. Recently, construction sector is one of the fields that used these innovative textile materials, and these textile materials are named "Buildtech." Buildtech products are defined as membrane, lightweight and massive construction materials, etc. for engineering and industrial buildings [1].

Construction is improving from the earliest times. At ancient times, people built kinds of buildings to shelter. To reduce cracking and to increase bearing capacity of buildings, they added hairs and vegetables in the mortar. Later with technology, construction materials changed. Different materials were used for each period. In the early period, mortar was used, but later various materials such as wood, stone, marble, and steel were used. Finally, with the discovery of concrete, its use in the construction industry has become widespread [2, 3].

Since 1800s, quick developments in construction materials technology have allowed civil engineers to achieve impressive gains in the safety, economy, and functionality of structures built to serve the common needs of the community [4]. Today, in the construction sector, steel-reinforced concrete is most important, and it is used widely for structural applications. For several decades, textile-reinforced concrete (TRC) has been innovative material for constructions. TRC can be used instead of conventional composite-building materials for many new applications [5]. TRC is a new composite material consisting of a fine-grained concrete matrix and corrosion-resistant high-performance multifilament yarns such as alkali-resistant glass, carbon, basalt, and polymer [6]. These raw materials are used to produce different textile forms for TRC. Those suited for the reinforcement of concrete matrix are yarns, warp knits (plain, circular or three-dimensional), multi-plies (plain or circular), and woven [7]. Thanks to these textile materials, production of TRC leads to thin structural elements with high strength, high durability, and corrosion resistance [8].

In the present architectures and constructions, there is a specific trend toward innovative structures of high-quality materials such as carbon, glass fibers, basalt, and aramid that continuously increase the requirements placed on the construction materials and that demand a continuous development of their properties. Using nonmetallic high-performance fibers as concrete reinforcement allows for the production of thin and light-weight elements with high durability and the potential of economic savings. These advantages together with the high scope of design options given to the architects have made glass-fiber–reinforced concrete (GFRC) a widespread construction material around the world. A disadvantage of the reinforcement with chopped strands (for example, AR-glass or PP) is the partial unorientated distribution of the fibers over the total cross-section, reducing their effectiveness. In contrast to steel reinforcement, AR glass or carbon fibers in the textile can be positioned in almost any direction and afterward nearly perfectly adopted to the orientation of the applied load. It is thus possible to create an extremely effective reinforcement. The use of corrosion-resistant technical textiles reduces concrete covers significantly and thus allows for light weight and thin concrete structures [9, 10].

Glass fiber is widely used in construction since 1950s. The cementitious matrix has highly alkaline environment (pH-value > 12.5), and glass fiber has poor corrosion resistance in a highly alkaline environment. Despite the glass fiber being considered as a reinforcement of cementitious materials for several decades, because of poor alkaline resistance, the limitation of structural applications still remains. For this reason, the alkali-resistant glass fiber (ARG) was preferably designed to reinforce cementitious matrices. Enhanced mainly by a high percentage of zirconia (ZrO_2 > 15% by weight) content, the alkali-resistant glass fiber (ARG) was designed to reinforce cementitious matrices that have been used in construction and civil engineering since the late 1960s. Despite ARG being more resistant to highly alkaline environment than normal glass fiber, many attempts have been made to modify either matrix or fiber by adding fillers or by surface coatings of polymer and carbon layers. In the composites sector, a coating is known as a widespread method of providing corrosion protection in order to improve the durability of engineering structures. For alkali-resistant glass fibers (ARGs), multifunctional sizings are required to provide the surface protection, abrasion resistance, and strength maintenance in the concrete. However, both durability in alkali environment and economic considerations have limited the commercial use of these materials. For enhancing the long-term resistance of glass-fiber–reinforced cement products, it is thus very important to develop an inexpensive and applicable coating to modify the ARG fiber surface and examine how the coatings interact with the surrounding cementitious matrix [11–13].

In the production of TRC, the application of polymer coating to the textile materials improves the utilization of mechanical performance and handling properties as well. Generally, thermoset resin is used as a coating material. Coating process is very long term and tough, and also, cost of coating materials is very expensive. Hence, researchers have searched for alternative methods to provide the surface protection, abrasion resistance, and strength maintenance for textile materials in the concrete.

In this study, hybrid yarn was used in the production of TRC as a reinforcement material. First of all, hybrid yarn was produced with braiding technology. In the hybrid yarn production, while alkali-resistant glass fibers (ARGs) were used as a reinforcement material, polypropylene filament (PP) is used as a matrix material. After production of hybrid yarn was heated in the oven, melted polymer covered alkali-resistant glass fibers (ARGs). Melted hybrid yarns were used in the production of TRC. All TRC samples were tested to see the effect of the hybrid yarn on the strength of the concrete. Information about hybrid yarn technology and braiding technology will be mentioned in the following sections.

2. Hybrid yarn technology

For many years, thermoset and thermoplastic composites are often used to produce advanced lightweight structures. Thermoset prepregs, which were previously a combination of resin and fiber, as preformed materials must be kept refrigerated to interrupt the ongoing curing reaction [14]. Generally, in many composite processing methods, the matrix is added to the fibers at the time of manufacture. By using additives and fillers, it improves the surface quality of the part, alleviates some processing problems and reduces the total cost [15, 16].

In the thermoset composite materials, the curing process requires heat and pressure. To improve its usability, the prepreg is typically stored in a freezer and then cured at a high temperature. In the 1960s prepregs produced necessitated quite high cure temperatures and had very short out-lives; nowadays, it is possible for them to be stored in a freezer, have an out-life of possibly several months and a cure temperature of between 50 and 200°C. Prepregs tend to be epoxy resins reinforced with carbon, glass, or aramid fibers. High-temperature polymer composites are occasionally used as other matrix materials [16].

In thermoplastic polymers, the organic molecular units are bonded together. They differ from thermoset polymers in this way. Thermoplastic composites are preferred for their ability to be formed by heat at low pressure. While thermosets are produced with manual production, the thermoplastic processing techniques are machine friendly and enable short processing cycles [14, 17].

Thermoplastic polymers are divided into two groups of amorphous and semicrystalline polymers according to their molecular structure. While amorphous polymers have randomly oriented molecular chains, in semicrystalline polymers, the chains are symmetric enough to be fitted into an ordered crystalline state. Because of the large molecular chain length, complete crystallinity cannot be obtained, and both crystalline and amorphous regions coexist in the solid. When a polymer is cooled, there are some changes. From above, the melt temperature (T_m) to glass transition (T_g), the chains lose mobility and begin to interact segment by segment. At T_g, the molecular chains are locked in place, and the polymer assumes a "glassy" state. In this case, the molecular orientation is random due to the irregular chain structure. At semicrystalline state, polymers crystallize at temperatures so high from the T_g, regions of amorphous, and crystalline phase coexist [17].

When a thermoplastic polymer is heated above the T_m, polymer melt viscosity reduces, and it melts. Many thermoplastic polymers can be used as matrix materials in textile composites. Generally, polymers having a lower melting temperature are preferred due to their ease in processing and drying. There are many approaches such as co-woven yarns, inter-dispersed fabrics, plied matrix, powder impregnation, and commingled materials, and they are used to impregnate the reinforcing textile structure with the matrix polymer [17]. In this techniques, the formability is improved. The first three techniques require direct flow of the polymer melt into the fiber bed. Powder impregnated and commingled materials have fiber and matrix constituents intimately mingled before the melt impregnation process. This reduces the distance that the resin must flow to achieve impregnation, offering the possibility of a fast wet-out during consolidation [18]. Nevertheless, the production of thermoplastic composites has some technical processing problems. For example, as thermoplastic melt has much higher viscosity than common thermoset resins, it is difficult to obtain homogeneous composite structure. There are some methods to produce homogeneous composite structure, and they are the following:

- Polymer direct melt extrusion or pultrusion that reinforce fibers are pulled through a resin;
- Solvents or low viscosity precursors; and
- Close contact between the fiber and the matrix [14].

Although there are novel thermoplastic composite manufacturing methods such as co-compression molding of textile preforms with a flowable core and overinjection molding of stamped preforms, close contact between the fiber and the matrix is widely used in textile composites [17].

2.1. Types of hybrid yarns

Hybrid yarns can be produced with many methods such as ring spinning, commingling, and braiding. These methods provide uniform distribution of matrix and reinforcement fibers as well as reduce the damage to reinforcing fibers [19]. The information on methods will be given following sections.

2.1.1. Ring spinning

Classical ring spinning machines have a pair of rollers for the drafting, twisting, and winding mechanism and can use only one type of fiber to manufacture spun yarns. Present ring spinning machines need some modifications to manufacture hybrid yarns. Generally, core ring spinning on a modified ring frame needs with it special guiding devices and feeder rollers (**Figure 1**). With slight modifications and little investment, this technique can be used to produce hybrid yarns by the core spinning system, and these machines are most commonly used to produce core-spun yarns containing elastane as a core. While one type of fiber is used in classical ring spinning, staple fibers and filaments can be used in core ring spinning [19–21].

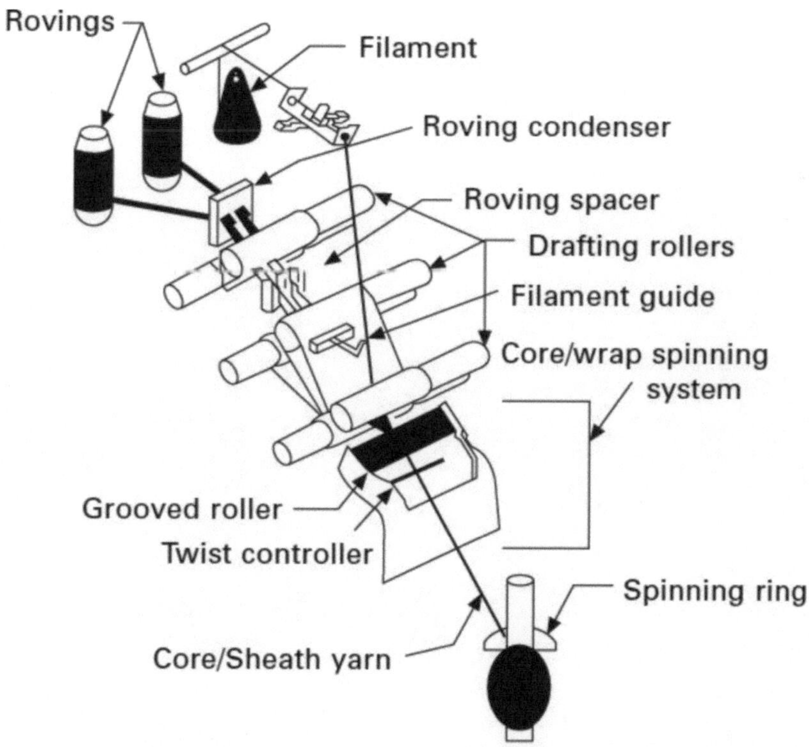

Figure 1. Modified core spinning system [21].

2.1.2. Rotor spinning

As in ring spinning, rotor spinning needs some modifications to manufacture hybrid yarns. In the modified rotor spinning machine, hybrid yarns can be produced by combining staple fibers with filament yarns under varying filament overfeeds. **Figure 2** shows the modified rotor and the diagram of spinning process. In this machine, the staple fiber and the continuous filament were fed into the rotor to produce hybrid yarn. This technology has a tendency to introduce filament misalignment in the core yarn, which is not desirable for composite application [19, 22].

2.1.3. DREF spinning

This spinning system can be used to manufacture core-spun hybrid yarns for thermoplastic composites. In classical DREF spinning, the completely opened fiber strand is brought into engagement with the rotating open end of the yarn by a perforated drum. Attachment and strengthening of the fibers are accomplished by continuously rotating the yarn in the converging region of the two drums. The spinning of the yarn takes place with the help of the rotational movement of the two cylinders, and the friction on the cylinder surface contributes to this [19].

This system can be used to fabricate a hybrid yarn consisting of reinforcement filaments of high-performance fibers in the core and staple fibers of the thermoplastic matrix material such as polyester, poly-ether-etherketone or liquid crystal polymers in the sheath. Hasan et al. studied the application of carbon filament yarn (CFY)–based conductive hybrid yarn as the heating element in a textile-reinforced concrete structure. In order to manufacture this hybrid yarn, they used DREF-2000 spinning technique. Carbon fiber was used as a core, and a mixture of short glass and polypropylene fibers was used as a sheath. **Figure 3** shows friction spinning machine and cross-section of hybrid yarn [19, 23].

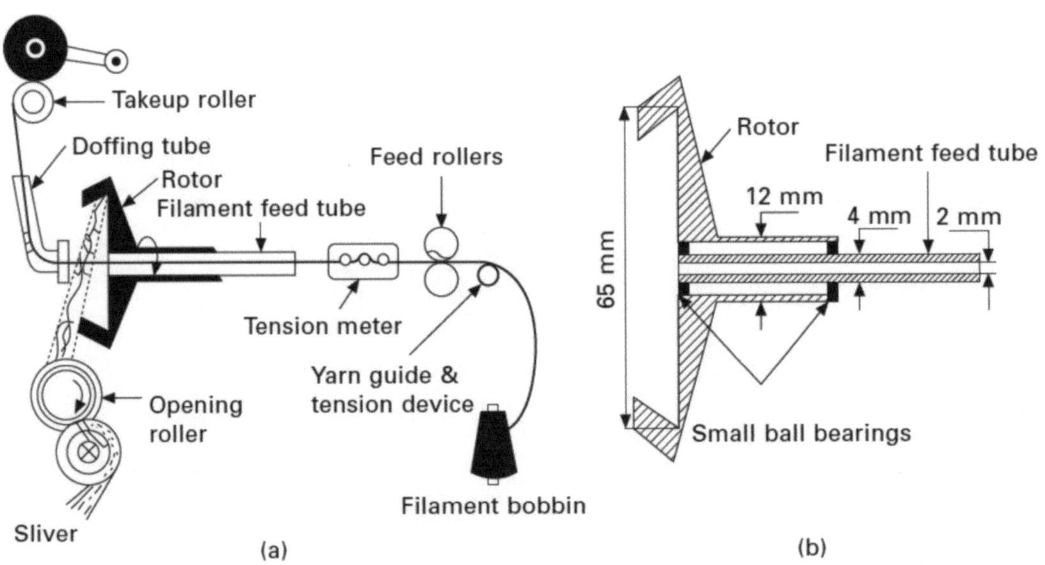

Figure 2. (a) Diagram of spinning process (b) Modified rotor [22].

2.1.4. *Wrap spinning*

In this system, a roving or a sliver feedstock passes through a hollow spindle without receiving true twist. The continuous filament yarn, which is mounted on the hollow shaft, is passed through the hollow spindle as shown and is wound on the central component (**Figure 4**). Hence, this filament strand is wound around the twistless strand in the core. With some modifications, this system can manufacture hybrid yarn using filament for both core and wrapping. When thermoplastic filament is used as a wrap yarn, during the consolidation process, this component will melt and it becomes part of the matrix [19, 24].

2.1.5. *Air-jet texturing*

Air-jet texturing is completely a mechanical process, and reinforcing and matrix filaments are combined by air. In this process, the air nozzle is the heart of the air-jet texturing process. Supply yarn is overfed into the turbulent zone, and here, compressed air is guided mainly parallel to the yarn path. This compressed air flow opens up filament bundles and then builds mingling sections (**Figure 5**), as a result, mixed filament yarn is manufactured, but, while filaments mix, some loops occur [19, 25].

2.1.6. *Commingling*

Like air-jet texturing, in the commingling process, compressed air is used to generate entanglements in and among filaments. In this process, two or more yarns, such as carbon or glass, and a thermoplastic matrix yarn are converted to a single strand. Commingled yarn consists

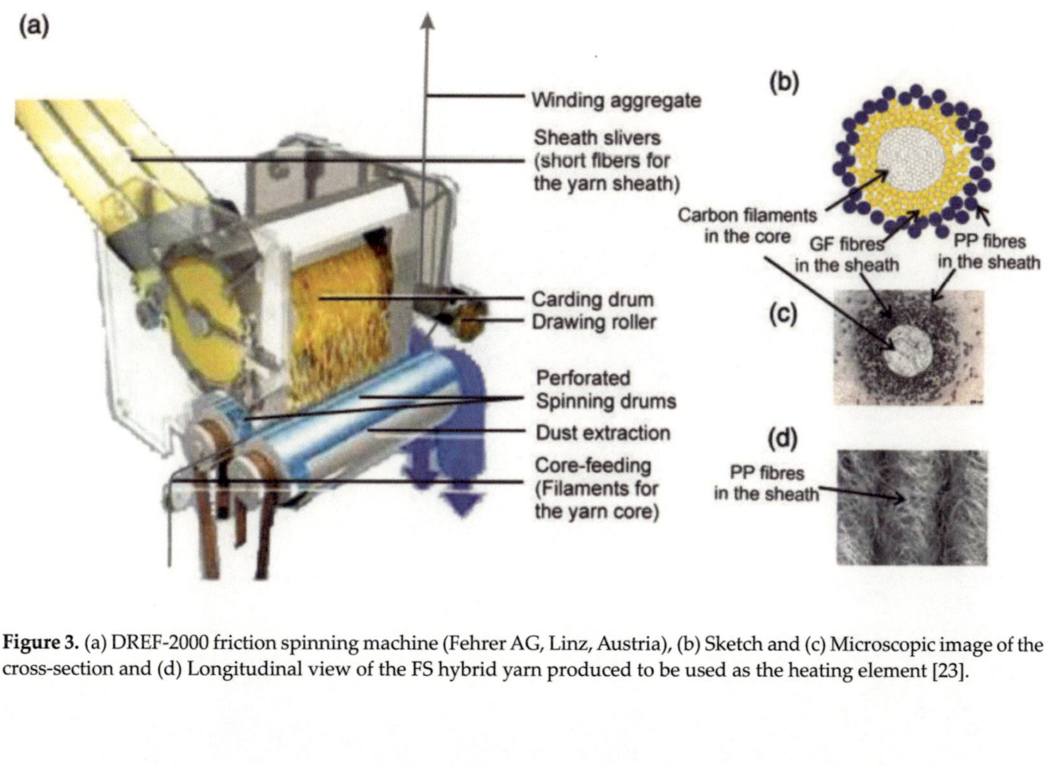

Figure 3. (a) DREF-2000 friction spinning machine (Fehrer AG, Linz, Austria), (b) Sketch and (c) Microscopic image of the cross-section and (d) Longitudinal view of the FS hybrid yarn produced to be used as the heating element [23].

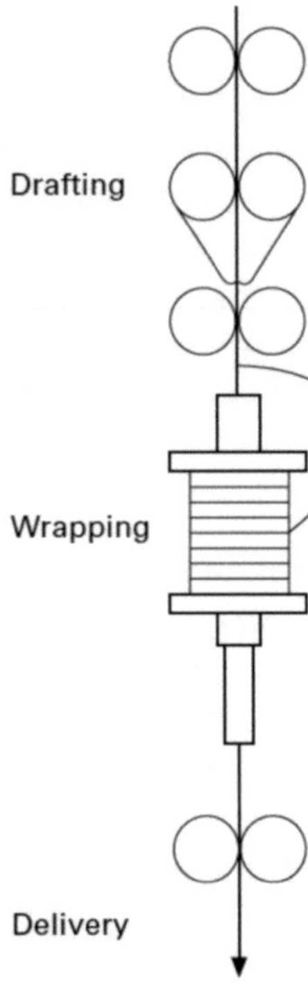

Figure 4. Filament wrap spinning [19].

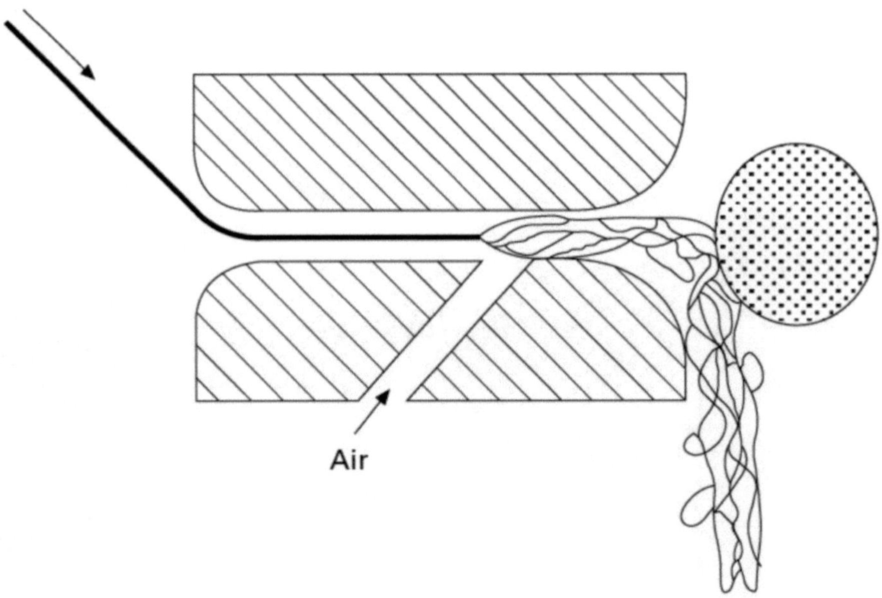

Figure 5. Air-jet yarn texturing [25].

of blended two parts combination, reinforcing filament yarn and filament yarn spun from thermoplastic polymers. In the beginning, the multifilament yarns are scattered by compressed air, and then they are often mixed with each other in parallel. Unlike air-jet texturing, loops do not occur, and generally, filaments are parallel to each other (**Figure 6**). Thanks to the changes in main production parameters such as degree of overfeeding, production speed, and air pressure, it can be manufactured as a hybrid yarn in the desired filament distribution. It is possible to obtain a homogenous distribution of reinforcement and matrix, and this reduces the mass transfer distance of the matrix during processing. In this way, a fast and complete impregnation of the reinforcement filaments is possible [19, 25, 26].

2.1.7. Parallel winding

In this simple process, the two components of the hybrid yarns are brought together side by side, as shown in **Figure 7**. Parallel winding is also called tape winding or the side-by-side technique. In this process, the continuous reinforcement filaments and the thermoplastic filaments are combined in the form of a band. Then, after the heating process, this tape is converted to a composite material [19, 24].

2.1.8. Stretch breaking

In this process, hybrid yarns consist of discontinuous thermoplastic fibers and filaments. They are brought together into a well-oriented coherent bundle by the insertion of a degree of twist. Thanks to this technology, a broken fiber feed on one or two tows of high-modulus filaments such as carbon or glass and also produces highly consistent yarns with minimal fiber damage. When heating is applied, the composite structure is formed by melting the thermoplastic structure [19, 24].

2.1.9. KEMAFIL technology

This technology has been developed in Germany for geotextiles. In this technology, a type of circular knitting machine is used to manufacture hybrid yarn. In this machine, a parallel

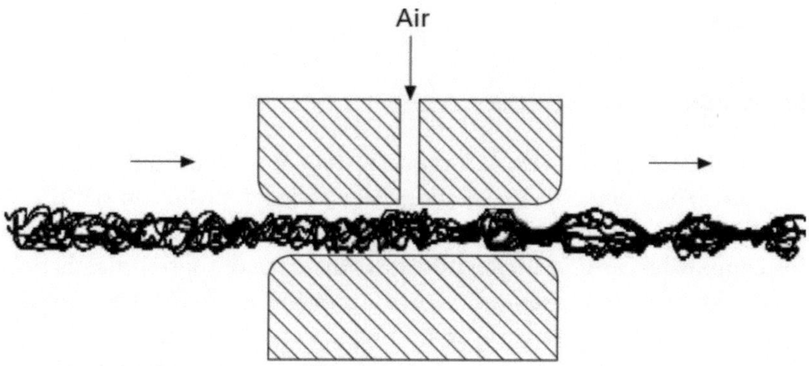

Figure 6. Hybrid yarn by commingling process [19].

Figure 7. Structure of side-by-side hybrid yarn [19].

arrangement of matrix fibers is surrounded by parallel reinforcing filaments. In this hybrid yarn, while the core consists of matrix and reinforcing filaments, the knitted sheath consists of only matrix fibers [19, 24].

2.1.10. Schappe technology

Schappe technology is same as stretch breaking. This technology is used to obtain more bulky and higher tenacity hybrid yarn from long fibers. These types of hybrid yarns are composed of a mixture of discontinuous reinforcing and matrix filaments surrounded by continuous matrix filaments. As this consists of transforming the continuous filaments put on top of long fibers, this technique removes the weak points of the fibers [19, 24].

2.1.11. Braided yarn

Braiding has been used since 1800s to produce textile fabrics and is generally used for producing narrow rope-like materials. In this technology, three or more strands of filaments or yarns are interlaced diagonally like in a Maypole dance. The filament bundles forming the braid are combined in a manner similar to the formation of the ribbon to form the braided yarn. In this way, tubular woven structure occurs. Researchers have focused braiding technology to meet new demands in composite production [27, 28]. Details of braided yarn and its production are discussed in Section 3.

2.2. Fiber distribution in hybrid yarns

The type of hybrid yarn production technique is very important on the fiber distribution in the hybrid yarn. The homogeneity of component fiber distribution within the matrix is strongly dependent on the hybrid yarn structure. Fiber distribution is influential in the quality of the composites, as it will affect the flow distance of thermoplastic polymer when heat is applied. When they are ranked according to the flow distance, the flow distance increased in the order of Schappe yarn, commingled yarn, Kemafil yarn, side-by-side yarn, and lastly, friction spun yarn. This means that the best degree of mixing of reinforcing and matrix are Schappe and commingled hybrid yarns fibers. Hybrid yarn structures are shown in **Table 1** [19].

Hybrid yarn structure	Position of performance filaments
	**Ring core-spun yarn (RS)** Performance filaments in core and covered spun fibers in a twisted form
	**Rotor core-spun yarn (ROS)** Performance filaments in core and covered spun fibers in a wrapped form
	**DREF core-spun yarn (DF)** Parallel arrangement of performance filaments in core and covered spun fibers
	**Wrap yarn (WS)** Parallel laid fibers in core and covered by performance filaments
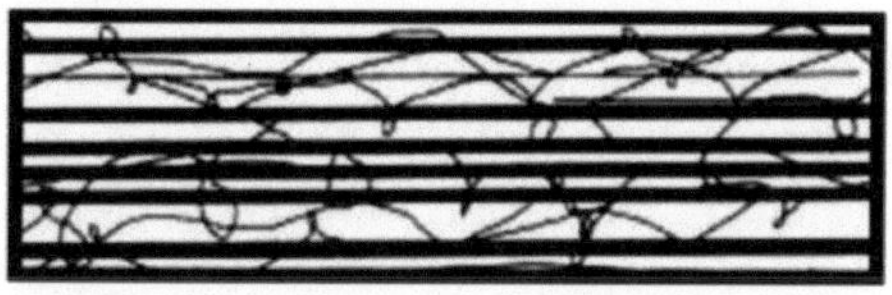	**Air-jet textured yarn (AT)** Thermoplastic filament and performance filaments are interlaced forming loop structure
	**Commingled (COM)** Thermoplastic filaments and performance filaments in a mingled form
	**Parallel winding (SBS)** Parallel arrangement of performance filaments and thermoplastic filaments
	**Stretch break (SB)** Stretched break thermoplastic filaments covered by performance filaments
	**Kemafil technology (KEM)** Parallel arrangement of thermoplastic filaments surrounded by parallel performance filaments in the core, sheath in continuous filament in a knitted form

Hybrid yarn structure	Position of performance filaments
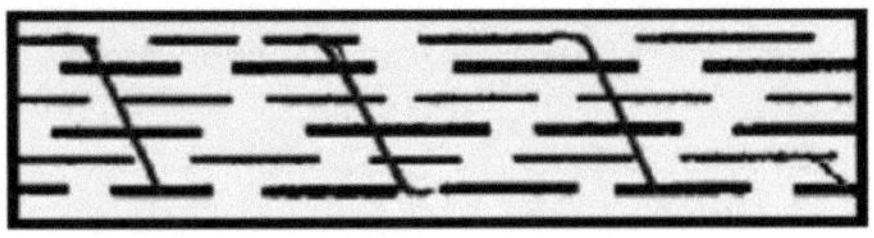	**Schappe technology (SCH)** Mixture of discontinuous performance material and fibers surrounded by thermoplastic filaments
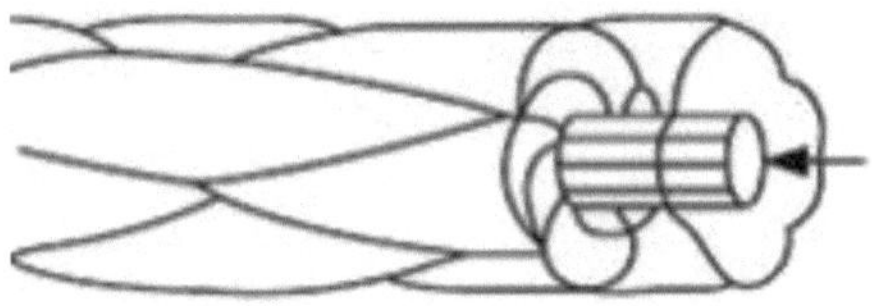	**Braiding technology** Filament or yarns in core and covered by performance filaments

Table 1. Hybrid yarn structures based on production method [19, 24].

3. Braiding technology

In composites manufacturing, the molten thermoplastic must flow through the capillaries between the reinforcement filaments. Theoretically, this can be achieved with a commingling method. In our previous studies, this method was tried, and some results were realized. It produced hybrid yarns approximately 1100 tex from 640 tex AR-glass roving and 400 tex polypropylene (PP) filament. In textile reinforcement concrete, generally 2400 tex or more linear density is used as a reinforcement material. In our study with approximately 1100 tex, it could not get a good homogeneity of fiber distribution. A further development of higher linear density with homogeneous fiber content is necessary. Therefore, the braiding yarn was used in this study as a new approach in order to obtain hybrid yarn for TRC production. Also, braiding technology offers minimum or no damage to the reinforcement fiber bundles, when compared to using commingled yarns.

3.1. Introduction to braiding technology

Braiding has been used for 200 years to produce textile fabrics. In this process, three or more threads are interlaced diagonally to the product axis [28, 29].

As the threads pass diagonally, they make an angle between 1 and 89° with the product axis of the yarns, usually within the range of 30–80° (**Figure 8**). This angle is called the braiding angle and is the most important geometric parameter of braided structures [29].

It is possible to produce a thicker, wider, or stronger product or cover some profiles in braiding technology. These products can be linear products (ropes), curved or plane shell, or solid structures (one, two, or three-dimensional (3D) fabrics) with constant or variable cross-section and of closed or open appearance. Researchers have focused braiding technology to meet new demands in composite production [28, 29].

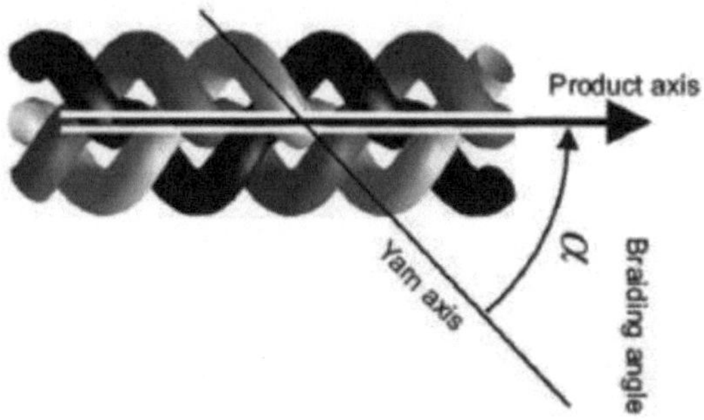

Figure 8. Structure of braided yarn [29].

3.2. Machine and product classification

There are several different classifications of the braided processes, machines, and products. Generally, braided products of maypole braiding machines are divided into two main groups: Braids with constant cross-section and variable cross-section form. While there is flat braiding, tubular braiding, and form braiding for braids with constant cross-section, there are only 3D Braids for variable cross-section form. Each group has normal or biaxial braiding and triaxial braiding types. The classical and mainly used braiding machines are known as maypole braiding machines. This name will always be used where these machines have to be distinguished from the other (spiral, lace) braiding machines. Details of these classification are mentioned in the following sections. In the future, it is expected that with the increasing speed of electromechanical drives, maypole braiding machines with switches (3D braiding) will be able to work continuously [29].

3.3. Basic principle

Braiding has been used for many years in different application areas such as braiding of yarns, flowers, and hair. The application areas for braided products are widespread such as medical items, electric cables, ropes, laces, the huge ropes, and tubes used in the marine oilfield sector. Also, use of braids is growing in the production of fiber-reinforced composites. All the products are manufactured in the same way, i.e., by the interlacing of yarns at an angle of about 40–60° to the main product axis. **Figure 9** shows the principle of hand braiding for a simple braid of three yarns. Initial structure (**Figure 9(a)**) is transformed to the interlacement in two steps:

- In the left two yarns, the outer left yarn goes over the next one (**Figure 9(b)**);
- In the right two yarns, the right yarn goes over the left next nearest yarn (**Figure 9(c)**), and it repeats in the same way [29].

3.4. Flat braiding

Flat braiding structures can be obtained when basic braiding is applied to more yarns, e.g. 5, 7. Flat braiding principle is seen in **Figure 10(a,b)**. In this case, at the first step, all left pairs interlace

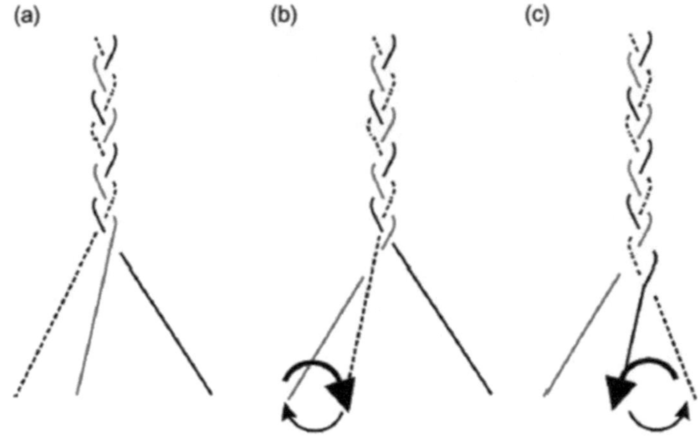

Figure 9. Principle of hand braiding with three yarns [29].

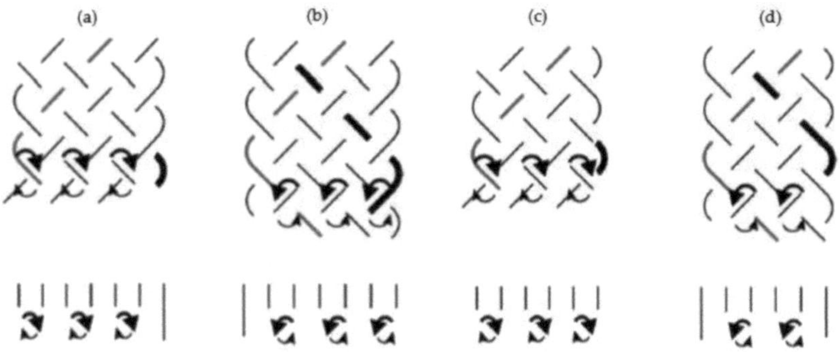

Figure 10. Hand braiding of odd (here seven) yarns (a,b) and of even (here six) yarns(c,d) [29].

so that the left goes over the right, and in the second step, all right pairs interlace so that the right yarn goes over the left [29].

Flat braiding is possible for both odd and even numbers of yarns. **Figure 10(c,d)** shows the sequence in the case of six yarns: at every second step, yarns from both the left and right sides stay and wait, while when using an odd number of yarns, only one yarn per cycle stays unused during the interlacement [29].

3.5. Tubular braiding

In this system, even number of yarns arranged around a circle interlace for production of tubular braids or ropes (**Figure 11**). The sequence of steps in this case is the same as for the flat yarn. In order to produce a regular structure, all yarns have to be kept under constant tension during the braiding process [29].

The interlacing sequence in the case of tubular braiding is same in flat braiding, but unlike flat braiding, all the yarn ends are located around a circle (**Figure 12**). The process can be described as follows:

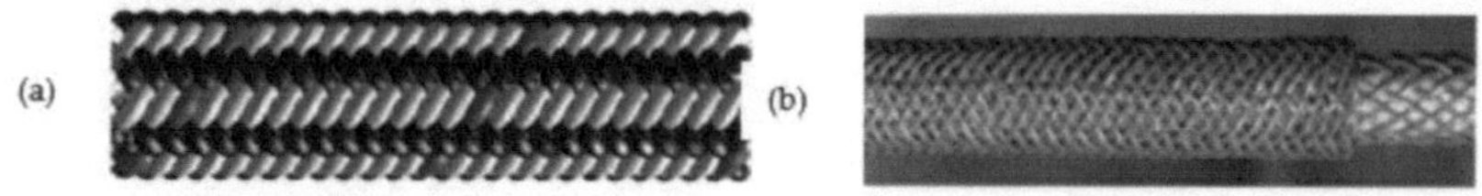

Figure 11. Tubular braids (rope). (a) Geometrical model and (b) Single (upper part) and Tripple (over-) braided part [29].

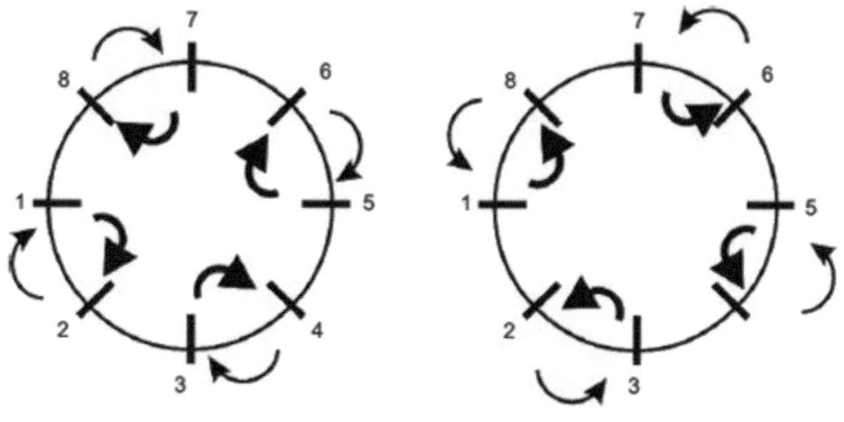

Figure 12. Sequence for tubular braiding [29].

- Interlacement of the left yarn over the right yarn in each pair (the yarn at position 1 goes over the yarn at position 2 and the yarn from position 2 goes to position 1, but interlacing under the first yarn; the pairs are 1–2; 3–4; 5–6; 7–8).
- After shifting the positions (the pairs become 8–1; 2–3; 4–5; 6–7), the right yarn goes over the left yarn in each pair [29].

In tubular braiding, the beam is slowly covered from the top down. In braiding terminology, the beam is called the core (if it had a regular form) or the mandrel (if it had a more complex geometry), and this braiding system is used today for overbraiding of profiles with carbon or glass fiber composites for aerospace, cables, high pressure ropes, etc., for automotive and other applications [29].

3.6. Biaxial and triaxial braids, inlay yarns

Inlay yarns can be put in between the core and the yarns. These are named differently in each every different application areas, e.g., middle ends, inlays, and triaxial braids (**Figure 13**) [29].

Inlay yarns can be inserted into all kinds of braided structures; however, the core or the mandrel requires a hollow structure and is mainly used in tubular braids (**Figure 14**) [29].

3.7. Industrial maypole braiding

In industrial braiding, the machine dancer element is called a carrier, and the path of the carrier is called the track. If the motion of the dancers/carriers is analyzed carefully, it can be seen that all dancers/carriers in one direction are following the same track, while the dancers/

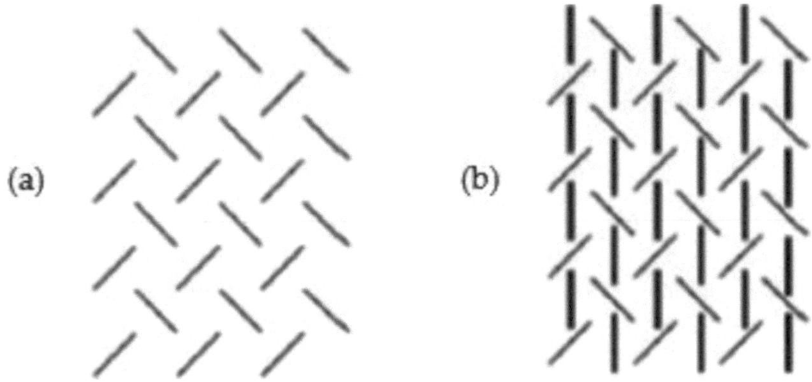

Figure 13. (a) Normal (biaxial) braids and (b) Braids with inlay yarns (triaxial) [29].

Figure 14. Tubular braid with core and inlay yarns (triaxial) [29].

carriers in the opposite direction follow the opposite track. Hence, for tubular braiding, there are two tracks and two systems of carriers [29].

The path of the motion is determined by the track, but the carriers are moved forward by the horn gears (**Figure 15**). At the beginning of the process, all outer ends of the yarns are connected together at the braiding point. When the carriers start their motions, the yarns interlace together. With this move, the next piece of braid is built at the braiding point. Finally, the braid produced is finally pulled out with the help of a take-off system [29].

3.8. Form braiding and 3D braiding

For different applications, such as thicker braided products of square or other cross-sectional type is required. For example, gaskets or fiber-reinforced composites can be manufactured with these products. In this case, the braiding process is called packing braiding, or 2.5D braiding (2.5D means more than the two dimensions of length and width, but less than

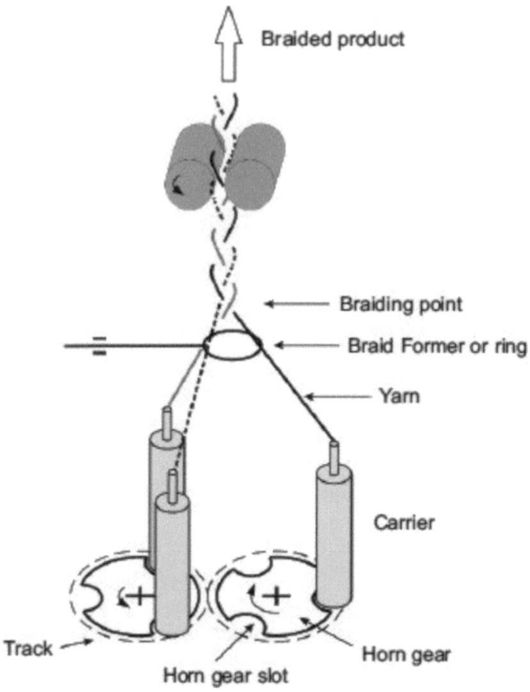

Figure 15. Principal construction of maypole braiding machine [29].

three dimensions), or alternatively, 3D braiding (due to the braided product having thickness in all three directions). However, according to some classifications, they may also be named 3D, 4D, etc., braiding, but D means "diagonal" here. In the new generation of braiding machines, they are controlled by a computer, and the track is not constant. The carrier can change its motion and can travel along a set of connected curves, and the selection of the next curve can change dynamically during the braiding process. Thanks to diversity, the braids do not have a constant cross-section, they are called 3D braids, and the process itself is 3D braiding [3, 29].

3.9. Applications of braiding yarns

The application of braids is various from medical items to electric cable. Also, use of braids is growing in the production of fiber-reinforced composites [3, 29].

If we summarize, we can classify the application fields as follows:

- Clothing: women's and men's wear, outwear, underwear, shoes (laces).
- Sports: aerial sports (starting rope for glider pilots), sailing (anchor ropes, sailcloth), mountain climbing (ropes), camping (tent ropes), fishing, tennis.
- Home textiles: curtains (laces, cords), decoration threads.
- Medical: surgery (blood veins, sewing threads, catheter), prostheses, tapes for orthopedics.

- Machine engineering: fiber-reinforced materials, package seals.
- Civil engineering: fiber-reinforced concrete.
- Traffic: fiber-reinforced materials for aircraft and car construction, track vehicles, and shipping.
- Electrical terotechnology: cables, insulation.
- Household: packaging, clothesline, do-it-yourself requirements [3, 29, 30].

3.10. Braiding technology in future

- Replacements of mechanical controls of braiding machines by electronic controls will reduce setup times.
- In future decades, even if braiding machine by electronic control will be used, mechanical carriers will surely continue to be used in a large number of applications.
- Braiding machines for the production of 3-D braids to obtain larger and more complex cross sections will be developed, and they can be used for nearly every application.
- Mathematical models between machine controls and the position of threads in the braid will be defined [3].

4. Experimental

Experimental study is focused on use of braiding yarn for textile-reinforced concrete. Hybrid yarn was produced with using braiding method. Hybrid yarn to be used as reinforcement has a unique combination of reinforcement and matrix component that was produced using a tubular braiding machine consisting of 16 spindles (**Figure 16**). Hybrid yarn can be thought of as a single yarn, although it is composed of two components. Continuous AR-Glass roving was used as the straight inserted axial fibers, and matrix fibers (PP fibers) were braided around the reinforcing AR-Glass roving.

AR-Glass fiber roving (Cem-FIL® 5325 from OCV Reinforcements) and polypropylene (PP) filament yarn (Aker Textile Yarn) were used to produce hybrid yarn with braiding method. The linear density of AR-Glass is 2400 tex, and the linear density of PP filament is 666.6 dtex. To produce hybrid yarn with braiding method, 1-filament AR-Glass and 16-filament PP were used. Yarn linear density was measured according to ISO 2060. The linear density of hybrid yarn, its components, and other reinforcing components are given in **Table 2**.

After the production of hybrid yarn, a hot compression molder was used to fabricate continuous fiber-reinforced thermoplastic composites. Matrix fibers were melted by heating at appropriate molding temperatures and become the matrices for the fiber-reinforced composites that easily wet out the reinforcing AR-glass roving. **Figure 17** shows the view of hybrid yarn before and after heating.

Figure 16. Braiding machine.

Yarn	Linear density (tex)
AR-Glass roving	2400
AR-Glass roving (coated with epoxy resin)	3840
Polypropylene (PP) filament	66.66
Hybrid yarn	3430
Carbon	1600
Carbon (coated with epoxy resin)	2560

Table 2. Linear density of hybrid yarn and its component.

Before and after heating, the cross-section of hybrid yarn was investigated on a microscope with 40× (**Figure 18**). It was observed that while structure of hybrid yarn was a bit loose before heating (a, b), it turned into compact structure after heating (c, d).

After hybrid yarns were prepared, coated AR-glass fibers were prepared to compare with hybrid yarn and uncoated AR-glass fibers. Epoxy resin (SR 8500/ SD 8605 from Sicomin) was applied to AR-glass fibers with roller coater. After coating, all samples were preheated and fixed at appropriate temperature and time. **Figure 19** shows the view of AR-glass fibers before and after coating.

(a)

(b)

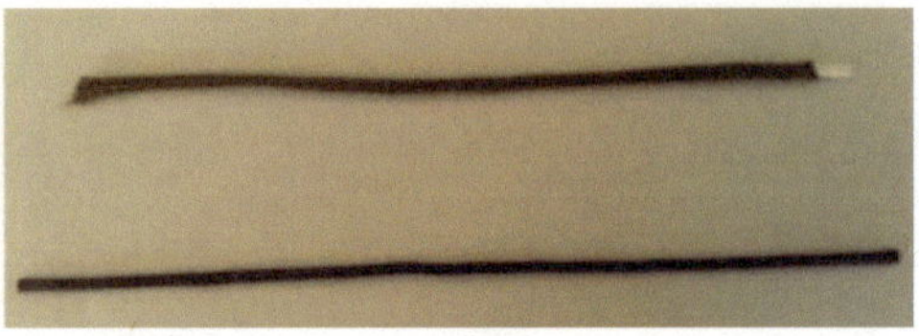

Figure 17. (a) Hybrid yarn before heating (b) Hybrid yarn after heating.

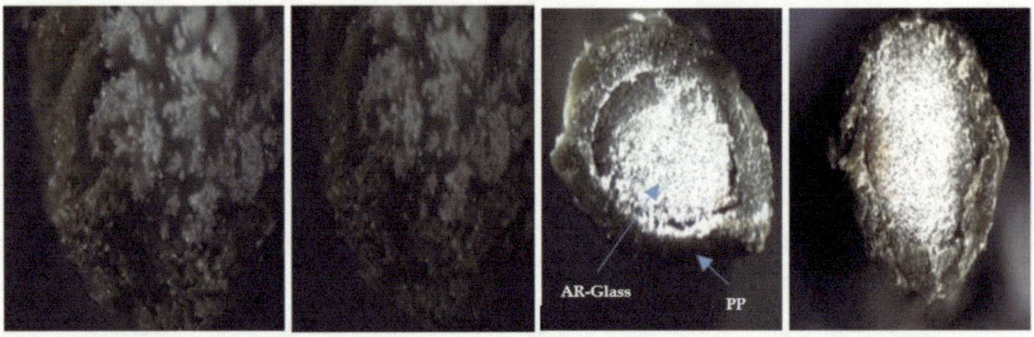

(a) (b) (c) (d)

Figure 18. Cross-section of hybrid yarn before heating (a, b), Cross-section of hybrid yarn after heating (c, d).

(a)

(b)

Figure 19. (a) Uncoated AR-Glass fibers (b) Coated AR-Glass fibers.

Prepared textile materials were placed in the mold for flexural test. For each sample, five yarns were placed in the mold along the long edge, and the distance between each yarn was set to 1 cm. The yarns were placed at a distance of 3 mm from bottom of the sample (**Figure 20**).

After all the textile components were prepared, second component for TRC which is concrete was prepared as shown in the mix proportions in **Table 3**.

The mixtures were batched in the vertical axis concrete mixer (**Figure 21**). The cement and fly ash were dry mixed for 1 min. This was followed by the addition of fine-grained aggregate, water, and the superplasticiser, with a mixing time of 5 min. After pouring the mix into oiled molds, a vibrator was used to decrease the amount of air bubbles. The specimens were demolded after 1 day and then placed in a curing room in the special pool which its water temperature is 20°C for 27 days of curing according to TS EN 12390-2 standard. For 12 h prior to the tests, the specimens were allowed to air dry in the laboratory.

Figure 20. Yarn position in the TRC sample.

Mix proportions	
Cement CEM I 42.5 R (c)	480 kg/m^3
Fly ash (f)	240 kg/m^3
Water (w)	284 kg/m^3
w/b*	0.39
Superplasticizer	1.5% b. m. of binder
Siliceous fines (0–0.3 mm)	642 kg/m^3
Siliceous sand (0.2–0.5 mm)	503 kg/m^3

*w/b = w/(c + f)

Table 3. Mix proportions of fine grained concrete.

Figure 21. Concrete mixer.

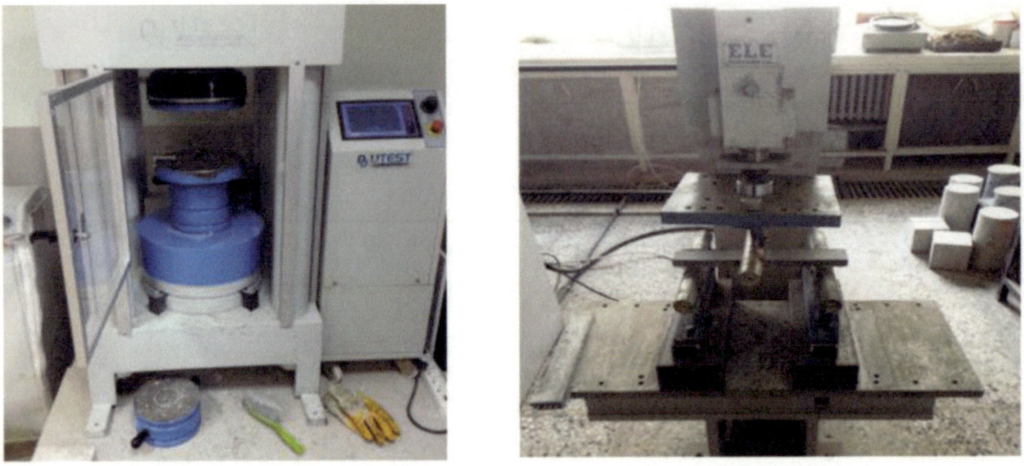

(a) (b)

Figure 22. (a) Compression test machine (b) Flexural strenght test machine.

For the prepared mixture, eight specimens (three 150 × 150 × 150 mm cubes for compression test and five 325 × 50 × 20 mm beams for three-point loading flexural test) were prepared. The compression tests were carried out in the UTEST compression test machine (Type UTC-5750) (**Figure 22(a)**) at a loading rate of 4 kN/s on the 150 × 150 × 150 mm cubes according to the requirements of TS EN 12390-3 standard. The three-point loading flexural tests were carried out in ELE flexural test machine (Model 37-6330; **Figure 22(b)**) at a loading rate of 0.4 kN/s on the 325 × 50 × 20 mm beams according to the requirements of TS EN 12390-5 standard. The mean values of the test results were shown in table at next section.

5. Results and discussion

The mechanical properties of the concrete were measured with a compression test to see the strength of concrete used in the TRC production. The average compressive strength (28 days) of a cube specimen (150 × 150 × 150 mm) was taken as 52.3 MPa. This result is good for conventional concrete. This concrete mix was used in the production of all TRC samples, and different materials were used as a reinforcing component.

In this study, only AR-glass roving was used to manufacture hybrid yarn with braiding technology. There is only AR-glass roving inside the braided yarn, and it is covered with PP filaments. To see the effectiveness of coating, carbon fiber (1600 tex from DOWAKSA) was used to produce TRC. The results of 3-point loading flexural test can be seen in **Table 4**.

As seen in **Table 4**, each reinforcement material contributes to the flexural strength of the concrete. Commonly used AR-glass roving and carbon filaments in concrete reinforcement have contributed to flexural strength, 41.19 and 146.57%, respectively, according to without reinforcement. As expected, the epoxy resin coating also has contributed to flexural strength

TRC samples	F (MPa)
Pure concrete	10.05
Reinforced with uncoated AR-Glass roving	14.19
Reinforced with coated AR-Glass roving	25.32
Reinforced with heated hybrid yarn	15.85
Reinforced with uncoated carbon fiber	24.78
Reinforced with coated carbon fiber	40.81

Table 4. Results of 3-point loading flexural test.

according to reinforcement with AR-glass roving and carbon filament, respectively, 78.44 and 64.69%. Heat-treated hybrid yarn reinforcement has contributed to a flexural strength of 57.71% according to without reinforcement. While reinforcement with epoxy resin coating has contributed to the flexural strength of 78.44%, reinforcement with heated hybrid yarn has contributed to flexural strength of 11.70% according to reinforced with AR-Glass roving. In this study, as it is known, it was seen that the epoxy resin coating provides a significant contribution. It is possible to evaluate the use of heated hybrid yarns produced by braiding technology in the production of TRC though it has contributed less than epoxy resin, since the epoxy resin cost is high and epoxy resin coating process is long and difficult.

6. Conclusion

In the experimental part of this study, the braiding method for using of TRC was investigated. Tubular braiding technique was applied to produce hybrid yarns using AR-glass roving as the core reinforcement fibers and PP fibers as the matrices around AR-glass roving. At the next step, these hybrid yarns were heated, and they were used for TRC production. All prepared samples used flexural strength test. When all test results were examined:

- Reinforcing materials such as AR-glass roving, carbon filament, and heated hybrid yarn have been found to increase the flexural strength.
- Carbon filament is better than AR-glass roving to reinforce the concrete for higher flexural strength.
- As expected, the epoxy resin coating also has contributed to high flexural strength according to reinforcement without coating.
- Also, the application of epoxy resin coating to the textile yarn improves the utilization of mechanical performance and handling properties as well.
- Although the contribution of the heated hybrid yarn is limited, it is expected that the desired results will be obtained by changes in braiding yarn production and yarn composition ratios.

Acknowledgements

This research is supported by "Scientific Research Projects Governing Unit of Çukurova University" with project number FDK-2016-6682. We would also like to thank KORD Industrial Rope and Yarn Industry and Trade Inc.—for hybrid yarn production; ÇİMSA Cement Industry and Trade Inc.—for their materials and laboratory; and Çukurova University Civil Engineering Department—for their laboratory.

Author details

Mutlu Kurban[1*], Osman Babaarslan[2] and İsmail Hakkı Çağatay[3]

*Address all correspondence to: mkurban@cu.edu.tr

1 Çukurova University, Vocational School of Technical Sciences, Textile Technology, Sarıçam, Adana, Turkey

2 Çukurova University, Faculty of Engineering and Architecture, Department of Textile Engineering, Balcalı, Sarıçam, Adana, Turkey

3 Çukurova University, Faculty of Engineering and Architecture, Department of Civil Engineering, Balcalı, Sarıçam, Adana, Turkey

References

[1] McCarthy BJ. An overview of the technical textiles sector. In: Horrocks AR, Subhash C, editors. Handbook of Technical Textiles. Volume 1: Technical Textile Processes. 2nd ed. Woodhead Publishing: The Textile Institute; 2016. pp. 1-20. ISBN: 978-1-78242-458-1, Cambridge, England

[2] Picon A. Construction history: Between technological and cultural history. Construction History. 2006;**21**:5-19

[3] Gries T, Veit D, Wulfhorst B. Textile Technology. 2nd ed. Hanser Publishers; 2015. 441 p. ISBN 978-1-56990-565-4, München, Germany

[4] Bakis CE, Bank LC, Brown VL, Cosenza E, Davalos JF, Lesko JJ, Machida A, Rizkalla SH, Triantafillou TC. Fiber-reinforced polymer composites for construction—State-of-the-art review. Journal of Composites for Construction. 2002;**6**(2):73-87. DOI: 10.1061/(ASCE)1090-0268(2002)6:2-(73)

[5] Holler S, Butenweg C, Noh SY, Meskouris K. Computational model of textile-reinforced concrete structures. Computers and Structures. 2004;**82**:1971-1979. DOI: 10.1016/j.compstruc.2004.03.076

[6] Brameshuber W, editor. Textile Reinforced Concrete. 1st ed. RILEM Publications: State-of-the-Art Report of RILEM Technical Committee 201-TRC. RILEM Report 36; 2006. 273 p. ISBN: 2-912413-99-3, Bagneux, France

[7] Hausding J, Engler T, Franzke G, Köckritz U, Cherif C. Plain stitch-bonded multi-plies for textile reinforced concrete. AUTEX Research Journal. June 2006:**6**(2):81-90

[8] Häußler-Combe U, Hartig J. Bond and failure mechanisms of textile reinforced concrete (TRC) under uniaxial tensile loading. Cement & Concrete Composites. 2007;**29**:279-289. DOI: 10.1016/j.cemconcomp.2006.12.012

[9] Hegger J, Voss S. Investigations on the bearing behaviour and application potential of textile reinforced concrete. Engineering Structures. 2008;**30**:2050-2056. DOI: 10.1016/j.engstruct.2008.01.006

[10] Brückner A, Ortlepp R, Curbach M. Textile reinforced concrete for strengthening in bending and shear. Materials and Structures. 2006;**39**:741-748. DOI: 10.1617/s11527-005-9027-2

[11] Gao SL, Mäder E, Plonka R. Coatings for glass fibers in a cementitious matrix. Acta Materialia. 2004;**52**:4745-4755. DOI: 10.1016/j.actamat.2004.06.028

[12] Scheffler C, Gao SL., Plonka R, Mäder E, Hempel, S, Butler, M, Mechtcherine V. Interphase modification of alkali-resistant glass fibres and carbon fibres for textile reinforced concrete I: Fibre properties and durability. Composites Science and Technology. 2009;**69**:531-538. DOI: 10.1016/j.compscitech.2008.11.027

[13] Scheffler C, Gao SL, Plonka R, Mäder E, Hempel S, Butler M, Mechtcherine V. Interphase modification of alkali-resistant glass fibres and carbon fibres for textile reinforced concrete II: Water adsorption and composite interphases. Composites Science and Technology. 2009;**69**:905-912. DOI: 10.1016/j.compscitech.2008.12.020

[14] Golzar M, Brünig H, Mäder E, Commingled hybrid yarn diameter ratio in continuous fiber-reinforced thermoplastic composites. Journal of Thermoplastic Composite Materials. 2007;**20**:17-26. DOI: 10.1177/0892705707068069

[15] Boukhili R, Boukhili H, Ben Daly H, Gasmi A. Physical and mechanical properties of pultruded composites containing fillers and low profile additives. Polymer Composites. 2006;**27**:71-81. DOI 10.1002/pc.20127

[16] Dominy J, Rudd C. Manufacturing with thermosets. In: Long AC, editor. Design and Manufacture of Textile Composites. 1st ed. Woodhead Publishing Limited; 2005. pp. 181-196. ISBN: 10-1-85573-744-2, Cambridge, England

[17] Wakeman MD, Månson AE. Composites manufacturing—thermoplastics. In: Long AC, editor. Design and Manufacture of Textile Composites. 1st ed. Woodhead Publishing Limited; 2005. pp. 197-241. ISBN: 10-1-85573-744-2, Cambridge, England

[18] Wakeman MD, Cain TA, Rudd CD, Brooks, R, Long, AC. Compression moulding of glass and polypropylene composites for optimised macro and micro-mechanical

properties-1 commingled glass and polypropylene. Composites Science and Technology. 1998;**58**:1879-1898

[19] Alagirusamy R, Fangueiro R, Ogale V, Padaki N. Hybrid yarns preforming for thermoplastic composite. Textile Progress. 2006;**38**(4):1-71. DOI: 10.1533/tepr.2006.0004

[20] Qadir B, Hussain T, Malik M. Effect of elastane denier and draft ratio of core-spun cotton weft yarns on the mechanical properties of woven fabrics. Journal of Engineered Fibers and Fabrics. 2014;**9**(1):23-31

[21] Sawhney, APS, Ruppenicker GF, Kimmel LB, Robert, KQ. Comparison of filament-core spun yarns produced by new and conventional methods. Textile Research Journal. 1992;**62**(2):67-73

[22] Pouresfandiari F, Fushimi S, Sakaguchi A, Saito H, Toriumi K, Nishimatsu T, Shimizu Y, Shirai H, Matsumoto YI, Gong H. Spinning conditions and characteristics of open-end rotor spun hybrid yarns. Textile Research Journal. 2002;**72**(1):61-70

[23] Hasan MMB, Offermann M, Haupt M, Nocke A, Cherif C. Carbon filament yarn-based hybrid yarn for the heating of textile-reinforced concrete. Journal of Industrial Textiles. 2014;**44**(2):183-197. DOI: 10.1177/1528083713480380

[24] Mankodi HR. Developments in hybrid yarns. In: Gong RH, editor. Specialist Yarn and Fabric Structures. 1st ed. Woodhead Publishing Limited; 2011. pp. 21-55. ISBN 978-1-84569-757-0, Cambridge, England

[25] Alagirusamy R, Ogale V. Commingled and air jet-textured hybrid yarns for thermoplastic composites. Journal of Industrial Textiles. 2004;**33**(4):223-243. DOI: 10.1177/1528083704044360

[26] Kravaev P, Stolyarov O, Seide G, Gries T. Influence of process parameters on filament distribution and blending quality in commingled yarns used for thermoplastic composites. Journal of Thermoplastic Composite Materials. 2014;**27**(3):350-363. DOI: 10.1177/0892705712446167

[27] Rawal A, Saraswat H, Sibal A. Tensile response of braided structures: A review. Textile Research Journal. 2015;**85**(19):2083-2096. DOI: 10.1177/0040517515576331

[28] Ayranci C, Carey J. 2D braided composites: A review for stiffness critical applications. Composite Structures. 2008;**85**:43-58. DOI:10.1016/j.compstruct.2007.10.004

[29] Kyosev Y. Braiding Technology for Textiles. 1st ed. Woodhead Publishing Limited; 2015. 392 p. ISBN 978-0-85709-135-2, Cambridge, England

[30] Khondker OA, Ishiaku, US, Nakai A, Hamada H. A novel processing technique for thermoplastic manufacturing of unidirectional composites reinforced with jute yarns. Composites: Part A. 2006;**37**:2274-2284. DOI:10.1016/j.compositesa.2005.12.030

Chapter 6

Three-Dimentional Textile Preform Using Advanced Textile Technologies for Composite Manufacturing

Peng Wang, Xavier Legrand and Damien Soulat

Additional information is available at the end of the chapter

http://dx.doi.org/10.5772/intechopen.68175

Abstract

Textile reinforcement structure plays an important role in the reinforcement/composite performances during the composite manufacturing and in-service life of the composite material. Structures with a three-dimensional (3D) fiber topology are desired due to their superior multiaxial performance and efforts have been made to modify 2D textile technologies to produce complex 3D shapes. Most of these 3D solutions are based on the principle of adding out-of-plane reinforcements to a planar 2D fabric. Well-established 3D textile methods such as braiding and knitting have also been demonstrated to directly produce near net-shape structures. To understand these potentialities, the first section of this chapter will present the several textile technologies with strengths and weaknesses of these processes to manufacture technical reinforcements for composite applications. In the following sections, several applications with specific textile architectures will be given, in particular, the applications of the through-the-thickness reinforcement and 3D textile ply during the composite manufacturing.

Keywords: textile reinforcement, 3D fabric, tufting, 3D surface weaving, forming

1. Introduction

The definition of textile reinforcement is an essential factor during composite manufacturing and for the performance of composite material. The step of the impregnation of the reinforcement is governed by its porosity (size and distribution) [1]. Porosities can be controlled by the permeability defined as the ability of the reinforcement to transmit fluids [2]. Permeability behavior of reinforcing textiles is a strong function of the textile's complex architecture [3] and depends on the fiber volume fraction. For the specifications of composite materials, load transfer from the matrix to the reinforcement is governed by the fiber orientation, which plays a paramount role in the composite stiffness [4]. Fiber direction and fiber volume fraction must be managed during the manufacturing process of composite materials. Resin transfer molding (RTM) [5] is one of the main manufacturing processes to produce composite parts for the transport industries [6].

A lot of experimental and numerical studies [7–11] concerning the draping stage of dry reinforcement, the first step of RTM process, were carried out to analyze the deformability of different reinforcements on specific tools. Research works have been published about the textile composite forming of hemispherical [12, 13], double-dome [14], tetrahedral shapes [15], and also with square box [16]. On these complex shapes, this preforming stage modifies the main parameters as fiber direction and fiber volume fraction and consequently has a significant influence on the resin flow impregnation and on the characteristics of the final composite part. The behavior of highly aligned reinforcements (woven, braided knitted, etc.) during the preforming step is characterized by complex coupled tensile, in-plane shear, and bending, but also by compaction deformations. Criteria to define the feasibility to realize a particular shape are based on limits of these deformations, such as the locking angle [17] for the in-plane shear behavior. All these studies have shown that the reinforcement architecture is a main parameter to predict the feasibility forming conditions and to prevent the manufacturing defects.

During the manufacturing of composite parts, defects may be introduced in their structure. Potter et al. [18] have presented a taxonomy of defect states that may have adverse impacts on the performance of composite parts. During the first step of the RTM process (preforming stage), with the use of full-field optical measurements of strains, the detection of defects is possible, as wrinkles [19], nonhomogeneity of the fiber density, and if the discontinuity of the preform due to sliding of tows takes place. The quantification of different level of strains in tension or in-plane shear can be also analyzed for the mechanical state of the preform. At mesoscopic scale, tows misalignment in the plane of the fabric or tow buckles that are defined as out-of-plane misalignment [20, 21] can be investigated. This presentation of defects could not be complete without mentioning the difficulties associated to the preforming of multilayer dry textile reinforcement, or thick as preforms as those used in high-tech industrial domains, such as automotive and aerospace industries, for instance, fan blades developed by Snecma [22]. In the literature, the preforming of multilayer dry textile reinforcements has been one of the main subjects [10, 11, 23]. In the considered forming parameters of multilayers, the number and the orientation of each ply must be taken into account. The study presented in Ref. [8] shows that preforming of multilayer is not yet controlled and demonstrate the influence of the layer orientations especially on the draw-in and finally on the final shape. The interply behavior is used, for example, with the identification of friction laws, through numerical and experimental studies [24, 25]. However, these studies estimated criteria of forming only on the top or bottom ply and did not measure the quantities in each ply of the multilayer. When these defects manifest during forming applications (and also thermostamping [23, 26]), it can be controlled by the addition of lateral restraint (e.g., matched die) or modification of forming temperature, tooling velocity, contact-to-free edge distance (e.g., blank size), and pretensioning (e.g., binder pressure) [27, 28]. Parametric studies are developed to study independently the effect of each of these parameters on final part quality; especially, experimental approaches lead to well understand the influence of the processing conditions to fabricate defect-free preform, as for example, the effect of blank-holder [29]. The development of numerical tools, by finite element simulation, can be an effective approach to take into account whole coupling parameters, to avoid experimental trial and error, and finally define optimal configurations of these parameters. If research works were concerned the efficiency of the modification of forming process parameters to avoid/reduce these defects, the influence of specifically the nature and the architecture of dry preform during the preforming is less analyzed.

Studies concerning, separately, the preforming of carbon, glass, and also natural fiber textiles in woven fabrics [12], nonwoven fabrics [30], noncrimp fabrics [24, 31], three-dimensional (3D) interlock [12, 32], weft-knitted fabric [25], UD [33], or also braided reinforcements [34] can be found in the literature. These studies show that there is a very wide range of types of reinforcements from high performance fibers such as carbon, glass, or aramid fibers to natural or hybrid (as comingled [35]) fibers, which can be used to manufacture composite materials [36]. These reinforcements are developed from textile technologies, as weaving, knitting, breading, or nonwoven manufacturing processes, and have different mechanical properties, especially during the preforming step. The classification of these reinforcement architectures, in the textile literature, is based first on the technologies used but also on the fiber topology, e.g., the number of fiber directions [37, 38], but conventional laminated composite has fibers oriented only in the plane of the laminate and is therefore vulnerable to delamination. Due to the inclusion of the through-thickness yarns, three-dimensional (3D) fiber reinforced composites have several advantages over traditional two-dimensional (2D) composites such as elevated fracture toughness, better interlaminar fracture toughness, higher damage tolerance, impact resistance, and tensile strain-to-failure [39]. Due to the complexity and diversity [40] of these 3D composite architectures [41], the performance and efficacy of these 3D composite in service are predicted early from numerical computational work. In order to validate the computational results and gauge the influence of the through-thickness fibers on the in-plane and out-plane mechanical properties, numerous authors have worked on the identification of the mechanical properties of 3D reinforced composites [42]. Transverse cracking and propagation of delamination have been studied earlier with the help of digital image correlation (DIC), which maps the surface strain distribution [43]. Moreover, to describe the complex 3D preform architecture, X-ray microcomputed tomography has also been used [44]. Consequently, efforts have been made to modify 2D textile technologies to manufacture complex 3D preform [37, 38]. Most of these 3D solutions are based on the principle of adding out-of-plane reinforcements to a planar 2D fabrics, examples include *z*-pinning [45], interlock-weaving [46], stitching [47], or tufting [48].Well-established 3D textile methods such as braiding [49], weaving [40, 46], and knitting [50] have also been demonstrated to directly produce near net-shape structures. The control of these textile technologies that are able to manufacture 3D preform is absolutely necessary to avoid damage occurring by the insertion of the through-the-thickness reinforcement and consequently decreasing mechanical properties of 3D composites.

To describe potentialities of 3D preform reinforcement especially during manufacturing steps of composite materials, this chapter focused on applications. In the second section, the several textile technologies will be presented in fact to understand architecture of reinforcements and its influences on their properties. Strengths and weaknesses of these processes will be given. In the third section of this chapter, an example of through-the-thickness reinforcement by tufting will be described specifically for the preforming step of the RTM process. Experimental comparisons between untufted and tufted samples subjected to hemispherical preforming will be compared to minimize defects as interlayer sliding or excessive draw-in. In the latest section, concerning the manufacturing of a square box part, two methods are compared: the classical preforming realized experimentally by a specific preforming device and the surface 3D weaving able to deposit the warp and weft yarns perfectly perpendicular on all of the surfaces of the final 3D ply. Fiber orientations as well as wrinkles and in-plane shear angles are analyzed and compared for each preform obtained. Conclusions and prospects finish this chapter.

2. Classification of preform architecture

Several points need to be remembered to introduce textile reinforcement classifications. (1) In a composite material, the fiber contributes mainly to the mechanical performances of the final parts. The positions and orientations of the fibers are critical to the mechanics of final parts. (2) The textile structure and the composite part should be distinguished. The textile structure is so-called dry: it is a semiproduct. The composite part is composed of the impregnated textile structure of polymerized matrix or reconsolidated matrix. (3) Textile structures are fundamentally multiscale materials in which the macroscopic mechanical behavior of the structure is directly inherited behavior at lower scales. Three scales of observation of the reinforcement are generally recognized (**Figure 1**): microscopic scale (scale of fiber), mesoscopic scale (scale of the tow and the structure pattern), and macroscopic scale (scale of the whole part).

The textile structures are very diverse although traditionally classified into the four main technologies: weaving, knitting, braiding, and nonwoven. There are many ways of classifying textile structures:

- by construction technology: weaving, knitting, braiding, and nonwovens according to their dimensionality: 1D, 2D, 3D,… [51]
- by number of orientations: monoaxial, biaxial,…

Different classifications have been suggested for the fiber reinforcement architecture in order to help the designer to select the architecture that respond to the cost and performance requirements. These classifications are based on the structural integrity and fiber linearity and continuity, fiber orientation, or the textile manufacturing technology. The conventional textile technologies have significantly evolved over the last century, especially by using technical fibers as glass or carbon fibers. However, some limitations are associated with the employment of conventional textile processes in technical fiber manufacturing reinforcement preforms, as the high degradation of technical fiber/yarn on the conventional textile machineries are occurring, or either, the fact that conventional machineries allow only the production of planar (2D)

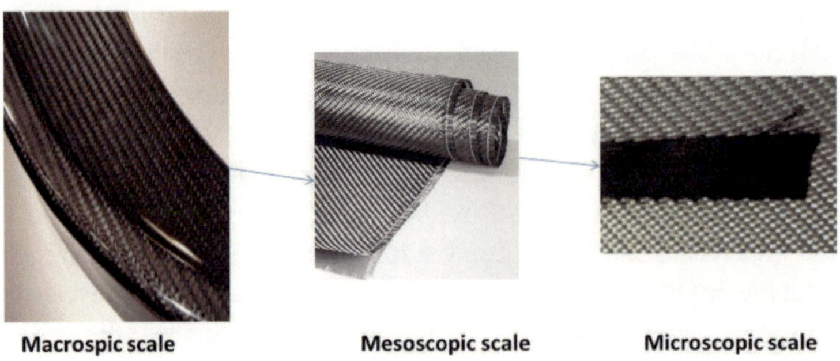

Figure 1. The three levels of study woven fibrous reinforcements.

or thin cylindrical preforms. It is possible to sort out the textile structures according to their size, i.e., size determined by fiber orientations.

The component of the textile structure is the fiber. It can be presented in the form of tape, roving, or yarn. The textile structure is a tangle of fibers. Let us agree that the twisted yarn or multifilament twisted or not is considered one-dimensional (1D), as in this longitudinal direction, the stiffness is predominant. Therefore, size of textile structure is drawn by mechanical stiffness orientations, i.e., along fibers. Similarly, the fabric that is obtained by nonwoven technology, by weaving, knitting, or by flat braiding is considered in 2D as the fibers are almost oriented into one plane, associated rigidities membrane. A so-called conventional 2D preform comprises fibers oriented in the main plane of the structure. For example, a conventional woven fabric, in the plane (*X*, *Y*), has fibers oriented along the *X* axe (said at 0° angle) and transverse fibers, along the *Y* axe (said at 90° angle). Therefore, none are along the *Z*-axis, i.e., out of the plane (*X*, *Y*). The major advantage of a so-called 3D preform is that, having fibers in the "third direction," the structure is generally more efficient because more mechanically efficient mechanically in this "third direction," as explained by Mouritz et al. [37, 47, 52–54].

Recall that the reinforcement into a composite material serves to strengthen, i.e., to provide the mechanical properties of the composite part. Moreover, one possible definition of the "size" of a textile structure is given by the orientation of the fiber bundles that constitute it. In fact, the comparison and classification of textile preforms is a difficult task because of the diversity of 3D structures as shown in **Figure 2**. Fukuta and Aoki have classified textile preforms according to different criteria such as the fiber orientation, the method of manufacture, or the geometry of the final preform (see **Figure 3**). Besides, a growing number of *Z*-reinforcement methods are available. Integral techniques (3D knitting, 3D weaving, or braiding) must be distinguished from insertion of the *Z*-reinforcement. For *Z*-reinforcement techniques, the insertion in *Z*-direction is separate manufacturing steps after the initial preform lay-up. It is the case for the structural stitching, the tufting, the *Z*-pinning or either the *Z*-anchoring. Besides, integral techniques are developed for the manufacture of large-scale 3D reinforced laminates, and *Z*-reinforcement techniques are ideally suited for the localized *Z*-reinforcement of sections in the composite with high out-of-plane stress state. A new class of reinforcements has to be introduced. Let us call it 2D+. A 2D+ structure is a structure where the thickness is greater than that of the 2D structures, but not as much as other dimensions. The 2D+ structures can be defined as structures where the thickness is greater than that of the 2D structures. For example, the woven structure called "interlock Aerotiss 2.5D" [55] is a bonding layer to chain interlock fabric layer having multiple layers in the thickness. A weave derived from the three-layer structure is carbon reinforcement, G1151, from Hexcel. Note that this structure is thicker than a simple fabric that tows bind the layers together. For interlock structures, the thickness is less negligible compared to the other dimensions and some are no longer just in the main plane (*X*, *Y*) of the structure.

In order to solve the problems associated with traditional laminated structures and especially their low interlaminar properties, much energy has been deployed in the past 50 years to develop composite structures with 3D reinforcements [37]. Many techniques can lead to the production of 3D textile architecture as 3D weaving, 3D braiding, and stitching. All these techniques are designed to obtain complex 3D shapes while providing improved mechanical performance. From all the classifications listed previously, a study of the typology of textile reinforcements is

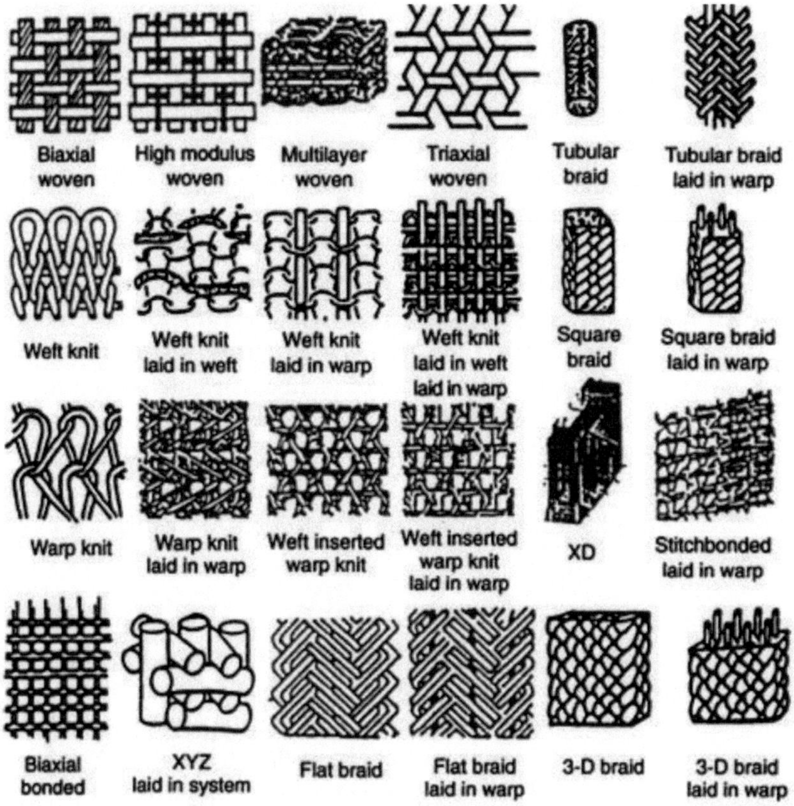

Figure 2. List of textile structures by Ko and Du [56].

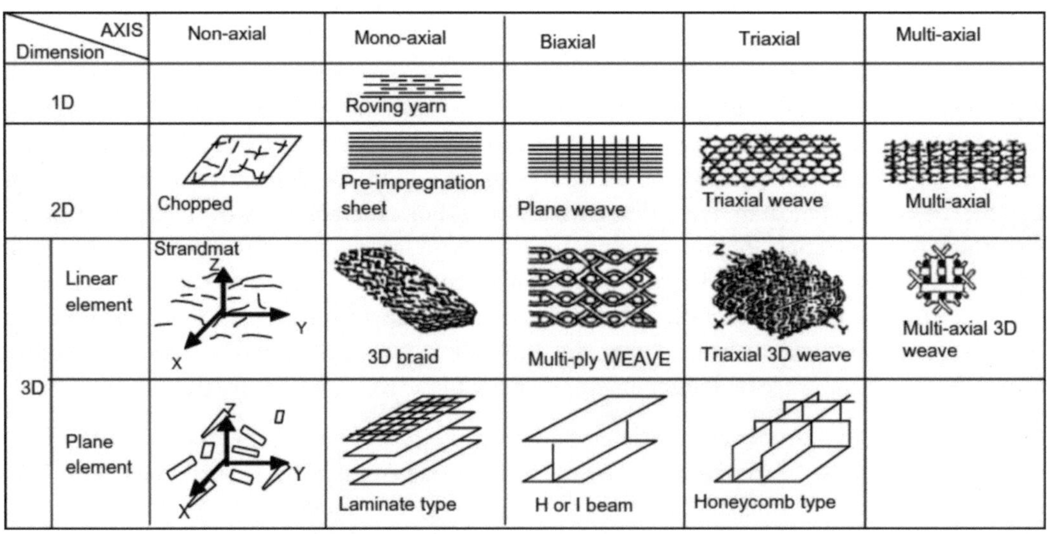

Figure 3. Classification by Fukuta and Aoki [57].

proposed in **Figure 5** to classify preforms during the various developments. The purpose of this classification with respect to all those present in the literature is to situate new textile reinforcement structures for composites. Moreover, the classification in **Figure 5** also proposes a class for flat thick preforms as woven interlock: the 2D+ class. It makes a case for assembly preforms. These structures are 3D structured using even strengthened by assembly, whether by stitching, but also by tufting or nailing. The classification in **Figure 4** explores unnamed geometric preforms. The main aim of the proposition presented in **Figure 5** is to clarify what is called 3D. It should also help researchers to define where are set their objects.

Dimensions \ Orientation		randomly	mono axial	bi axial	tri axial	multi axial
1D		fibre	tow, yarn			
2D		non woven	uni directional	Fabric: Woven, knitting, bi axial braiding	tri axial braiding: flat, over braid	NCF stitch bonded, NCF woven
2D+		thick non woven		woven interlock	multi axial multi-layer interlock	
3D	surface	shaped non woven		tetrahedron ply stamping	over-braiding, preforming, stamping of braid or/and multi axial	
3D	volume			assemblies by tufting, stitching, nailing/pinning	cross of shaped	BWS, cross of stiffeners (integral 3D)

Figure 4. Classification of textile preforms.

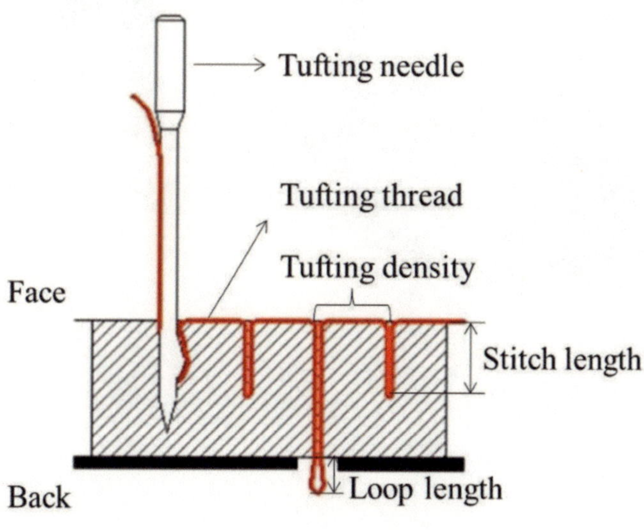

Figure 5. Tufting process.

3. Reinforcement by tufting technology

In this subdivision, an example of through-the-thickness reinforcement technology is applied and examined for its influences during the preforming stage of the RTM process. For thick composite parts, multilayer dry textile reinforcements are deformed on tools associated with the geometry of the composite part to fabricate. The interply sliding is identified as a major forming defect during this phase of the procedure. Moreover, the manufacturing defects will be presented and especially on the influence of the layer orientations on the sliding and wrinkles, which are not acceptable for the final part.

3.1. Tufting process and tufted 3D reinforcements

Tufting technology has been developed based on the stitching technique (see **Figure 5**). The dry reinforcements are tufted together by yarn loops. **Figure 5** shows a hollow needle passes through one side of dry preform without tension. When the hollow needle retracts, the tuft yarn is kept within the dry preform by friction and forms a loop. Tufting technology applies a formation of loops with a loose and almost stress-free presentation of the threading system that can bring down the stitching effect on the in-plane properties [55, 58]. Compared to the conventional stitching, tufting is much simpler as it does not require a second thread and lock the threads [58].

A specific tufting device developed in GEMTEX Laboratory is shown in **Figure 6**. **Figure 6a** shows the tufting equipment. Four principal parts of this equipment: tufting device,

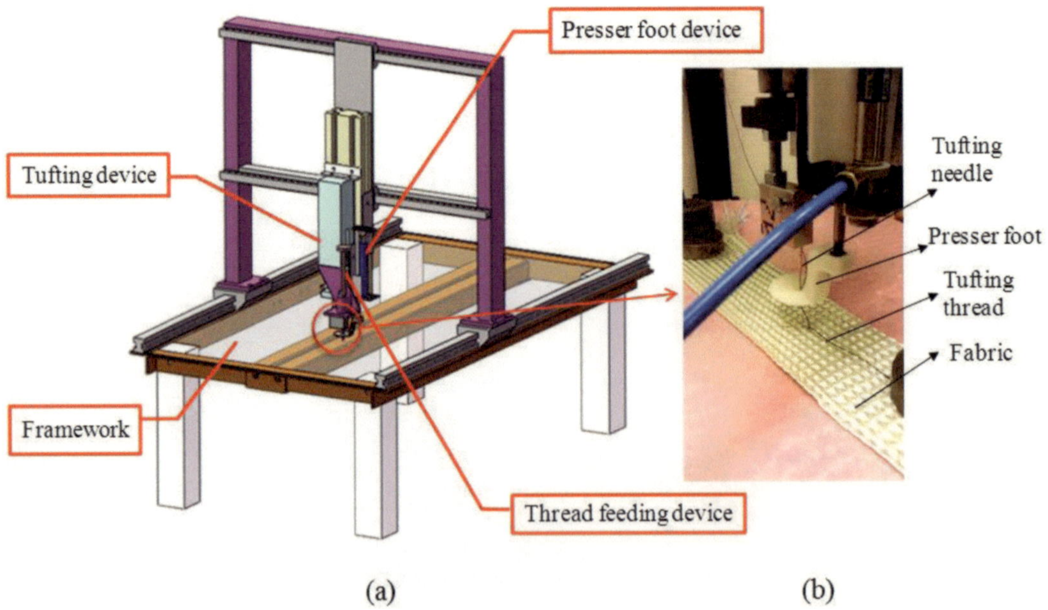

Figure 6. The tufting equipment (a) and zoom of the tufting head (b).

presser foot device, thread feeding device, and framework are illustrated. The tufting needle linked to a pneumatic jack in order to well-control the tuft length. A presser foot situated next to the needle is carried by another pneumatic jack to apply a force per unit area on the preform in tufting process. The thread feeding device provides tufting yarn with a certain length and tautness. These devices are installed along a mobile framework to form the tufting head (see **Figure 6b**). During the tufting, stitch length, tufting density, tufting direction, and pressing of the presser foot can be well-controlled. The maximum tuft length 50 mm is used in the present study, which is decided by the range of the pneumatic jack. A change of needles can be selected according to the yarn performance and the cloth of the preform.

For the presented case, the reinforcement fabric is E-glass plain weave with an area density of 157 ± 5 g/m^2. The tested preforms were first laminated with four plies [±45°, 0°/90°]^2. **Table 1** shows the preforms were tufted by TENAX® carbon yarn with different tufting densities corresponding to a tufting spacing. The tufting spacing is defined by the vertical distance of two neighbor tuft lines. One of the preforms after tufting is shown in **Figure 7**. The dimensions of the tested preforms are 280 × 280 × 1.0 mm^3. As presented in **Figure 7**, the preform is tufted in square spiral pattern to assure that the tufting thread is continuous and uninterrupted and inserted in two directions (warp and weft directions).

Ref. of samples	**Area density (g/m^2)**	**Tufting spacing (mm)**
Non-tufted	626.3	-
Tufted 2.0	639.0	20
Tufted 1.0	665.8	10
Tufted 0.5	707.9	5

Table 1. Main properties of the tufted 3D fabric specimens.

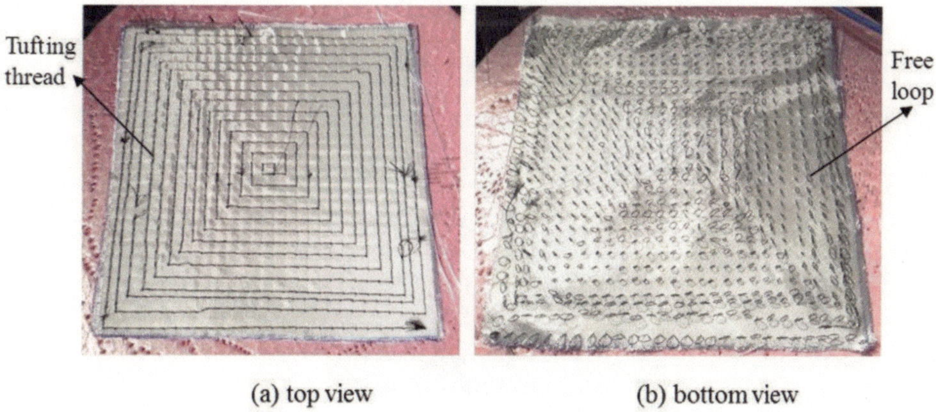

Figure 7. Tufted sample with a tufting spacing of 10 mm. (a) Top view; (b) bottom view.

3.2. Hemispherical forming

The hemispherical forming is performed on a specific preforming device shown in **Figure 8** [12]. The diameter of the hemispherical punch is 150 mm. The hemispherical preforming device was used to analyze the double-curved shape forming with a given textile reinforcement under different conditions (shape of punch, position, and pressure of blank-holder). The tufted 3D fabric is placed between the blank-holder and die. Four pneumatic jacks, connecting to the blank-holder, are used to set an adjustable pressure on the cloth. In order to measure the important forming parameters by optical measurement, such as the material draw-in and interlayer sliding, the open-die forming is used.

3.2.1. The multilayered reinforcement forming

In order to manufacture the thick composite parts, it is frequently used in multilayered forming [23, 59, 60]. **Figure 9** shows the shape of the E-glass plain weave preforms after the hemispherical preforming. The deformed preforms are symmetric, and there are no wrinkles that can be observed in the usefulness zone (the flat zone). The sliding is noted on the interface of plies as the plies are simply superimposed. The maximum intersliding can be measured by the difference of material draw-in between the top and the bottom plies.

3.2.2. The tufted 3D reinforcement forming

The preforming of four different tufted 3D fabrics is figured out. The forming conditions are identical to the multilayered forming presented previously (punch displacement 65 mm and blank-holder pressure 0.05 MPa). The deformed preforms are shown in **Figure 10**. The influence of tufting spacing can be followed on the different results. The tufting yarn reinforces

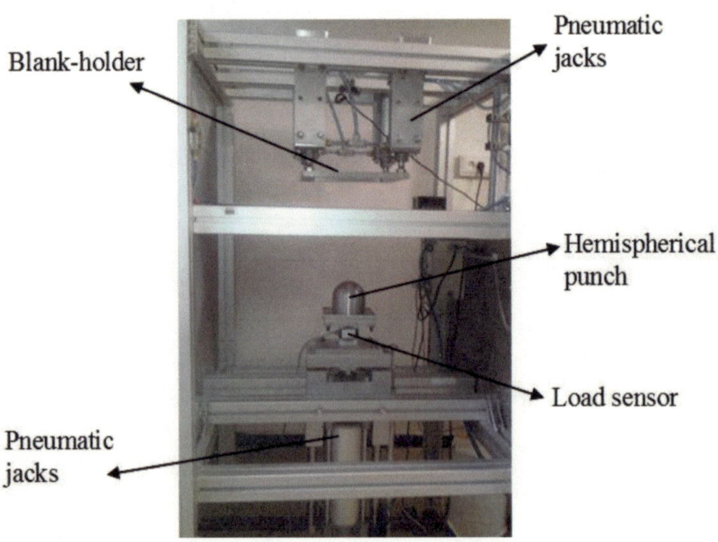

Figure 8. The hemispherical preforming device.

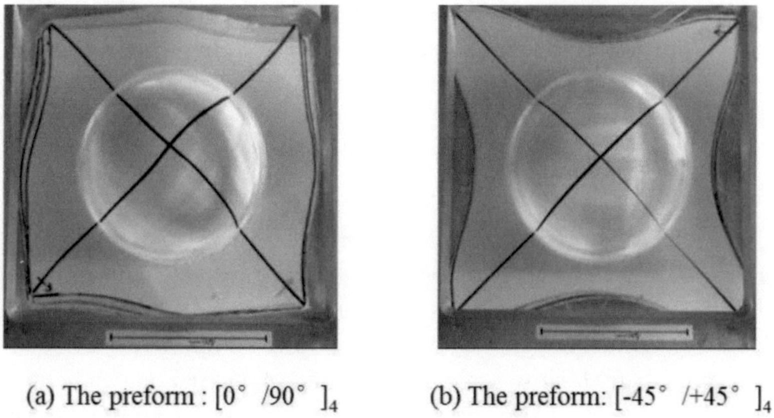

Figure 9. The multilayered forming. (a) The preform [0°/90°]^4; (b) the preform [−45°/+45°]^4.

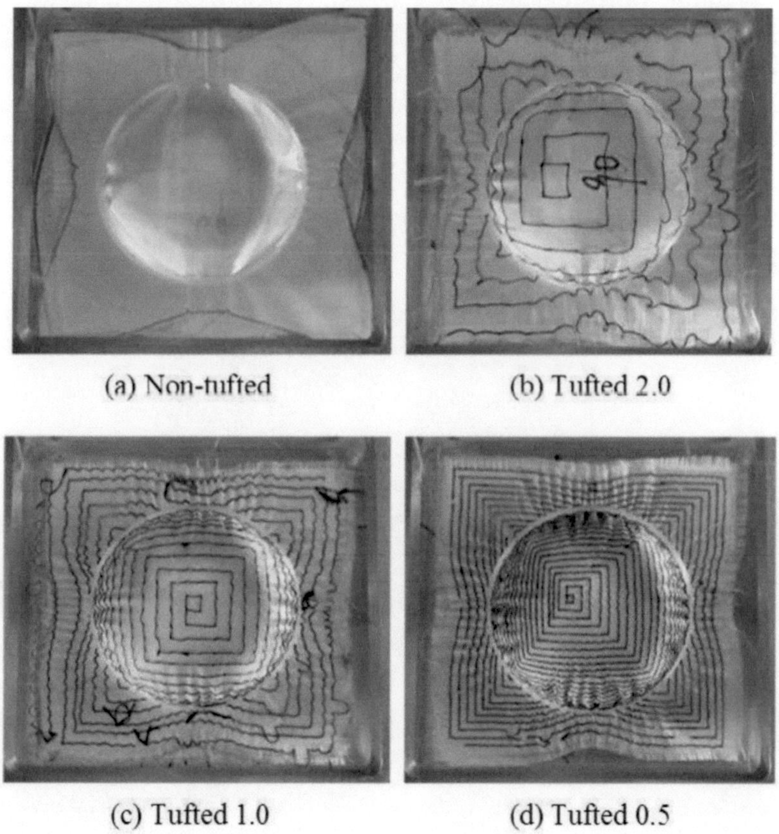

Figure 10. The different tufted 3D preforms after performing. (a) Nontufted; (b) tufted 2.0; (c) tufted 1.0; (d) tufted 0.5.

through the thickness of the preform and then the interply sliding can be minimized. It can be observed that the tuft yarns can be broken by the interply sliding in the weak tufting density cases (for example, tufting spacing of 10 and 20 mm). Consequently, it has better to choose a rather high tufting density to reinforce the "Z" direction (for example, tufting spacing of 5 mm), but a higher tufting density leads to a heavier and more rigid preform, and as a result the forming process will be difficult to achieve.

As one of the most common defects, wrinkling can be experienced frequently in textile reinforcement forming [23, 26, 29, 61]. The possible relative motion of fibers due to the internal composition of textile reinforcement leads to a weak bending stiffness [18, 62]. Wrinkling is a global phenomenon, which depends on the strain, force components, and boundary conditions of forming process [61]. However, the wrinkling phenomenon can be modified by the use of tufting technology. Wrinkling phenomena are shown in **Figure 11** for the forming with different preform. Big wrinkles with a nonregular shape can be noted in the forming of nontufted preform (**Figure 11a**). Compared to the deformed nontufted preform, in the forming of tufted 1.0 and tufted 0.5 preforms the wrinkles are regularly distributed and the size of wrinkle is much reduced.

The influence of tufting yarn orientations is analyzed. Two plies E-glass plain weave are superposed and tufted trough-the-thickness with the different orientation of the tufting yarn (see **Figure 12**). Mass of preform is not changed as the dimensions of ply and the tufting spacing are not changed. Comparing the $[0°/90°]^2$ preforms tufted in 0°/90° and in ±45° directions, the change of tufting yarn orientations does not alter the global deformation of preform (**Figure 12**).

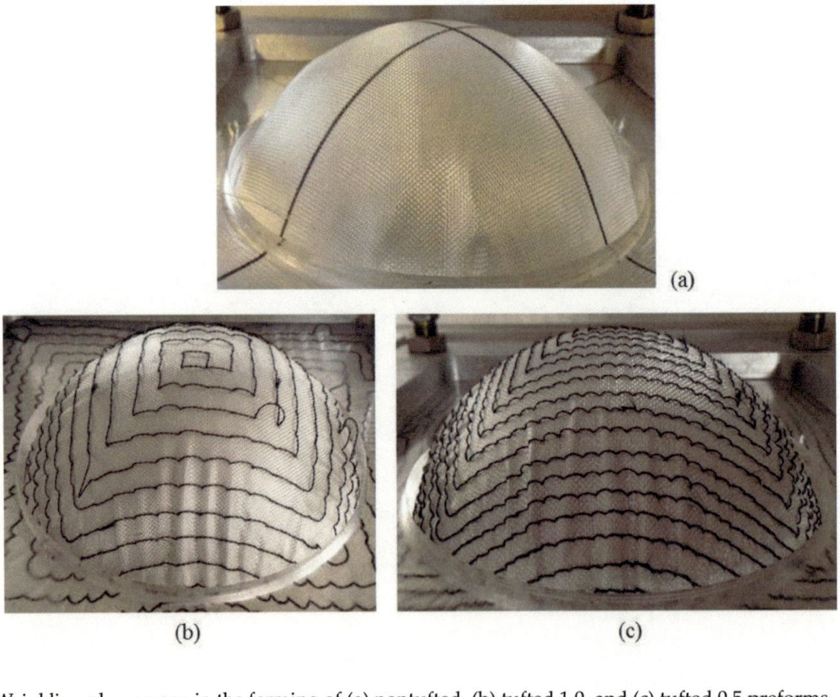

(a) (b) (c)

Figure 11. Wrinkling phenomena in the forming of (a) nontufted, (b) tufted 1.0, and (c) tufted 0.5 preforms.

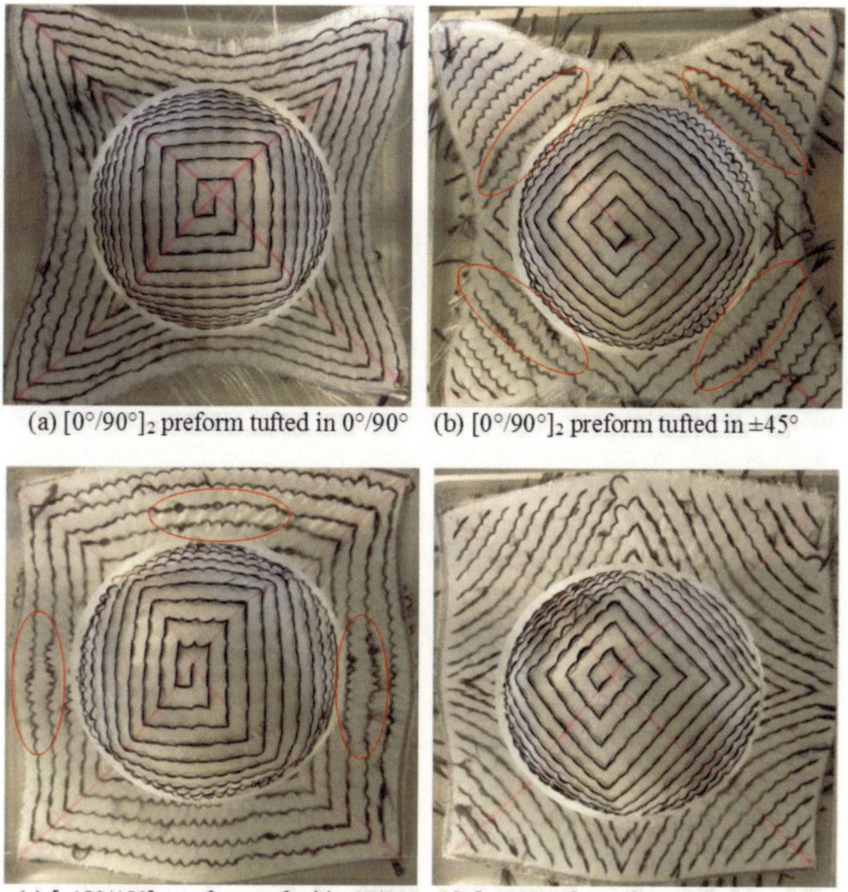

(a) $[0°/90°]_2$ preform tufted in 0°/90° (b) $[0°/90°]_2$ preform tufted in ±45°

(c) $[-45°/45°]_2$ preform tufted in 0°/90° (d) $[-45°/45°]_2$ preform tufted in ±45°

Figure 12. Hemispherical forming of different tufted 3D preforms. (a) $[0°/90°]^2$ preform tufted in 0°/90°; (b) $[0°/90°]^2$ preform tufted in ±45°; (c) $[-45°/+45°]^2$ preform tufted in 0°/90°; (d) $[-45°/+45°]^2$ preform tufted in ±45°.

As there is a strong in-plane shear effect in diagonal direction for 0°/90° ply (zones indicated on the figure), each segment of tufting thread between two tufting points is compressed and then the tufting yarns in these strong shear zones are misaligned. The same phenomenon is observed in the forming of $[-45°/45°]^2$ preform tufted in 0°/90° and in ±45°. As the strong shear zones are presented in longitudinal and transversal directions in [−45°/45°] ply forming, the misalignment is noted in the deformed $[-45°/45°]^2$ preform tufted in 0°/90°(**Figure 12c**).

The influence of tufting yarns during the tufted 3D textile reinforcements forming has been investigated. The tufted preform is more rigid than the multilayered preform due to the presence of tufting yarns through-the-thickness. Hence, it requires a bigger punch force in the tufted 3D reinforcement forming. When a quasiisotropic structure $[±45°, 0°/90°]^2$ is used in the multilayered forming, a significant intersliding can be observed. In tufting process, the slippage between the plies is reduced following the increasing of tufting density. Finally, the preform can be deformed as a single 3D ply when the tufting spacing 0.5 mm is used.

4. The manufacturing process of the composite corner fitting and square box parts

As the specific parts in aeronautical and automobile applications, it is very interesting to product the composite corner fitting and square box to replace the aluminum parts. As the classical manufacturing process, the composite forming is fast and more efficient, in particular to the manufacturing of multilayer laminates and interlock fabric-reinforced composites. Forming has been developed widely for dry textile reinforcements and prepregs. As presented previously, the double-curved shape forming can possibly lead to defects; it depends strongly on the forming parameters (tool loads, blank-holder forces, temperature). A novel manufacturing technology of composite materials called the surface 3D weaving seems very promising. It demonstrates directly the geometry of final composite part without the step of 2D product. The weaving in three directions is completely designed and the warp and weft yarns stay perfectly perpendicular on all of the surfaces of the final 3D ply. It is possible by using this technique to produce the 3D composite fabrics with geometry quite complex. Compared to the traditional forming process, the surface 3D weaving is a direct route to avoid certain manufacturing defects, such as wrinkling, porosities, slippage of the network, etc. On the contrary, the 3D weaving process is long and difficult to implement.

4.1. Textile reinforcements forming

4.1.1. Corner fitting

The experimental forming with the corner fitting punch is performed on a specific preforming device presented in Section 3. The punch dimensions are shown in **Figure 13**. A woven fabric with surface dimensions of 300 × 300 mm^2 was prepared for the forming tests. The corners fitting preforming results corresponding to a 65 mm displacement of punch are shown in **Figure 14**. The warp and weft directions on the deformed plies are figured out. The preforms

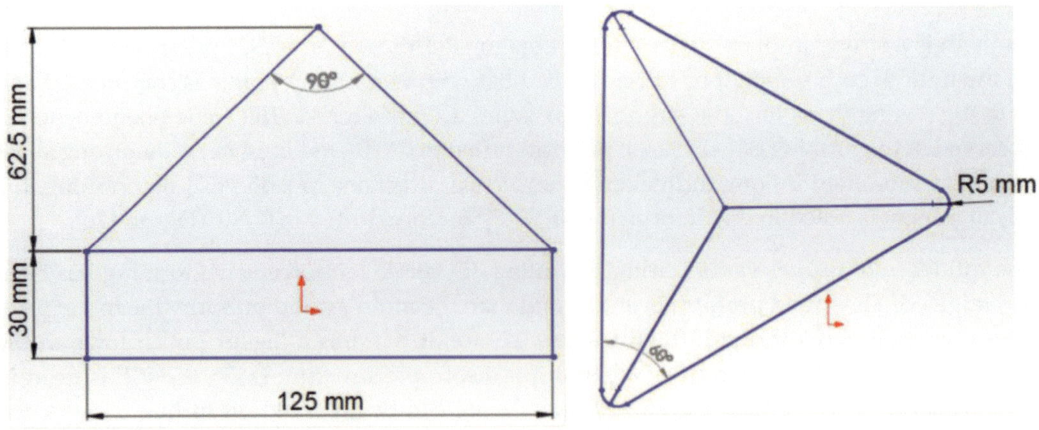

Figure 13. The punch dimensions.

http://dx.doi.org/10.5772/intechopen.68175

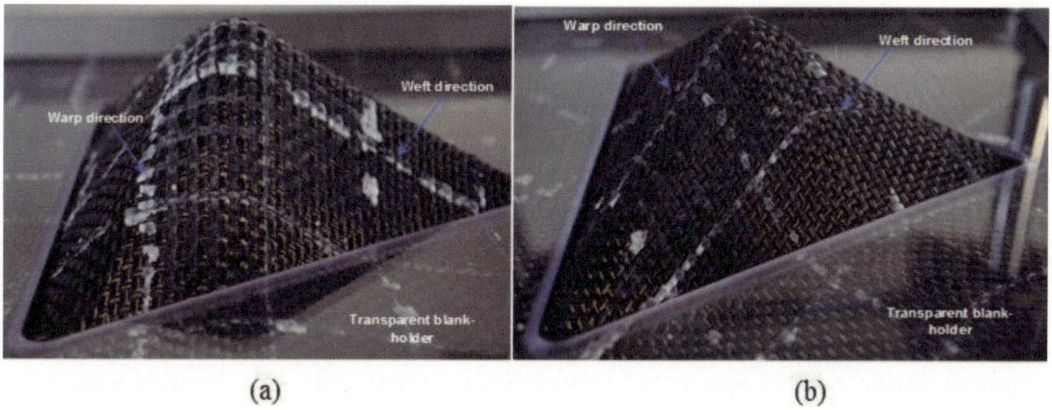

Figure 14. The experimental forming with 0°/90° (a) and −45°/45° and (b) plain weave fabrics.

are obtained without defects as the rupture or the buckle of the yarns. Moreover, no wrinkling is observed in these preforming tests.

Figure 15 shows the warp and weft directions of the deformed 0°/90° and −45°/45° plies. The angle between warp and weft yarns is not constant due to different local in-plane shearing effects. There is only a small zone where warp and weft yarns are (or quasi) perpendicular on each face of deformed ply (a small zone with zero in-plane shear angle around the middle line). Furthermore, it can be observed a positive in-plane shear angle (the angle between warp and weft yarns <90°) in the left zone and a negative in-plane shear angle (the angle between warp and weft yarns >90°) in the right zone on both 0°/90° and −45°/45° deformed plies. The in-plane shear angle can reach maximum 25° after forming in both left and right zones on the deformed −45°/45° ply. On the deformed 0°/90° ply, the maximum in-plane shear angle of 25° and 20° can be observed in the left and right zones, respectively. These changes in the angle between warp and weft yarns lead to a variation of fiber volume fraction and local permeability of the textile fabric. Subsequently,

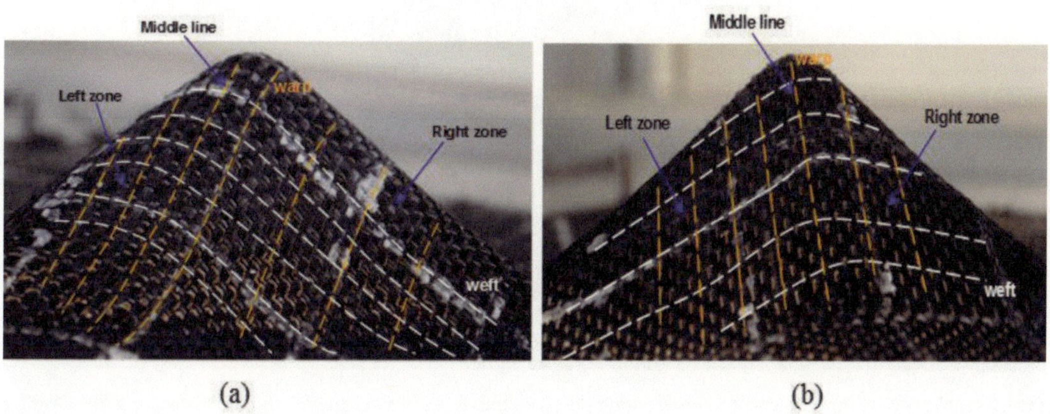

Figure 15. The yarns' directions of the deformed 0/90° ply (a) and ±45° ply (b).

it could result in a nonhomogenous flow during the resin infusion or injection stage. In this case, it will have a negative influence on mechanical performance of the final composite part. As an essential manufacturing parameter in liquid composite molding (LCM) processes, the permeability of the preform should be well controlled.

Another forming example with a tetrahedral punch performed through the numerical simulation analysis is presented in **Figure 16**. This forming simulation with a punch, as same geometry as the corner fitting part, is performed by using the approach semidiscrete [11, 23, 61]. **Figure 16a** shows the geometry of the tools (punch, die, and blank holders). The dimensions of a single ply are 425 × 425 × 0.3 mm^3. Six independent blank holders are used and apply a 0.01 MPa pressure during the forming stage. **Figures 16b** and **c** present the deformed blanks and the field of in-plane shear angle obtained by the forming simulation corresponding to a 95 mm displacement of the punch. The mechanical properties of the textile fabric used in forming simulation are given in **Table 2**. The friction coefficient on the interfaces of tool/ply and ply/ply is assumed to be equal to 0.2. In the major zone of the three surfaces of corner fitting part, the in-plane shear angle is between 15° and 25°. In addition, wrinkles can be observed in the useful zone in **Figure 16b** and **c**. Wrinkling is one of the most common flaws that experiences in both dry textile fabric [61] and prepreg composites forming [23] due to the possible relative motion of fibers, internal composition of textile reinforcement, causing a very weak bending stiffness [18, 61, 63]. Wrinkling depends strongly on the geometry of the final composite parts, on the fabric characteristics and on manufacturing parameters [23, 61].

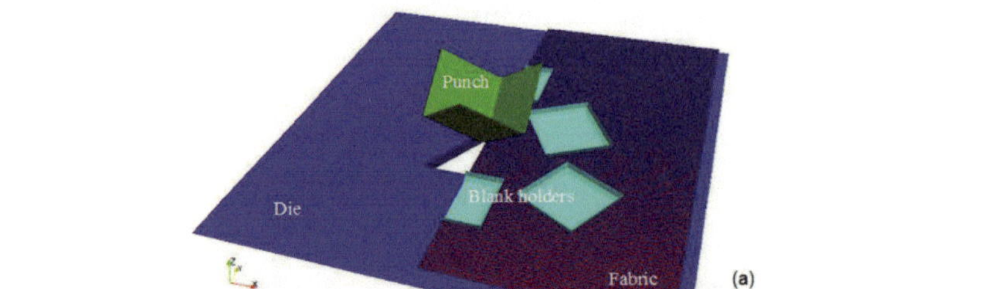

Figure 16. Numerical simulations of a corner fitting part forming. (a) Geometry model and mesh structure, (b) forming with a 0°/90° ply, and (c) forming with a −45°/+45° ply.

Tension in warp direction is assumed to be $T^{11} = C_1 \varepsilon_{11}$	Tensile stiffness in warp direction: $C_1 = 4600$ N/yarn
Tension in weft direction is assumed to be $T^{22} = C_2 \varepsilon_{22}$	Tensile stiffness in weft direction: $C_2 = 4600$ N/yarn
In-plane shear moment is assumed to be $M^s(\gamma) = k_1 \gamma + k_2 \gamma^3 + k_3 \gamma^5$	In-plane shear stiffness: $k_1 = 0.371$; $k_2 = -0.841$; $k_3 = 1.03$
Bending moment is assumed to be $M^1 = B_1 \chi_{11}$ in warp direction	Bending stiffness in warp direction: $B_1 = 0.5$ N mm
Bending moment is assumed to be $M^2 = B_2 \chi_{22}$ in weft direction	Bending stiffness in weft direction: $B_2 = 0.5$ N mm

Table 2. Mechanical behaviors of the textile composite fabric used in forming simulation.

4.1.2. Square box

Figures 17 and **18** present the square box forming with 0°/90° and −45°/45° plies by using blank-holder pressure of 0.05 MPa (a weak pressure). This pressure is applied homogenously on the blank-holder. The forming results correspond to a punch displacement of 85 mm. Due to the very weak intraply shearing effects, the in-plane shear angles inferior to 5° are measured on the upper surface and lateral surfaces of deformed 0°/90° ply (**Figure 17a**). On the contrary, the strong shear effects lead to an in-plane shear angle superior to 60°, which are observed on the corners of the square box and brings out the large wrinkles (**Figure 17b**). The in-plane shear angles between 25° and 35° are noted at two triangular zones shown in **Figure 18a**. Compared to these angles, the in-plane shear on the upper surface is negligible. As same as the forming of 0°/90° ply, the maximum in-plane shear angle can be observed on the corners (around 50°) and this strong in-plane shear effects lead to wrinkles (**Figure 18b**).

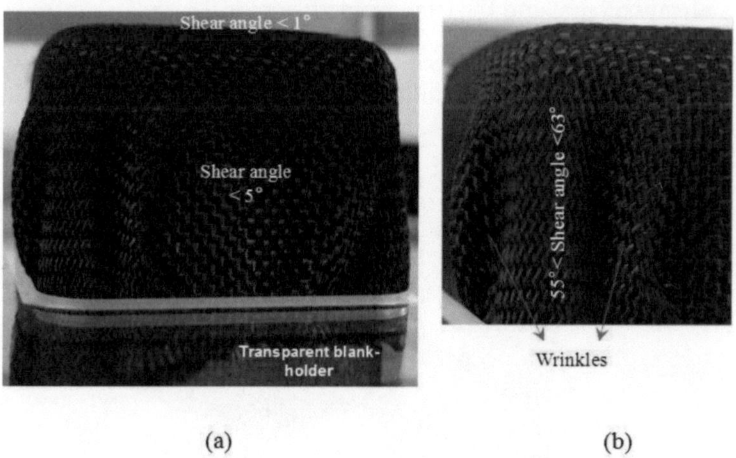

Figure 17. Forming with 0°/90° ply at 0.05 MPa (a) and zoom on the corner zone (b).

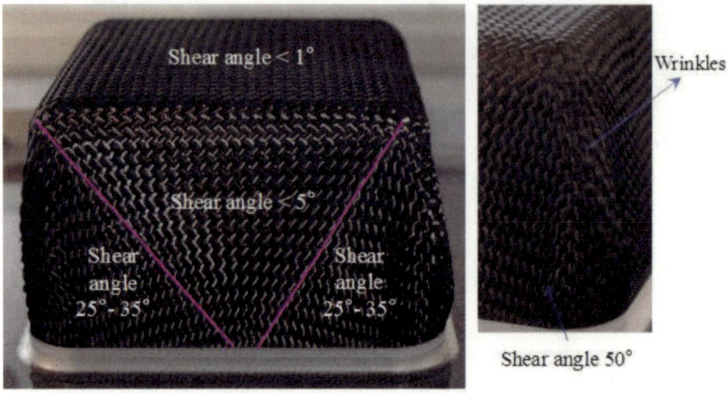

(a) (b)

Figure 18. Forming with −45°/45° ply at 0.05 MPa (a) and zoom on the corner zone (b).

The wrinkling depends strongly on the fabric characteristics. Therefore, the form and size of the wrinkles are different in the forming for 0°/90° and −45°/45° plies (**Figures 17b** and **18b**).

As one of the forming process parameters, the blank-holder force and the blank-holder shape and position can influence strongly on the wrinkling phenomena. A square box forming simulation with eight independent blank-holders is performed and shown in **Figure 19**. The same plain weave fabric presented in **Figures 17** and **18** is used in the forming simulation analysis. In this study, it is interesting to use the different pressure applied on different zones: a lower pressure in the corners or a higher pressure in the corners. As a higher pressure applied in the corners can lead to a more important tension and a more risk to create wrinkling and slippage of network, in the forming simulation a lower pressure of 0.05 MPa will be applied in the corners and a higher pressure of 0.2 MPa will be applied in the lateral sides (see **Figure 19**). The punch stroke is always 85 mm and the friction coefficient is 0.3.

The forming simulation results are shown in **Figures 20** and **21**. Wrinkles can be observed. Compared to the previous forming simulations, the distribution of in-plane shear angle is

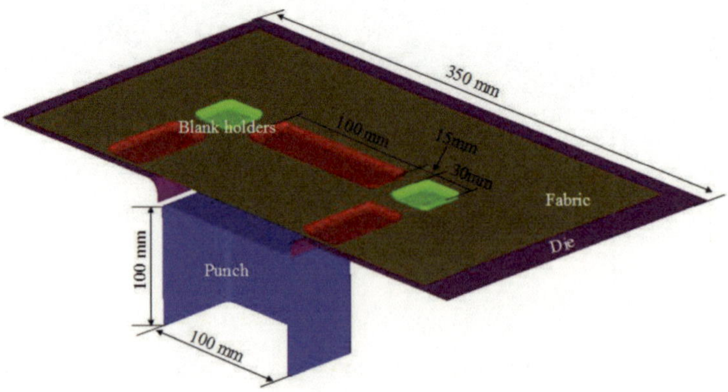

Figure 19. Geometry model and mesh structure of the square box forming with eight independent blank holders.

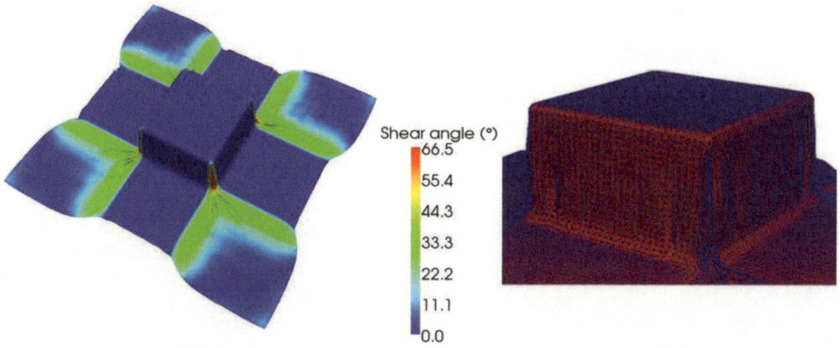

Figure 20. Square box forming simulation with eight independent blank holders for 0°/90° ply, (a) deformed ply and (b) zoom on the useful zone.

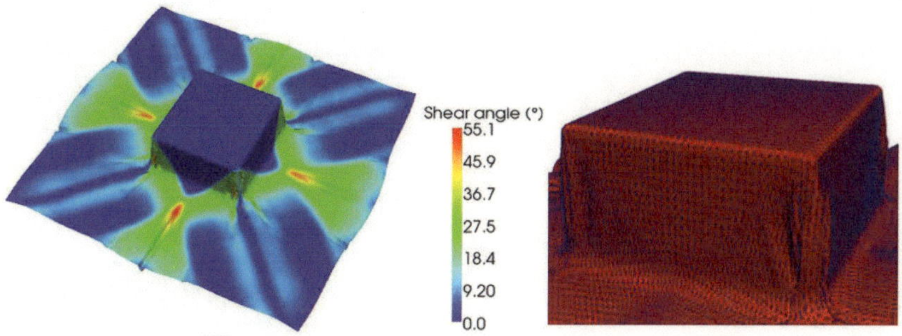

Figure 21. Square box forming simulation with eight independent blank holders for -45°/+45° ply, (a) deformed ply and (b) zoom on the useful zone.

very similar. Furthermore, the wrinkling phenomena are modified in the useful zone. Fewer wrinkles are noted in forming of 0°/90° ply but more on the corners of deformed −45°/45° ply. The forming simulation analyses emphasize that it is difficult to avoid the wrinkles in square box preforming with the considered textile reinforcement. Consequently, another manufacturing technology must be used to accomplish a product without wrinkles.

4.2. Surface 3D weaving

Compared to the bag weaving technique (the traditional weaving technique), two points should be developed in the surface 3D weaving method: assure the orientations and the continuity of fibers in each surface of corner fitting and square box.

4.2.1. 3D Corner fitting ply

The corner fitting 3D ply with the fiber direction of 0°/90° can be manufactured in the following main three steps. *The first step*: the first step concerns the yarns beaming and picking in the direction *X* as the traditional loom (**Figure 22a**). *The second step*: this step is still a traditional weaving loom process. The yarns in the direction *Y* are weaved with the ones in the direction *X* to build face 1 (**Figure 22b**). *The third step*: the Z yarns are generated automatically by the free yarns in the directions X and Y. Then X and Y yarns are weaved with Z yarns automatically as shown in **Figure 22c**. The insertion of X and Y yarns is not rectilinear but in the path of "L." In this way, the two faces (faces 2 and 3) can be constructed in the same time.

The desired orientations of fiber are set by controlling the tow tension during the whole structuring (interlacing) process, which is a traditional solution used in the textile industry. The X and Y yarns produced in steps 1 and 2 are under tension control. In order to guarantee the same tension force on each tow, a spring is employed for each loop of carbon tow during step 3. All the *X* direction springs are connected together by a transverse bar self-controlled on tension, to assure global tension applied on the X tows. Each "L" weft (inserted tow during the step 3) is also tensioned

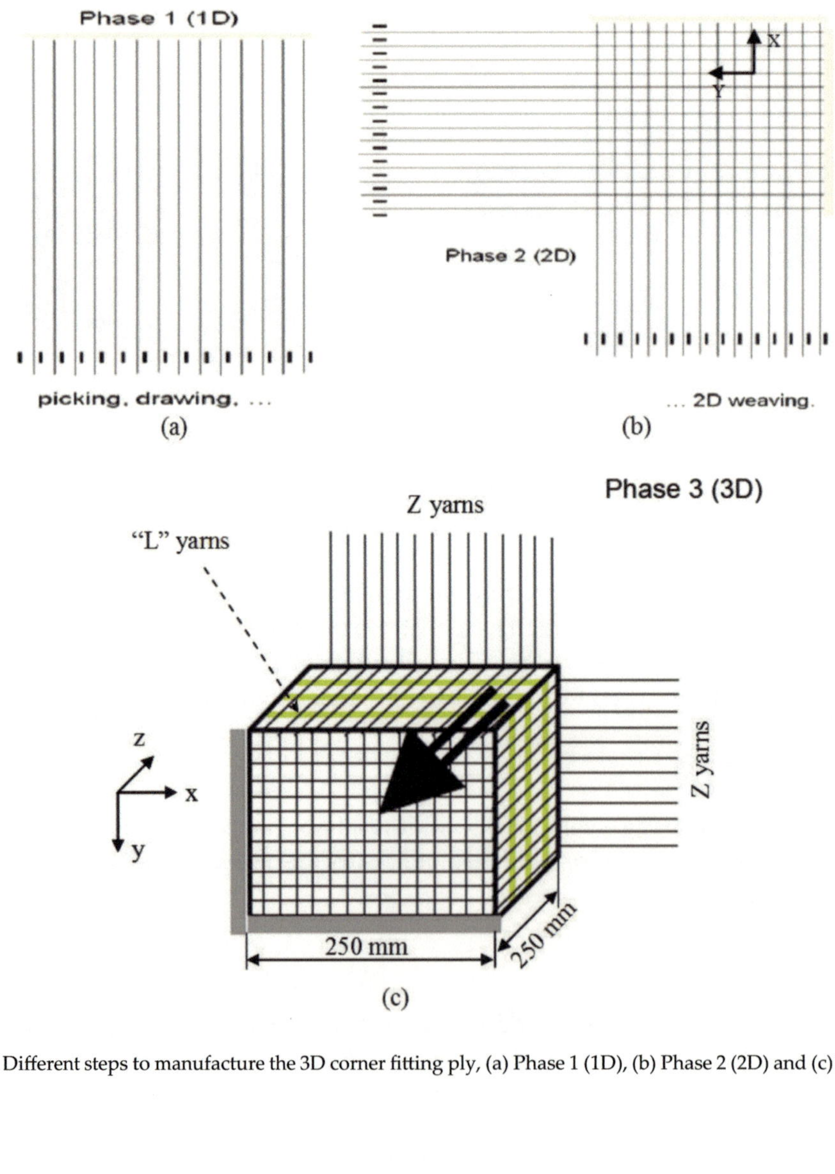

Figure 22. Different steps to manufacture the 3D corner fitting ply, (a) Phase 1 (1D), (b) Phase 2 (2D) and (c) Phase 3 (3D).

after its insertion between X and Y tows. Therefore, the tow is maintained in its right place and the desired orientations could be assured.

The −45°/45° corner fitting 3D ply can be produced in the similar way presented previously. The X and Y yarns are posed initially −45°/45°. Then the "L" yarns are weaved with XY yarns following −45° and 45° directions (**Figure 23**). The final 3D corner fitting plies with the fiber direction of 0°/90° and −45°/45° are shown in **Figure 24**. There are no wrinkles observed in these final 3D surface structures. The yarns in warp and weft directions stay perfectly perpendicular on the three faces and the continuity between faces is free from defects.

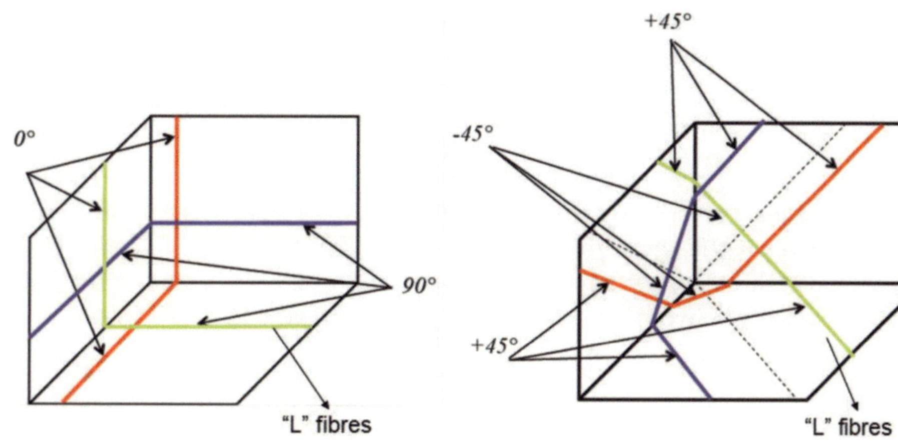

Figure 23. Directions of fibres in the 0°/90° and −45°/45° 3D corner fitting plies.

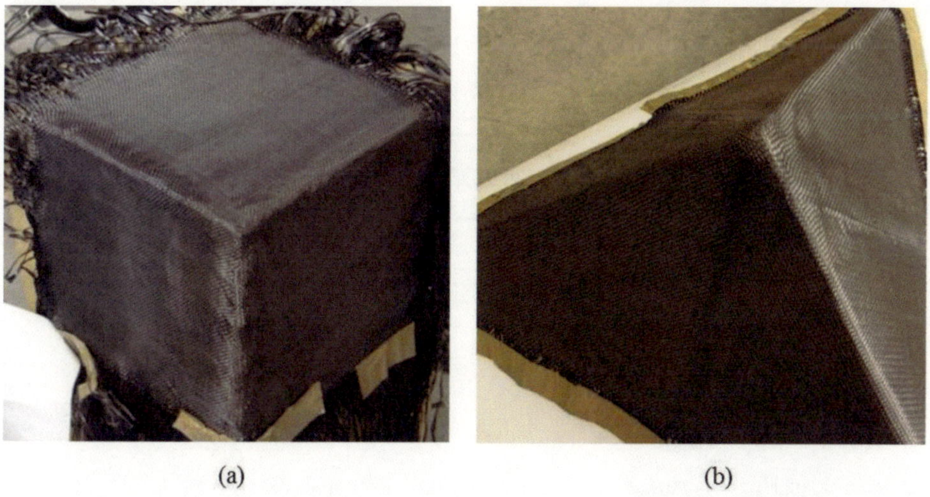

Figure 24. The 3D corner fitting plies obtained by surface 3D weaving process, (a) [0°/90°] and (b) [−45°/45°] plies.

4.2.2. 3D square box ply

Figures 25 and **26** present the surface 3D weaving process for manufacturing automatically the 3D square box ply with the fiber orientation of 0°/90°. In the manufacturing of the 3D corner-fitting ply, the first step concerns the yarns drawing and picking in the direction X (**Figure 25a**). The second step is called 2D weaving, through this step the Y yarns are performed. In this case, the X and Y yarns build automatically the face 1 (upper surface of the square box) (see **Figure 25b**). Thanks to a transversal forming frame in Z-direction, the free X and Y yarns can be considered as Z yarns (**Figure 26a**). To produce faces 2–5, the last X or Y yarn is used to weave with the Z yarns (see **Figure 26b**). The last two X or Y yarns are chosen on opposite faces and turn round to weave more rapidly with the Z yarns. The insertion of XY yarns is nonrectilinear but in a "rectangle" path (**Figure 26b** and **c**). In this way, the lateral faces (faces 2–5) can be constructed together. The warp and weft yarns on the lateral faces are not strictly perpendicular due to the yarn's width (**Figure 26c**). This deviation will be reduced by the increase of the dimensions of the box.

As for 3D corner fitting ply, the fibers' orientations must be well controlled. The manufactured 3D textile square box is shown in **Figure 27**. Some local defects due to manipulation of the ply are observed in the figure, these local defects can be absolutely avoided. No wrinkles on the faces and at the edges can be noted. The warp and weft yarns are perfectly under control and the continuity between faces is properly implemented.

The 0°/90° and −45°/45° 3D plies are produced and then laminated together. In this case, it should adapt the dimensions of final composite by changing warp and weft densities or by adding/removing one tow. The laminated corner fitting or square box plies can be manufactured by the RTM process to obtain a composite piece (**Figure 28**). Surface 3D weaving technique

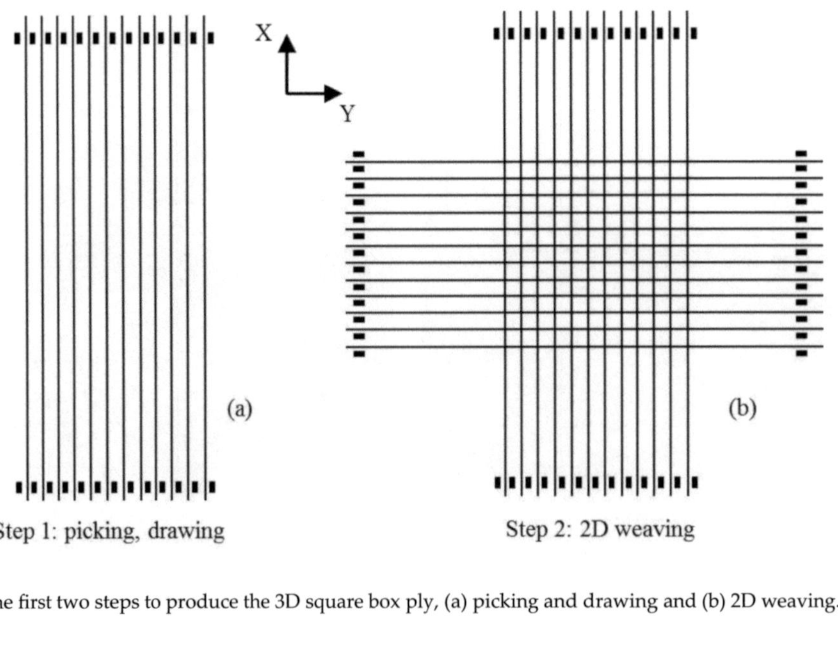

Figure 25. The first two steps to produce the 3D square box ply, (a) picking and drawing and (b) 2D weaving.

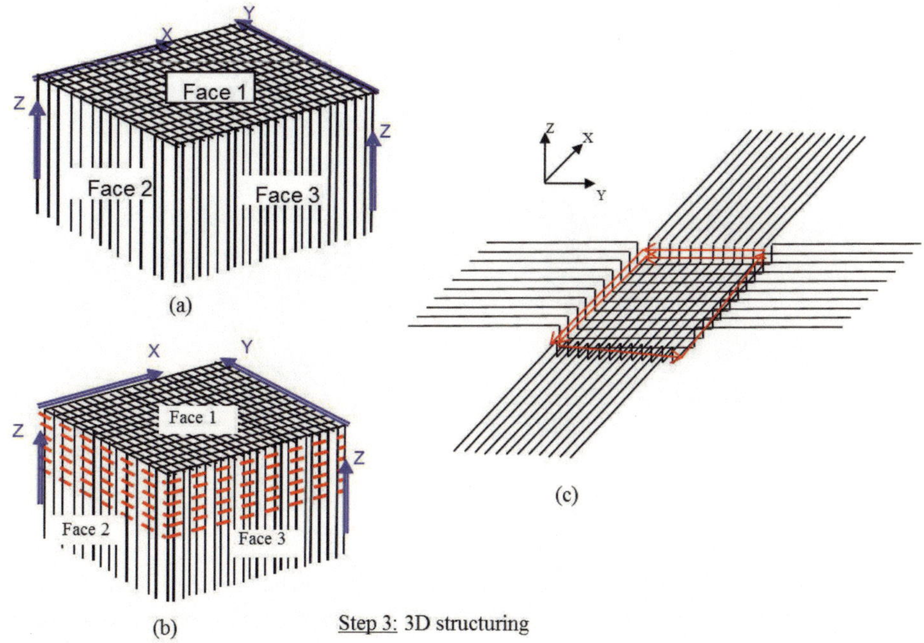

Figure 26. The 3D structuring step in surface 3D weaving process, (a) Z yarns, (b) face 2- face 5 and (c) weaving in a "rectangle" path.

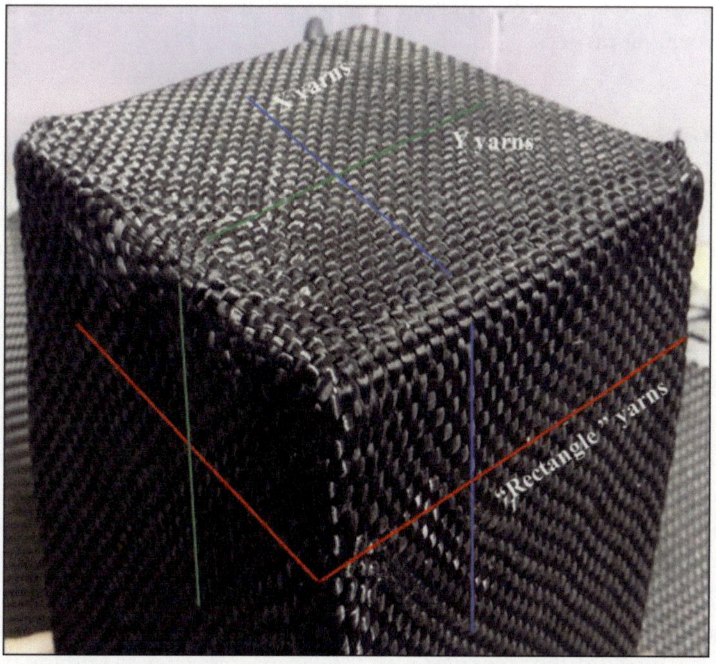

Figure 27. 3D square box ply obtained by surface 3D weaving.

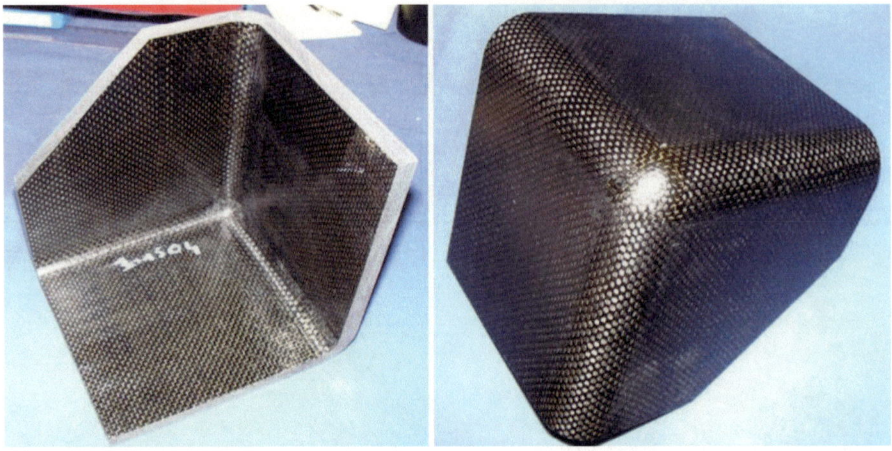

Figure 28. Final corner fitting composite part.

is defined in another way to design and produce the textile reinforcement for composite preforms when the shape is strongly double curved. Through this technique, it is possible to make a lighter composite part, nevertheless the conception and implementation are time-consuming. The main advantage of this new technique is the control throughout the process of the orientation of the carbon tow. The preform is free from any local or global misalignment. Moreover, the manufacture of the preform is made in a single operation (divided into three simple steps) very similar to traditional weaving. This is a direct process to obtain the 3D preform. Moreover, this new technique limits the tows' damage as they are less stressed than by the weaving technique and performing process.

5. Conclusions

The different promising textile reinforcements based on the advanced textile technologies for composite manufacturing have been presented in this chapter. The reinforcements (structure, manufacturing process, etc.) influence strongly on the properties of final composite materials. The innovation of the textile reinforcement structure is always a challenge in the composite research field. The 2D fabrics are employed frequently in the manufacturing of composite materials by forming/thermoforming processes. The double-curved shape forming can possibly lead to defects; it depends strongly on the forming parameters (tool loads, blank-holder forces, temperature, etc.). It is fast and more efficient, in particular for the manufacturing of multilayer laminates and interlock fabrics. The thick fabric can be used to produce thicker, lighter, and stronger composite parts. The through-the-thickness reinforcement of 3D structures is a promising attempt to overcome several problems. As one type of 3D textile reinforcements, the tufted 3D fabric is very interesting for the manufacturing of advanced composite part. Delamination can be reduced due to the existence of transverse reinforcement (reinforced yarns through-the-thickness), and the impact resistance and the damage tolerance are relatively strengthened. Compared to the stitch and interlock techniques, the tufting is user-friendly, more rapid, and low-cost process. The tufting yarns can minimize the interply sliding and the defects (e.g., wrinkles and misalignment) during the forming.

Compared to the textile reinforcement forming, the yarn directions can be well controlled for the surface 3D ply obtained by 3D surface weaving. The 3D surface weaving is a direct way to obtain 3D reinforcement. The wrinkling can be avoided completely and the local permeability of the 3D ply is homogenous, which is a big advantage for the resin injection/infusion stage during the liquid composite molding processes. It should be pointed out here that textile reinforcement forming is more developed. It is employed widely for both dry textile reinforcements and prepregs materials. To avoid the forming defects, it is important to predict the feasible conditions of composites forming, for example, by numerical simulation.

Author details

Peng Wang*, Xavier Legrand and Damien Soulat

*Address all correspondence to: peng.wang@ensait.fr

University of Lille, ENSAIT, GEMTEX, Roubaix, France

References

[1] Walther J, Simacek P, Advani SG. The effect of fabric and fiber tow shear on dual scale flow and fiber bundle saturation during liquid molding of textile composites. International Journal of Material Forming. 2012;**5**(1):83-97

[2] Trochu F, Ruiz E, Achim V, Soukane S. Advanced numerical simulation of liquid composite molding for process analysis and optimization. Composites Part A. 2006;**37**:890-902

[3] Swery EE, Allen T, Comas-Cardona S, Govignon Q, Hickey C, Timms J, et al. Efficient experimental characterisation of the permeability of fibrous textiles. Journal of Composite Materials. 2016;**50**(28):4023-4038

[4] Lomov S.V. Modelling the geometry of textile reinforcements for composites: WiseTex. In: Boisse P, editor. Composite Reinforcements for Optimum Performance. 1st ed. Woodhead Publishing; 2011. p. 200-238. DOI: http://doi.org/10.1533/9780857093714.2.200

[5] Long AC. Composites Forming Technologies. New York, NY: CRC Press; 2007

[6] Christos K. Design and Analysis of Composite Structures: With Application to Aerospace structures. New York, NY: Wiley; 2010

[7] Allaoui S, Hivet G, Soulat D, Wendling A, Ouagne P, Chatel S. Experimental preforming of highly double curved shapes with a case corner using an interlock reinforcement. International Journal of Material Forming. 2014;**7**(2):155-165

[8] Allaoui S, Cellard C, Hivet G. Effect of inter-ply sliding on the quality of multilayer interlock dry fabric preforms. Composites Part A. 2015;**68**:336-345

[9] Harrison P, Gomes R, Curado-Correia N. Press forming a 0/90 cross-ply advanced thermoplastic composite using the double-dome benchmark geometry. Composites Part A. 2013;**54**:56-69

[10] Boisse P, Aimène Y, Dogui A, Dridi S, et al. Hypoelastic, hyperelastic, discrete and semi-discrete approaches for textile composite reinforcement forming. International Journal of Material Forming. 2010;**3**:1229-1240

[11] Hamila N., Boisse P. Simulations of textile composite reinforcement draping using a new semi-discrete three node finite element. Composites Part B. 2008;**39**:999-1010

[12] Dufour C, Wang P, Boussu F, Soulat D. Experimental investigation about stamping behavior of 3D warp interlock composite preforms. Applied Composite Materials. 2014;**21**(5): 725-738

[13] Tephany C, Soulat D, Gillibert J, Ouagne P. Influence of the non-linearity of fabric tensile behavior for preforming modeling of a woven flax fabric. Textile Research Journal. 2016;**86**(6): 604-617

[14] Pazmino J, Carvelli V, Lomov SV. Formability of a non-crimp 3D orthogonal weave E-glass composite reinforcement. Composites Part A. 2014;**61**:76-83

[15] Allaoui S, Boisse P, Chatel S, Hamila N, Hivet G, Soulat D, et al. Experimental and numerical analyses of textile reinforcement forming of a tetrahedral shape. Composites Part A. 2011;**42**(6):612-622

[16] Boisse P, Zouari B, Gasser A. A mesoscopic approach for the simulation of woven fiber composite forming. Composites Science and Technology. 2005;**65**(3-4):429-436

[17] Syerko E, Comas-Cardona S, Binetruy C. Models for shear properties/behavior of dry fibrous materials at various scales: A review. International Journal of Material Forming. 2015;**8**:1-23

[18] Potter K, Khan B, Wisnom M, Bell T, Stevens J. Variability, fiber waviness and misalignment in the determination of the properties of composite materials and structures. Composites Part A. 2008;**39**:1343-1354

[19] Bloom LD, Wang J, Potter KD. Damage progression and defect sensitivity: An experimental study of representative wrinkles in tension. Composites Part. 2013;**45**:449-458

[20] Ouagne P, Soulat D, Moothoo J, Capelle E, Gueret S. Complex shape forming of a flax woven fabric; analysis of the tow buckling and misalignement defect. Composites Part A. 2013;**51**:1-10

[21] Ouagne P, Soulat D, Tephany C, Gillibert J. Measurement of the appearance and growth of tow buckling defect in the frame of complex shape manufacturing process by using fringe projection technique. Strain. 2016;**52**(6):559-569

[22] Warren KC, Lopez-Anido RA, Goering J. Experimental investigation of three-dimensional woven composite. Composites Part A. 2015;**73**:242-259

[23] Wang P, Hamila N, Boisse P. Thermoforming simulation of multilayer composites with continuous fibers and thermoplastic matrix. Composites Part B. 2013;**52**:127-136

[24] Bel S, Hamila N, Boisse P, Dumont F. Finite element model for NCF composite reinforcement preforming: Importance of inter-ply sliding. Composites Part A. 2012;**43**:2269-2277

[25] Mulvihill DM, Sutcliffe MPF. Effect of tool surface topography on friction with carbon fibre tows for composite fabric forming. Composites Part A. 2017;**93**:199-206

[26] Liang B, Hamila N, Peillon M, Boisse P. Analysis of thermoplastic prepreg bending stiffness during manufacturing and of its influence on wrinkling simulations. Composites Part A. 2014;**67**:111-122

[27] Bhattacharyya D, editor. Composite sheet Forming. 1st ed. Elsevier Science; 1977.

[28] Potter K., Ward C. Draping processes for composites manufacture. In: Boisse P, editor. Advances in Composites Manufacturing and Process Design. 1st ed. Woodhead Publishing; 2015. P. 93-109. DOI: http://doi.org/10.1016/B978-1-78242-307-2.00005-1

[29] Capelle E, Ouagne P, Soulat D, Duriatti D. Complex shape forming of flax woven fabrics: Design of specific blank-holder shapes to prevent defects. Composites Part B. 2014;**62**:29-36

[30] Omrani F, Wang P, Soulat D, Ferreira M, Ouagne P. Analysis of the deformability of flax-fibre nonwoven fabrics during manufacturing. Composites Part B. 2016. DOI: http://dx.doi.org/10.1016/j.compositesb.2016.11.003

[31] Duhovic M, Mitschang P, Bhattacharyya D. Modeling approach for the prediction of stitch influence during woven fabric draping. Composites Part A. 2011;**42**(8):968-978

[32] Mathieu S, Hamila N, Bouillon F, Boisse P. Enhanced modeling of 3D composite preform deformations taking into account local fiber bending stiffness. Composites Science and Technology. 2015;**117**:322-333

[33] Omrani F, Wang P, Soulat D, Ferreira M, Mechanical properties of flax-fibre reinforced preforms and composites: Influence of the type of yarns on multi-scale characterizations. Composites Part A. 2017;**93**:72-81

[34] Jacquot PB, Wang P, Soulat D, Legrand X. Analysis of the preforming behavior of the braided and woven flax/polyamide fabrics. Journal of Industrial Textiles. 2016;**46**(3): 698-718

[35] Risicato JV, Kelly F, Soulat D, Legrand X, Trumper W, Cochrane C, et al. A complex shaped reinforced thermoplastic composite part made of commingled yarns with integrated sensor. Applied Composite Material. 2015;**22**(1):81-98

[36] Kazemahvazi S, Khokar N, Hallstrom S, Wadley HNG, Deshpande VS. Confluent 3D-assembly of fibrous structures. Composites Science and Technology. 2016;**127**:95-105

[37] Mouritz AP, Bannister MK, Falzon PJ, Leong KH. Review of applications for advanced three-dimensional fiber textile composites. Composites Part A. 1999;**30:**1445-1461

[38] Kamiya R, Cheeseman BA, Popper P, Chou TW. Some recent advances in the fabrication and design of three-dimensional textile preforms: A review. Composite Science and Technology. 2000;**60**:33-47

[39] Potluri P, Hogg P, Arshad M, Jetavat D, Jamshidi P. Influence of fiber architecture on impact damage tolerance in 3D woven composites. Applied Composite Material. 2012;**19**(5): 799-812

[40] Boussu F, Cristian I, Nauman S. General definition of 3D warp interlock fabric architecture. Composites Part B. 2015;**81**:171-188

[41] Rahali Y, Goda I, Ganghoffer JF. Numerical identification of classical and nonclassical moduli of 3D woven textiles and analysis of scale effects. Composites Structures. 2016;**135**:122-139

[42] Saleh MN, Lubineau G, Potluri P, Withers PJ, Soutis C. Micro-mechanics based damage mechanics for 3D orthogonal woven composites: Experiment and numerical modeling. Composites Structures. 2016;**156**:115-124

[43] Gras R, Leclerc H, Roux S, Otin S, Schneider J, Périé JN. Identification of the out-of-plane shear modulus of a 3D woven composite. Experimental Mechanics. 2013;**53**(5):719-730

[44] Naouar N, Vidal-Salle E, Schneider J, Maire E, Boisse P. 3D composite reinforcement meso F.E. analyses based on X-ray computed tomography. Composites Structures. 2015;**132**:1094-1104

[45] Pegorin F, Pingkarawat K, Daynes S, Mouritz AP. Influence of z-pin length on the delamination fracture toughness and fatigue resistance of pinned composites. Composites Part B. 2015;**78**:298-307

[46] Chen X, Taylor WL, Tsai LJ. An overview on fabrication of three-dimensional woven textile preforms for composites. Textile Research Journal. 2011;**81**:932-944

[47] Mouritz AP, Leong KH, Herszberg I. A review of the effect of stitching on the in-plane mechanical properties of fiber-reinforced polymer composites. Composites Part A. 1997;**28A**: 979-991

[48] Dell'Anno G, Treiber J, Partridge I. Manufacturing of composite parts reinforced through-thickness by tufting. Robotics and Computer-Integrated Manufacturing. 2016;**37**:262-272

[49] Kyosev Y. Geometrical modeling and computational mechanics tools for braided structures. In: Kyosev Y, editor. Advances in Braiding Technology Specialized Techniques and Applications. 1st ed. Woodhead Publishing; 2016. P. 501-519. DOI: http://doi.org/10.1016/B978-0-08-100407-4.00021-1

[50] Abounaim M, Diestel O, Hoffmann G, Cherif C. High performance thermoplastic composite from flat knitted multi-layer textile preform using hybrid yarn. Composites Science and Technology. 2011;**71**(4):511-519

[51] NASA Technical Reports Server (NTRS) 19970017583: Handbook of Analytical Methods for Textile Composites, 1997.

[52] Mouritz AP. Flexural properties of stitched GRP laminates. Composites Part A. 1996;**27**(7): 525-530

[53] Mouritz AP, Cox BN. A mechanistic approach to the properties of stitched laminates. Composites Part A. 2000;**31**(1):1-27

[54] Koh T, Isa M, Chang P, Mouritz AP. Improving the structural properties and damage tolerance of bonded composite joints using z-pins. Journal of Composite Materials. 2012;**46**(26):3255-3265

[55] Pérès P., Desmars B., Léard J.P. Composite behavior of assemblies with AEROTISS® 03S technology. In: 16th International Conference on Composite Materials; July 8-13; Kyoto, Japan. 2007.

[56] Ko F.K., Du G.W. Textile Preforming. In: Peters S.T., editor. Handbook of Composites. Springer; 1998. p. 397-424.

[57] Fukuta K, Aoki E, Nagatsuka Y. 3-D fabrics for structural composites. In: 15th Textile Research Symposium; Philadelphia, PA. 1986

[58] Henao A, Carrera M, Miravete A, Castejon L. Mechanical performance of through-thickness tufted sandwich structures. Composite Structures. 2010;**92**:2052-2059

[59] Vanclooster K., Lomov S.V., Verpoest I. On the formability of multi-layered fabric. In: 17th International Conference on Composite Materials; July 27-31; Edinburgh, UK. 2009.

[60] ten Thije RHW, Akkerman R. A multi-layer triangular membrane finite element for the forming simulation of laminated composites. Composites Part A. 2009;**40**:739-753

[61] Boisse P, Hamila N, Vidal-Sallé E, Dumont F. Simulation of wrinkling during textile composite reinforcement forming. Influence of tensile in-plane shear and bending stiffnesses. Composites Science and Technology. 2011;**71**(5):683-692

[62] Prodromou AG, Chen J. On the relationship between shear angle and wrinkling of textile composite performs. Composites Part A. 1997;**28A**:491-503

[63] Kawabata S, Niwa M, Kawai H The finite deformation theory of plain weave fabrics part I: The biaxial deformation theory. Journal of the Textle Institute. 1973;**64**(1):21-46

Chapter 7

Textile Composites for Seat Upholstery

Stana Kovačević, Jacqueline Domjanić,
Snježana Brnada and Ivana Schwarz

Additional information is available at the end of the chapter

http://dx.doi.org/10.5772/intechopen.68984

Abstract

Textile for seat upholstery is a part of wide field of Mobiltech. It is very important segment of the visual identity of a vehicle, so its design, along with other high set requirements, is of the utmost importance. Textiles for seat upholstery are subject to many challenges; therefore, some of the required properties are resistance to stress, abrasion, UV radiation, external temperature and humidity, static electricity and peeling, as well as offering safety and comfort. For cladding car seat construction with textile, individual cutting parts of the fabric are joined by seam, and this part usually has the lowest mechanical properties. Therefore, in designing seat upholstery, the focus needs to be on these material segments. The function of the seam is to provide uniform transmission of loads between two joined materials and keep their integrity, which is not completely possible with stitched seam. Therefore, the material behaviour in places of sewn seam, the impact of the needlepoint and needle type on the strength, the appearance and seam quality will be discussed. Since the textile composites for seat upholstery are exposed to the multi-cyclical stress on certain areas, a part of this chapter will be focused precisely on that field.

Keywords: textiles in transportation, seat upholstery, physical-mechanical properties, biaxial cyclic stress, seam

1. Introduction: textiles in transportation

Transport vehicles imply all types of conveyance for people and cargo, such as cars, trucks, buses, trains, ships, submarines, airplanes, spacecraft and various devices for sport and recreation. The industry of transport vehicles is one of the largest users of technical textiles. Mobiltech covers all fields of transportation on land, air, space or water. The usage of Mobiltech textile is wide (**Figure 1**), where some of these materials visible (upholstery, carpets, seat belts,

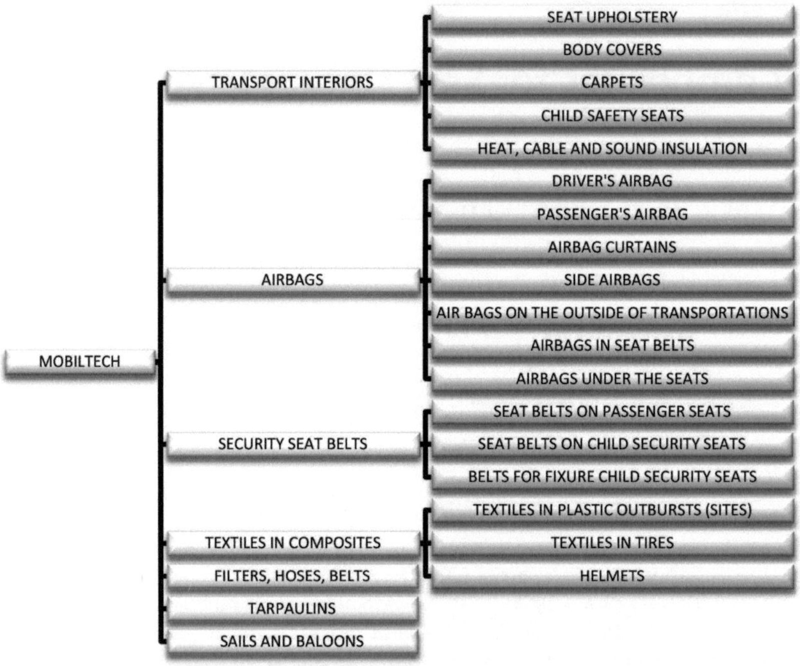

Figure 1. Distribution of technical textiles used in transport vehicles.

headliners, etc.), while other are concealed (tyre cords, airbags, hoses, belts, air and fuel filters, noise and vibration dampening, body panel reinforcement in composites, etc.).

Today's life style leads to the fact that people spend much more time in the vehicles that are sometimes used as places to work, eat, sleep, etc. Therefore, the safety and comfort of a conveyance are of paramount importance, which contributes to the design, functionality and cost-effectiveness of the vehicle interior.

Textiles in transport vehicles have multiple functions and can be summarized as follows:

- they provide a pleasant sensation during the long sitting in the same position. Textile is used for filling spaces and cladding of seat constructions (made of a metal, plastic and wood) with composite materials (woven fabric + polyurethane foam + knitting fabric) and thus contributes to its ergonomic design
- they provide passenger safety (seat belts and airbags)
- they ensure the protection of transport vehicles (shields and reinforcements for tyres, reinforcement in the walls of the transport vehicles, air and fluids filters, external airbags)
- they provide noise and vibration reduction in vehicles (multilayer materials for coating the interior).

The use of raw materials in transport vehicles is various in order to achieve the best possible application properties and maximum weight reduction (**Table 1**).

Application area	Fibres
Upholstery	Polyester, wool, cotton, nylon, acrylic
Carpets and coverings	Nylon, polyester, polypropylene
Airbags	Nylon 6, 6, nylon 4, 6
Seat belts	HT polyester
Composites	Glass, carbon, aramid, HT polyester, HT polypropylene
Tyres	Polyester, nylon, HT rayon, steel and aramid

Table 1. Application of textile raw materials in transport vehicles.

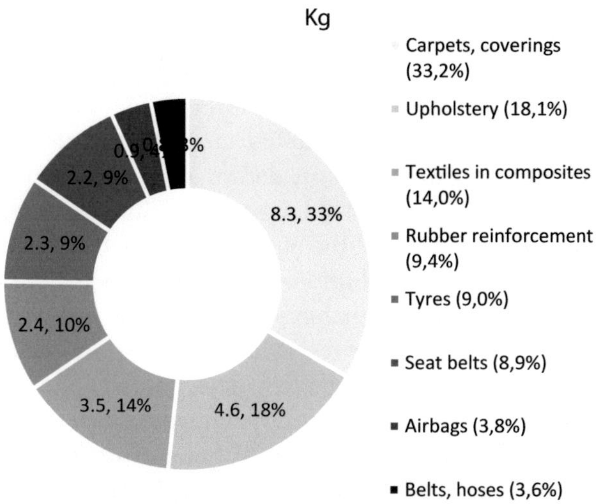

Figure 2. Textiles used in car.

Textiles in the transport vehicles, beside the basic physical-mechanical (strength, abrasion resistance, pilling) and thermo-physiological properties (comfort), must meet a number of other specific properties (resistance to sunlight and UV radiation, reduced flammability, odour free, antistatic, soil resistance) and at the same time be stable under the external temperature and humidity conditions (temperature of − 20 to + 100°C and humidity of 0–100%) through the entire vehicle life time. Textiles must also meet the high requirements for attractiveness following global trends in design.

Nevertheless, the largest consumers of textiles are still personal cars. Around 25 kg of textiles (**Figure 2**) is incorporated in the car, of which visible components cover approx. 45 m^2. In order to achieve lighter and more effective vehicle, metal parts are increasingly replaced by textile composites. It also increases the interest in the production and use of fabrics and composites made from natural fibres (cotton, flax, jute, sisal, kapok, wool, etc.) as a replacement for synthetic fibres, but these changes are still in infancy [1, 2].

2. Car seat upholstery

Almost all seats in the conveyance of passengers, that is, cars, aeroplanes and boats intended for luxury tourism are ergonomically designed with the possibility to control the height and position of the body. The shift of the seats in the vertical and horizontal direction and rotation of the backrest at desired angle enables adjustment of the best body position while driving. The gaps of the metal seat construction are filled with a variety of fibres, non-woven textiles and polyurethane foam in order to improve comfort during prolonged sitting and protect car seat cover from damage caused by direct contact with the metal frame of the seat (**Figure 3**).

The softness and breathability of the car seat upholstery material contribute to the feeling of comfort when driving, especially during a long stay in vehicle when the human body is in the passive sitting position. Long-term pressure of the same part of the body to the car seat reduces blood circulation and causes pain in muscles and joints. Functional properties and aesthetic appearance of the interior play a significant role in the vehicle selection.

The first materials used for car seats were single-layered, made from leather or rough and solid fabrics. Leather is characterized by durability, strength, abrasion resistance and appealing looks. Disadvantages of the leather seats are deformations that occur in short time of usage on sitting places and backrest as a result of long-term multidirectional stress. In the beginning of leather application, the surface of the material was not processed in order to increase resistance and protection. Therefore, additional disadvantages (of the leather material) are difficult maintenance, inadequate breathability and poor resistance to UV radiation and the impact of various weather conditions (especially in transport vehicles with unprotected seats). By its mechanical properties, woven fabric does not have the firmness of the leather, but its elastic properties ensure material stability. However, the woven fabrics have some advantages over the leather, such as a selection of various structures and designs, reasonable price, better thermal properties and softness to the touch. Today's development of textile for car seat upholstery is oriented to its structure and surface treatment. With the advent of more and more advanced textile materials and the possibility of various surface finishes, car seat upholstery is able to meet the growing stringent requirements of the market. The largest increase in the quality and functionality of car seat upholstery was followed by development of textile composite materials.

There are number of technical, design and purchase demands and end-use requirements placed on this type of materials. Exceptional strength, but also sufficient elasticity, breathability,

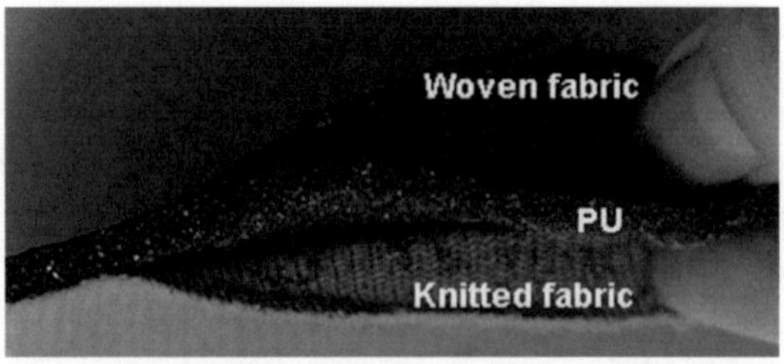

Figure 3. Composite material for car seat upholstery.

flammability and resistance to abrasion, UV radiation, extreme temperatures, humidity and microorganisms are some of the most important properties, which are achieved by using appropriate materials and structural parameters of materials and finishing process. Attractive appearance, soft touch, ergonomic design, comfort and easy maintenance will result in the feeling of pleasant stay in vehicle. Those performance properties and design are among the most important criteria for customer satisfaction. There are also some special properties, like reduction of electrostatic charge in seat upholstery (that can be achieved). A person exiting a car and leaving the seat can generate a body voltage of up to 30.000 V, which may affect passengers comfort, endanger sensitive electronic components or even cause an explosion during refuelling. Body voltage control can be achieved by interweaving antistatic yarn into woven fabric structure [3].

Targeted properties of the final composite and ultimately car seat upholstery can be easily achieved by a combination of various types of materials with predetermined properties. Car seat upholstery is exposed to multiple, long-term and cyclical stresses on the seat and backrest. Therefore, it is necessary to explore the flow of fatigue and product life by material testing with stress simulation. One of the main demands on car seat upholstery is that it should last longer than the transport vehicle. An important role in the car seat upholstery quality assessment has a sewn seam deformation. Since the sewn seam is inelastic and is usually located on the folding areas, it makes the weak point of the car seat upholstery. Prior to perform a sewing process, it is necessary to select an optimum combination of needle geometry, sewing thread, stitch class, and sewing machine type [1].

To study textile composite materials for seat upholstery, six samples with different layer thickness and lamination speed were selected. Total number of sample involves 18 composite materials made by flame lamination method with three different laminating speeds (30, 34 and 39 m/min) as a combination of woven fabric on the face of the material, polyurethane foam in the middle, and knitted fabric on the reverse. The woven fabric was produced from 100% polyester (PES) multi filament, weave: dobby, density warp/weft: 290/205 (filament/10 cm), count warp/weft: 620 f 144 dtex/167 f 48 × 3 dtex, mass 316.1 g/m^2, using weaving machine Dornier type: S 220 cm, weft insertion with rapier (electronical dobby).

The foam layer was made from polyurethane with 2 mm thickness and weight of 76.5 g/m^2 for the sample F2 and 4 mm polyurethane thickness with weight 144 g/m^2 for the sample F4 using flame lamination Schmid, model: 1281/2200.

The lower layer used as lining fabric knitted from 100% polyester multifilament, knitted fabric 1 + 1, density wales/course: 130/110 (per 10 cm), yarn fineness: 75–84 f 36 dtex, mass 51.4 g/m^2, using Terrot machine type: S296–1; E28 30''.

The results obtained by these tests are shown in the different sections of this chapter.

3. Composite material features

The first composite materials that have appeared on the market were used for making car seat upholstery, and even nowadays, they account for the largest share of the total amount of technical fabric production. In the last few decades, intensive development of materials has resulted in the emergence of new composites that meet the most stringent demands of car manufacturers. A wide range of new materials with outstanding properties and added value has been developed due to large funds investment in the car industry in order to increase market competitiveness. Developed new composites have higher level of quality and functionality. A use and qualitative value of the composite are usually determined by its physical-mechanical properties, which are directly correlated with the properties of individual components (woven fabric + polyurethane foam + knitted fabric) (**Figure 4**). Characteristics of the composites are given by the properties of the components from which the composite is made. This means that composite characteristics can be modified with combining certain properties of each component.

Woven fabric that provides the composite reinforcement is usually placed on the face of the composite material, while the polyurethane foam, in the middle, and knitting fabric, on the reverse, contribute to the composite comfort. Each component in the composite has its own function. Woven fabric with its targeted properties should provide adequate strength, aesthetics appearance, and affordable price. The second layer of the composite polyurethane foam is sandwiched between knitted and woven fabric to provide comfort when sitting. Therefore, it has to have certain elasticity, good thermal adhesion to the woven and knitted fabric in the joining process. In the same time, materials used for the production of car seat upholstery require good breathability, porosity and high elasticity. Knitted fabric used on the back of the material should protect polyurethane foam but also to improve durability, stability and elasticity of the composite material. With the implementation of new technologies, the life time of the composite has become longer. Contemporary composites have relatively higher strength, high resistance to abrasion and delamination, excellent resistance to UV radiation, good thermostability and seam quality, to achieve the best functionality and aesthetic appearance [4–6].

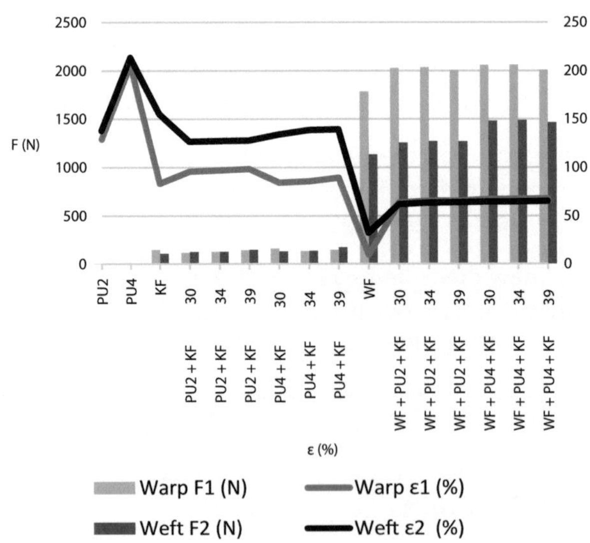

Figure 4. Tensile properties of tested materials in warp and weft direction of: each separate components in the composite (WF, PU, KF), semi-composites (PU + KF) and composites (WF + PU + KF); where WF is the woven fabric; KF is the knitted fabric; PU is the polyurethane foam (2 or 4 mm); speed of thermal joining (30, 34 or 39 m/min).

Beside the abovementioned composite, the materials for car seats develop in the direction of production of 3D spacer textile structures. Spacer fabric is a three-dimensional network, which has two outer layer structures connected with spacer yarn, which acts as a linear spring offering more energy absorption during compression and excellent recovery at the time of deceleration. Such fabrics can be used as substitutes for polyurethane foam, which beside its specific characteristics, thanks to which is the key element of the multilayer fabric in terms of comfort and mechanical behaviour, has great number of problems. Polyurethane foam, which generates toxic gases during its manufacturing process and during laminating processes, has a problem with flammability as well as recycling issues. All these problems lead to a need for its replacement. This new product has to be environmentally friendly and has to answer to the automotive specifications in terms of weight, formability and cost [7].

4. Composite material fibre content

In the beginning of the automobile production, products for the car seat upholstery were made by single-textile materials from natural fibres (fabric) or leather. The materials had a high value of surface mass, especially in the wet state, so their durability caused problems during the usage. Former car seat upholstery had a short product life due to the following disadvantages: long water retention, slow drying, low dimensional stability in wet state and others. With the advent of synthetic fibres, natural fibres are almost abandoned from the application in the production of car seat upholstery. On exceptional basis, natural fibres such as jute, sisal and kapok, with high strength values, could be used as the alternative to polyurethane foam. Nowadays, the woven and knitted fabrics are made from multifilament, synthetic yarns in various structures, fineness and twist count.

Composites from woven and knitted fabrics made of polyester multifilament fibres are most commonly used (**Figure 5**) due to the relatively low cost and good properties. The most important features of polyester fibres are good durability and dimensional stability, high strength, easy maintenance, fast drying, low moisture adsorption, resistance to the most chemicals, resistance to high temperatures and compatibility with cotton. There are many kinds of polyester fibres on the market, of which the following is used for woven fabrics intended for car seat upholstery: FR Trevira CS: flame-resistant polyester fibres, high strength Trevira and particularly a modified polyester fibre with antimicrobial activity (Bactekiller). Rapid technological development, interdisciplinary research and cooperation between science and industry have enabled the development of new materials in composites for car seat upholstery.

Since the woven fabric is usually located on the face of the composite, good durability properties and comfort of car seat upholstery depend mostly on it. The trend of using wool-based materials for furnishing luxury vehicles is increasing. Woollen fabrics enhanced by surface treatment achieve better properties of breathability, comfort, insulation and thermal properties, compared to synthetic fabrics. Still, the woollen fabric has some disadvantages such as higher price, shorter durability, peeling tendency (in mixtures) and poor resistance to UV radiation and abrasion.

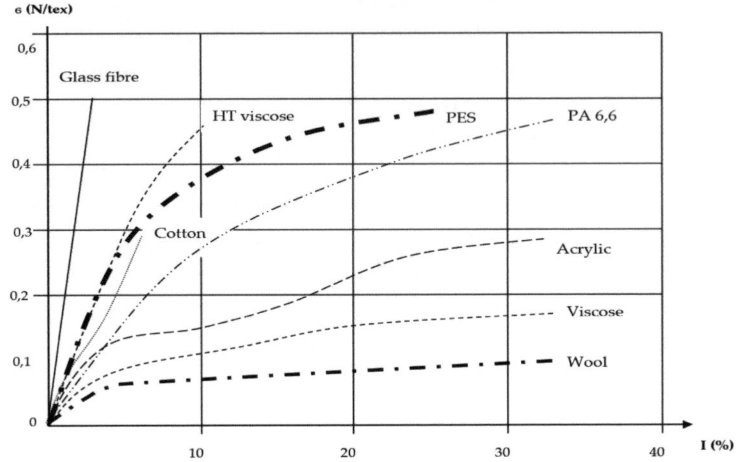

Figure 5. Tensile properties (strength and elongation) by fibre types.

Polyurethane foam has a low strength and low wear resistance but provides the car seat upholstery a very comfortable softness while sitting. One of the important properties of the polyurethane foam is its rigidity that prevents bending, wrinkling, creasing and stretching of the sitting and backrest areas during extended use. There are different thicknesses of polyurethane foams (1–11 mm) that are used in car seat upholstery composites (most often 2–5 mm). On the car seat, there are areas that are exposed to higher values of pressure. Therefore, those parts are reinforced with thicker polyurethane foam layer.

5. Thermal joining of composite

The components in the composite are thermally joined at the certain temperature and speed of the material, which has a direct impact on the final quality of the composite. Good quality of thermal joining of the components (first: PU + knitting, and then: PU and knit fabric + woven fabric) is achieved by good and solid adhesion between the components, while keeping elasticity properties.

The quality of adhesion between components can be estimated by testing the required force for composite components separation. In purpose of finding the optimal solution that will give the most solid thermal contact between the components, the separation force of composites joined with different bonding speeds (30, 34 and 39 m/min) and thicknesses of polyurethane (2 and 4 mm) was tested. The separation force between the knitted fabric and polyurethane + woven fabric composite and woven fabric and polyurethane + knitted fabric composite was conducted. Testing was performed on samples in longitudinal (warpwise) and transversal (weftwise) directions on the tensile tester Pellizzato/Tinius, C. Olsen type. H5KS, according to standard DIN EN ISO 13934–1.

Insufficient temperature in production process or an excessively high speed of the material in the process of thermal joining will result in insufficient adhesion between composite layers

(**Figures 6–9**). In addition, higher temperatures or lower velocity of the material will provide better adhesion of the components, but there is a risk that the mixture of melted adhesive agent and polyurethane foam penetrates into the structure of woven and knitted fabrics and stiffens the material. By this, the quality of material will be degraded in terms of bending characteristics and sewability.

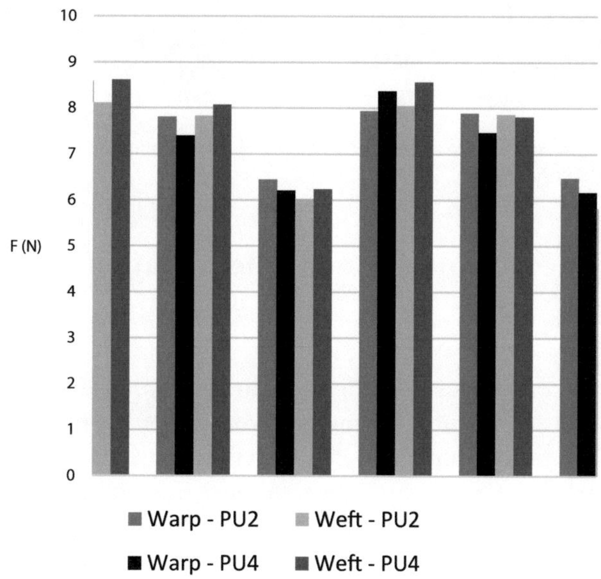

Figure 6. The separating force of the composite components, where WF is the woven fabric; KF is the knitted fabric; PU is the polyurethane foam (2 or 4 mm); speed of thermal joining (30, 34 or 39 m/min).

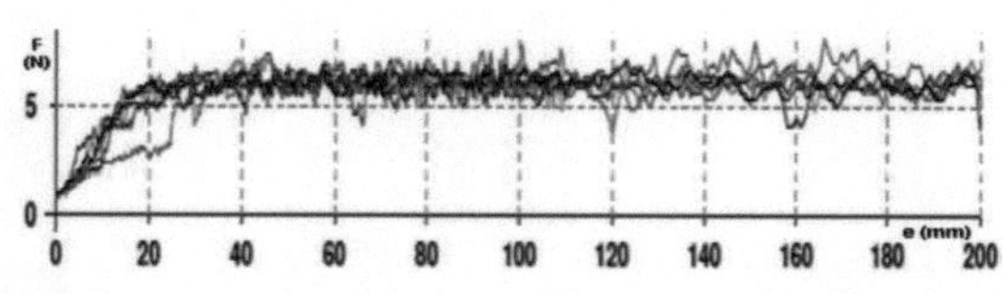

Figure 7. The movement of force and elongation at separation of the PU2 from the WF, in the warp direction, speed of thermal joining 39 m/min.

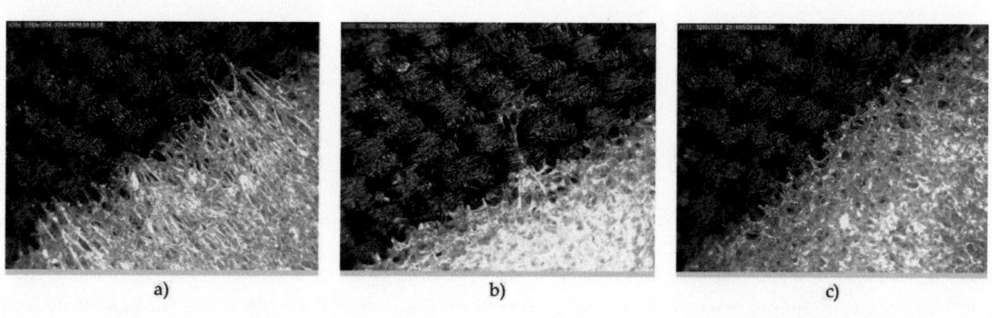

Figure 8. Separation of the woven fabric from the PU: (a) speed of thermal joining 30 m/min; (b) speed of thermal joining 34 m/min; and (c) speed of thermal joining 39 m/min.

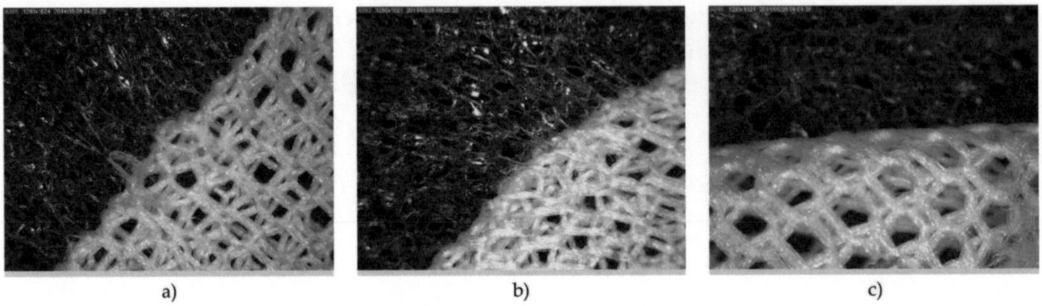

Figure 9. Separation of the knitted fabric from the PU: (a) speed of thermal joining 30 m/min; (b) speed of thermal joining 34 m/min; and (c) speed of thermal joining 39 m/min.

6. Material resistance to abrasion and determination of mass loss after abrasion

Abrasion resistance of composites, intended for car seat upholstery, is of great importance since it affects durability and quality. The test is carried out on the face of the composite, or the woven component. The tested composite (fabric) has an extremely high abrasion resistance. This claim can be supported by a relatively small loss of weight of the samples (1.2–2.5%) after 100,000 cycles of abrasion (**Figure 10**). It is important to mention the relatively big difference between the composite with a polyurethane (PU) foam thickness of 2 and 4 mm. Samples with thicker PU foam are softened, and the pressure between the fabric and the

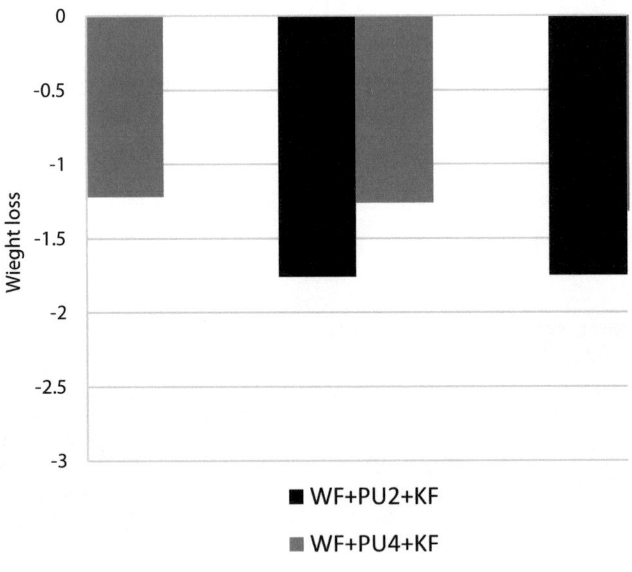

Figure 10. Mass loss of samples after 100, 000 cycles of abrasion testing, where WF is the woven fabric; KF is the knitted fabric; and PU is the polyurethane foam (2 or 4 mm).

material, subjected to abrasion, is slightly lower, but the surface force acting during the testing also has lower values.

7. Spherical straining of material

Car seat upholstery testing of strength at a stress in the material subjected to a spherical load is of great significance. In this example, test results of various composites bursted with different diameters of the balls (60, 40, 20, 10 mm) were performed with the diameter of the ring (100 mm). The circular shaped sample with 120 mm in diameter is fixed between two rings (clamp). The sphere is placed above the sample and fixed to the movable clamp of the tensile tester. Sphere moves towards the sample at a uniform speed rate of 100 mm/min till the testing material is bursted when it stops. The results of the testing are bursting force (N) and sphere displacement (mm). The testing results on **Figure 11** show the force and the depth of sphere penetration in the moment of bursting for four different composite samples and with the four different sphere diameters. The greater penetration force is in correlation with the diameter of the sphere. Also, the higher values of bursting forces are followed by lower values of sphere penetration depth at the bursting. Considering that bursting sphere depth penetration is in correlation with bursting elongation, it can be seen that thickness of polyurethane layer has small influence in composite elastic properties in a way that thicker the polyurethane foam is, the composite is slightly stiffer. Regarding the sphere diameter, greater sphere dimensions means the greater sample area affected in testing which results in wider stress distribution ($\sigma = F/A$, where A is area which is increasing along with sphere diameter) and delayed bursting in the terms of sphere penetration depth.

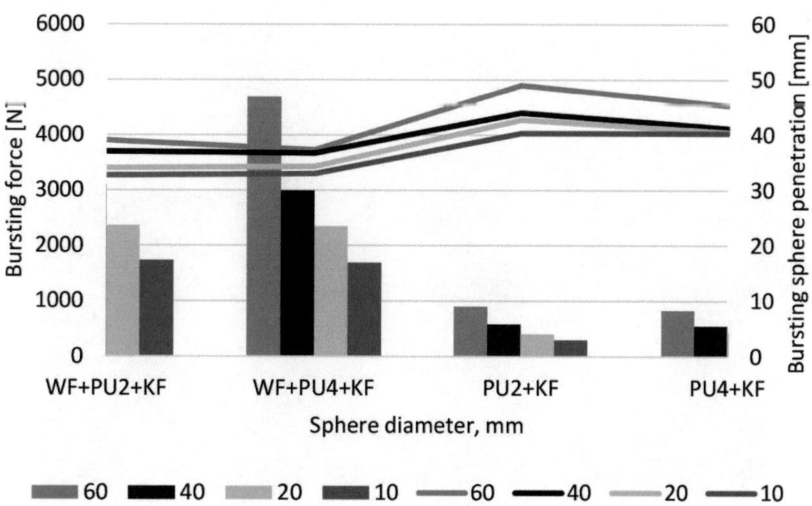

Figure 11. Bursting force and bursting depth of sphere penetration for composite samples (WF + PU + KF) and semi-composites (PU + KF), where WF is the woven fabric; KF is the knitted fabric; PU is the polyurethane foam (2 or 4 mm); and Ø is the diameter of the ball (10, 20, 40 and 60 mm).

8. Biaxial cyclic strain

A material fatigue could be determined by exposing the material to biaxial strain caused by the action of certain forces in two mutually perpendicular directions that stretch the material through defined number of cycles. It is therefore of utmost importance to define the change of rheological and usage properties of material, and its resistance to biaxial cyclic stresses. This test is particularly important on materials for car seat upholstery, in order to define their durability to multiple stresses.

8.1. The innovative device for determining material fatigue after cyclic stress

The device for measuring the resistance of fabrics to the biaxial cyclic stress is an innovative device (**Figure 12**). With this device, the material could be subjected to the cyclic stress in one or two directions. Regulation of pretension is the one of the most important actions in testing preparation, and it could be achieved with this device. Testing samples are prepared according to the method of testing (biaxial or uniaxial). The surface of the testing sample is always

Figure 12. The device for measuring uniaxial and biaxial cyclic stress.

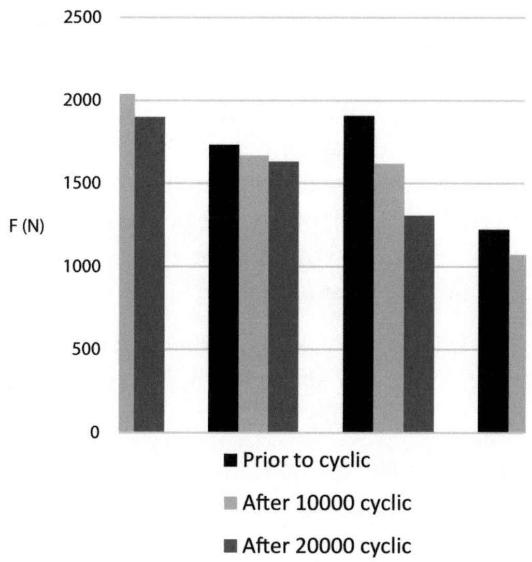

Figure 13. Breaking forces of the composite before and after exposure to the cyclic stress.

200 × 200 mm with the addition of fixing preclamps and clamps (100 mm). Sample pretension is adjusted by putting weights on two or four sides of material (depends of testing type) and fastening the preclamps to it. After a uniform prestressing, the sample is fastened by clamps that will hold it through entire testing.

The board with the side roller moves vertically in the up and down direction. At the upward move, the sample is tensed, while at the downward move, the material is relieved of stress. Cyclic strain of the sample is done by subjecting it to the certain force. The force and the speed of the board can be regulated. The device stops automatically after a specified number of cycles. The obtained samples are tested by one of the standard test methods for breaking strength and elongation of textile fabrics. The results of tests gave promising knowledge in the field of the durability of textile materials.

Figure 13 shows an example of test results for car seat upholstery materials that are cyclically burdened with a known force and then tested according to ISO 13934–1:2013 (Textiles—Tensile properties of fabrics—Part 1: Determination of maximum force and elongation at maximum force using the strip method). Due to high extensibility of individual composites components (knitted fabric, PU foam), was not possible to inflict such fatigue with cyclic stresses, by which differences in the decrease of mechanical properties could be determined. The results reveal that the material that was subjected to the action of the biaxial cyclic strain has lower values of breaking properties. A large number of cycles increase fatigue and permanent deformation.

9. The sewing of vehicle seat covers

Sewing has still a prime role in joining textiles in the automotive industry. The manufacturing of vehicle upholstery must fulfil high demands on product functionality, quality and sustainability. Sewing pattern, needles, threads and stitch density has to be carefully chosen to deliver the best possible seam quality to ensure performance standards and durability throughout the vehicle's life span [8].

Sewability is defined as the sewing thread's performance evaluated after the sewing process where the functional behaviour of the seam is described by seam properties such as seam stability (seam breaking strength and seam slippage), seam elasticity, abrasion resistance, individual product-related criteria [9, 10]. Behaviour properties of seams used for the production of vehicle seats represent the most significant problem, since the seam is the place of the weakest link in the motor vehicle seat, **Figure 14** [11].

During seam formation, the triple composite fabric is being mechanical damaged. Immediately after sewing, there are visible signs of damage on the fabric and sewing needle surface (**Figure 15**).

Needle penetration force is defined as the quantitative measure to determine the damage of sewn fabrics that has negative consequences on seam performance. The quantitative value

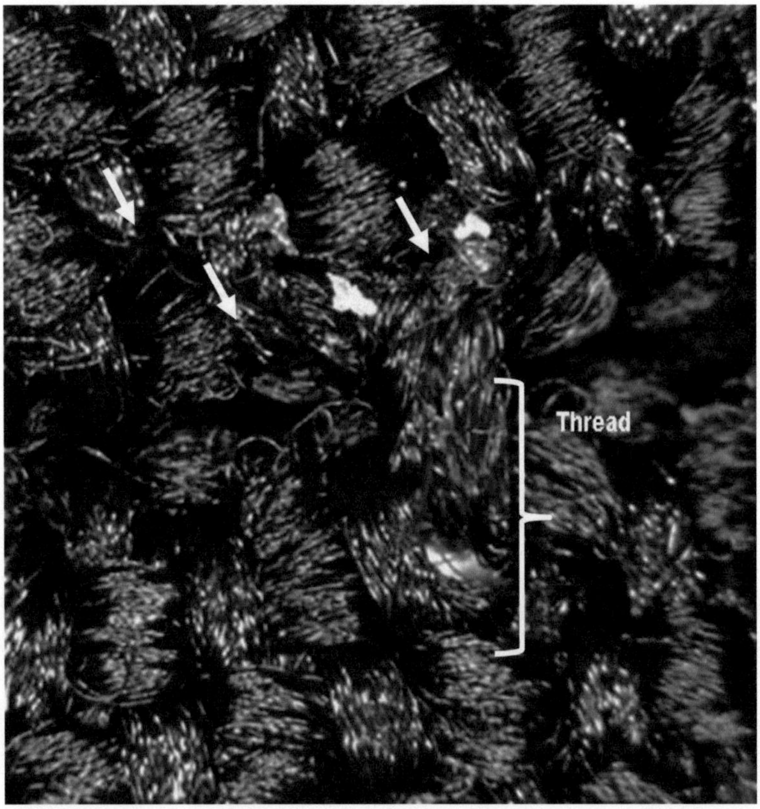

Figure 14. Mechanical damage appeared during seam formation: sewing needle penetrates through composite and drags the polyurethane foam on the front side of the composite.

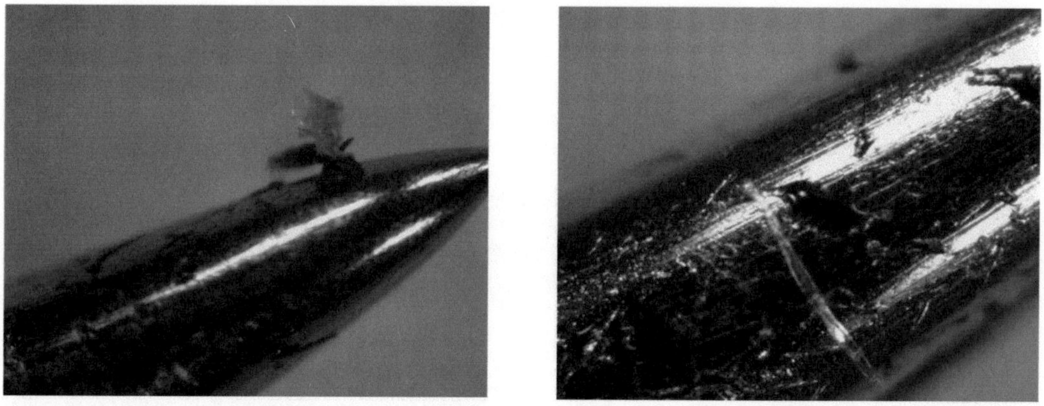

a) b)

Figure 15. Mechanical damage on needle surface: (a) sewing needle heating causes adhesion of polyurethane foam to the needle surface and (b) microscopic enlarged needle surface image with polyurethane foam.

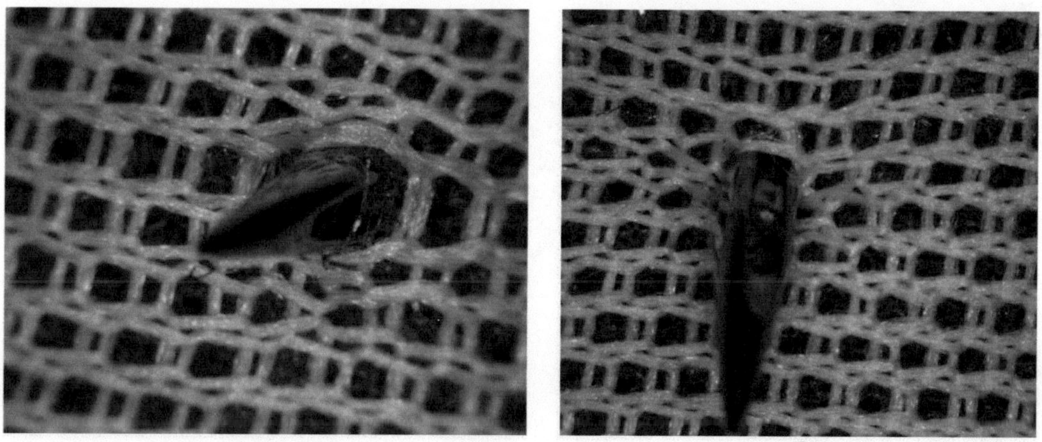

a) b)

Figure 16. Sewing needle penetrates through composite fabric: (a) sewing needle NM 100 RG and (b) sewing needle NM 100 RG SAN6.

of force that occurs when the needle penetrates through fabric (**Figure 16**) is influenced by fabric construction, needle geometry, sewing thread construction and sewing machine settings [12–14].

If the sewn fabric has high values of needle penetration force, than there is a high risk of fabric damage because the fabric has high resistance to the penetration of sewing needle. Damage that appears after sewing process is linked to seam quality and to the quality of the final product. **Figure 17** shows needle penetration force measured on composite fabric for producing vehicle seats [13].

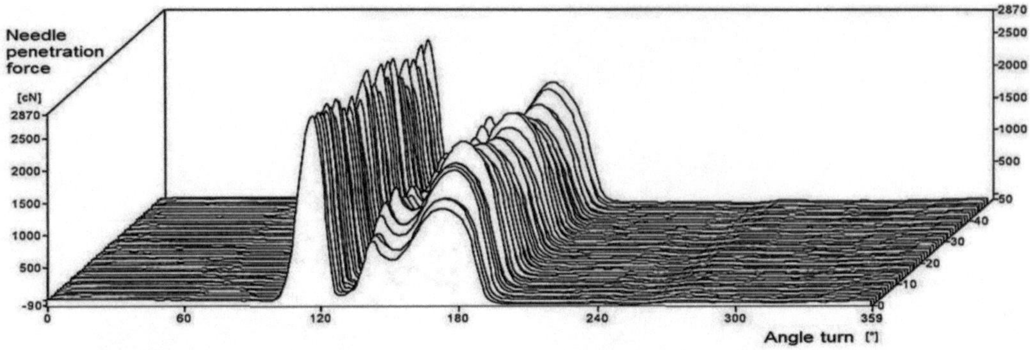

Figure 17. Graphical representation of measured needle penetration force based on 50 stitches without the use of sewing thread.

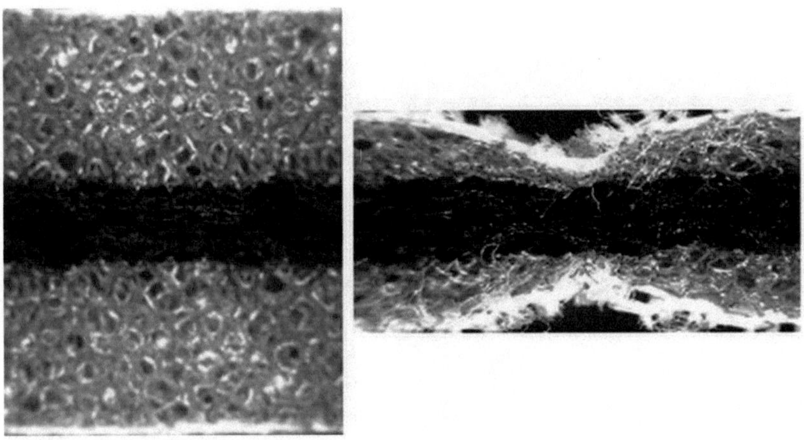

Figure 18. Presentation of composite fabric: (a) prior seam formation and (b) after seam formation.

The function of the seam is to provide uniform load transmission of two or more connected fabric layers to achieve their integrity (**Figure 18**). At the seam position, flat composite material becomes thicker, uncomfortable in the end use, it is subjected to higher wear and forces that further weak the joint place. An additional problem in sewing ergonomically shaped seats is insufficient elasticity at the seat folding place and in the area of multi-directional stresses [6, 7, 15].

Deformation of composite fabric and sewing needle damage can be explored by a systematic analysis of the seam. Based on the result, the best possible sewing needle will be selected in the production process to achieve minimum damage to sewing needle and fabric [16–19].

The samples were sewn using 100% PES sewing thread as the upper and under thread at single needle lockstitch (ISO 4915) with stitch density of 4.5 stitches per cm. The properties of the sewing thread were as follows: thread finesses 97 tex, breaking force 58 N, elongation at break 18, 47% determined according to the ISO 2062 and using the Statimat M tensile tester produced by Textechno [20].

An industrial high-speed sewing machine was used to join two layers of material PFAFF 1053 in warp direction, with Groz-Beckert needles of NM 90 and NM 100, standard round point (R).

In its performance, a seam is exposed to longitudinal and transversal loadings, and thus, certain discrepancies are visible when considering breaking force and elongation at break of the seam made using needle size NM 90 and NM 100, **Figure 19**.

The seam strength was higher when using a thinner needle size for the selected material, and vice versa.

Seam quality contributes to functional and aesthetic performance of automotive upholstery in the end use. A sewn seam has a good quality if its features such as strength, elasticity, durability, security and appearance are properly balanced with the fabric properties in order to be joined together [21]. Fabrics used for the production of car seats upholstery are made as multi-layered composite materials where each fabric layer has different properties.

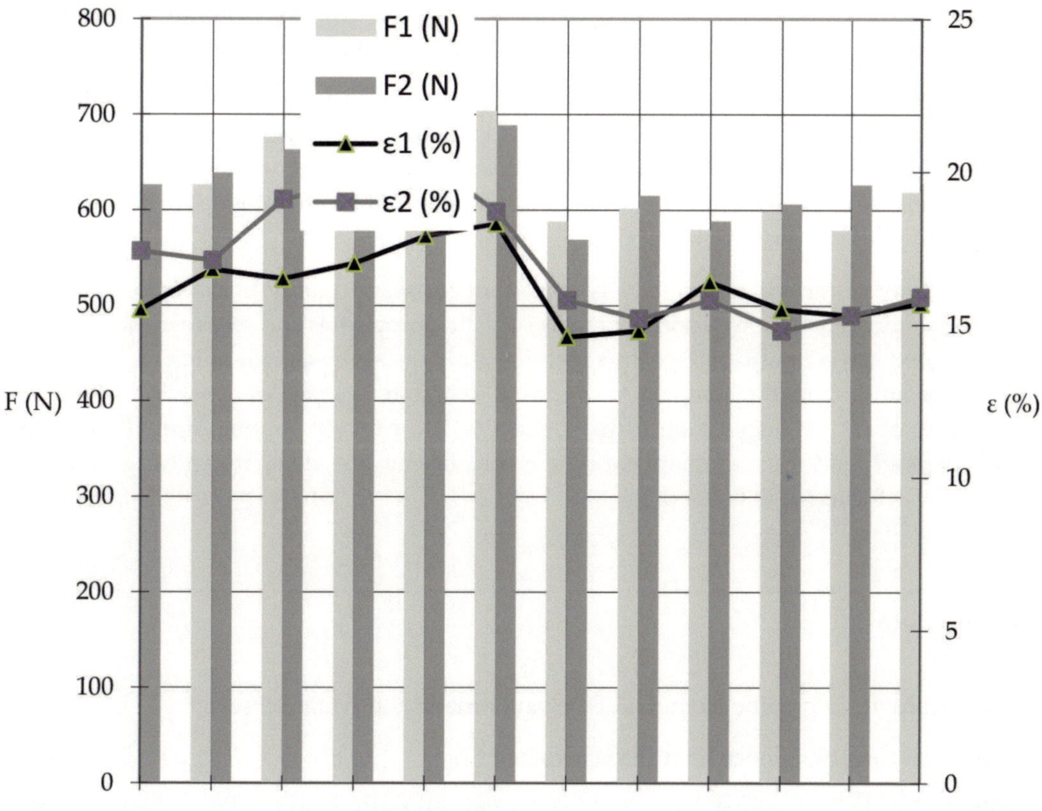

Figure 19. Breaking force and elongation at break of the seamed composite materials joined using: (a) Groz-Beckert needle of NM 90 and (b) Groz-Beckert needle of NM 100.

10. Review of the test results

The breaking strength of the composite is higher than the total amount of components breaking strengths prior to the thermal joining, at both (longitudinal and transversal) directions, at all thermal joining speeds and with both polyurethane foam thickness. Breaking forces of composites depend on the joining process speed. There is no linear trend that connects joining speed values with breaking forces of the composite. In the given example, the maximum tensile forces are at a medium speed of 34 m/min, which can be considered optimal for the tested composite.

The resistance of the composite to abrasion is essential for assessing the sustainability of the tested materials since materials in the vehicles are in the constant contact with the human body. Composite car seat upholstery is highly resistant to abrasion, where even after 100, 000 cycles, there was no appearance of the hole, and a mass loss is minimal (1–3%).

The composite with thicker polyurethane foam has a lower bursting force. This suggests that higher stiffness and thickness of the material provide less resistance to spherical stress. The greater diameter of the punching sphere is proportional to higher punching force.

Increase of the cycle number in cyclic stress setting leads to a growing material fatigue, which is reflected in the reduction of breaking force of materials tested on dynamometer.

The separation force between polyurethane foam and woven fabric is usually greater than the separation forces of polyurethane foam and knitted fabric. Lower joining process speed results in larger separation forces. Thinner polyurethane foam in the longitudinal direction results in larger separation forces, whereas thicker polyurethane foam gives higher values of separation forces in transversal direction.

The sewability of car seat upholstery is important for seam quality but also to insure cost effective production process. Good interaction of fabric properties and sewing conditions will result in good sewing performance. In a real-life situation, sewn seam is made by high-speed sewing machine where the composite is exposed to high temperatures of sewing needle and high values of needle penetration force. A seam is also exposed to longitudinal and transversal loadings, and thus, an inappropriate choice of any one element can cause failure of functional performance in the end use.

Author details

Stana Kovačević, Jacqueline Domjanić, Snježana Brnada and Ivana Schwarz*

*Address all correspondence to: ivana.schwarz@ttf.hr

Faculty of Textile Technology, University of Zagreb, Zagreb, Croatia

http://dx.doi.org/10.5772/intechopen.68984

References

[1] Pačavar S. The Influence of sewing parameters on the production quality of car seat covers [dissertation]. Zagreb: University of Zagreb, Faculty of Textile Technology; 2015. P. 192

[2] Kovačević S, Ujević D, Schwarz I, Brlobašić Šajatović B, Brnada S. Analysis of Motor vehicle fabrics. Fibres & Textiles in Eastern Europe. 2009;**71**: 32-38

[3] Shishoo R, editor. Textile advances in the automotive industry. Cambridge, England: Wodhed publishing Limited; 2008

[4] Mukhopadhyay SK, Patridge JE. Automotive Textiles. Textile progress. 1999;**29**(1/2):97-107

[5] J. Mcloughlin S. Hayes. Types of fabric and their joining requirements. In: Jones I, Stylios GK, editors. Joining Textiles, Principles and applications. UK: Woodhead publishing series in textiles; 2013. pp. 1-44

[6] Hayes S, Mcloughlin J. The sewing of textiles. In: Jones I, Stylios GK, editors. Joining Textiles, Principles and applications. UK: Woodhead publishing series in textiles; 2013. pp. 62-122

[7] Njeugna N, Schacher L, Adolphe DC, Schaffhauser JB, Strehle P. Development of a new 3D nonwoven for automotive trim applications. In: Chiaberge M, editor. New Trends and Developments in Automotive Industry. Rijeka, Croatia: InTech Publisher; 2011. pp. 323-346

[8] Gries T, et al. Textile structures for load-bearing applications in automobiles in Textile advances in the automotive industry. UK: Woodhead Publishing Limited; 2008

[9] Amann Group, editor. Quality, In: Focus Sewing and Embroidery Threads. Vol. 1. Basic Knowledge. Boennigheim: Amann Group; 2010

[10] Carvalho M, Carvalho H, Silva LF. Problems relating to sewing. In: Joining Textiles. Cambridge, UK: Woodhead Publishing; 2013. pp. 149-171

[11] Kovačević S, Ujević D. Seams in car seat coverings: Properties and performance. In: Jones I, Stylios GK, editors. Joining Textiles, Principles and Applications. UK: Woodhead publishing Series in Textiles; 2013. pp. 478-506

[12] Gotlih K. Sewing needle penetration force study. International Journal of Clothing Science and Technology. 1997;**9**(3):241-248

[13] Gurarda A, Meric B. Sewing needle penetration forces and elastane fiber damage during the sewing of cotton/elastane woven fabrics. Textile Research Journal. 2005;**75**(8):628-633

[14] Haghighat E, Najar SS, Etrati SM. The prediction of needle penetration force in woven denim fabrics using soft computing model. Journal of Engineered Fibers and Fabrics. 2014;**9**(4):45-55

[15] Geršak J, Knez B. Određivanje probodnih sila šivaćih igala u procesu šivanja odjeće. Tekstil. 1985;**34**(10):759-768

[16] Rudolf A, Geršak J. Influence of sewing speed on the changes of mechanical properites of differently twisted and lubricated threads during the process of sewing. Tekstil. 2007;**56**:271-277

[17] Ujević D. Utjecaj probodnih sila šivaćih igala u procesu šivanja pletene odjeće [dissertation]. Zagreb: University of Zagreb, Faculty of Textile Technology; 1998. p. 506

[18] Robak D. A model of fabric transport in a sewing machine using a feed dog covered with a suplle material of increased friction. Fibres & Textiles in Easten Europe. 2002;**7**(9):59-62

[19] Mukhopadhy A, Midha VK. The Quality and performance of sewn seams. In: Jones I, Stylios GK, editors. Joining Textiles, Principles and Applications. UK: Woodhead Publishing Series in Textiles; 2013. pp. 175-207

[20] Pačavar S, Ujević D, Domjanić J. Sewability of composite materials for car seats, 5th International Textile, Clothing & Design Conference—Magic World of Textiles, 5-8 October 2014, Dubrovnik, Croatia, 312-315; ISBN 978-953-7105-54-9

[21] Akter M, Khan MR. The effect of stitch types and sewing thread types on seam strength for cotton apparel. International Journal of Scientific & Engineering Research. 2015;**6**(7):198-205

Section 4

Textiles for Protection and Filtration

Chapter 8

Contemporary Personal Ballistic Protection (PBP)

Izabela Luiza Ciesielska-Wróbel

Additional information is available at the end of the chapter

http://dx.doi.org/10.5772/intechopen.69085

Abstract

The review concerns existing contemporary protective equipment and their components serving against ballistic and non-ballistic threats of different sorts. The main focus, however, is on the personal ballistic protection (PBP) based on textile components and their role in the protective elements. Soft ballistic protections are crucial Soft ballistic protections are crucial elements of PBP, for example in military and law enforcements. Although the subject of PBP was limited in this chapter to soft ballistic protection, other elements, e.g. hard ballistic protection, stab-resistant vests, dual threat, so-called in-conjunction protective elements, modern helmets, were also mentioned in this chapter to demonstrate positioning of the soft ballistic protection and other elements in the global personal protection approach. Apart from it, the chapter contains selected information concerning high-performance polymers and fibres as well as a brief notes about their application in protective panels being basic elements of any protective element. The final remarks concern the most up-to-date approach in relation to ballistic protection, which is immersing high-performance fibres into non-Newtonian liquid substances having the ability of ordering their chemical particles and changing into a high concentration and high segregation lattice under the influence of kinetic energy impact.

Keywords: personal ballistic protection, soft body armour, hard body armour, dual threat, stab-resistant vests, bullet-resistant vests, protective panels, high-performance fibres

1. Introduction

This chapter provides selected and the most up-to-date information concerning personal ballistic protection (PBP). The reason for writing this chapter is that the world of textiles is developing rapidly and this development includes technical textiles, protective clothing and equipment. Therefore, the novelty of these aspects is worth presenting. In addition, some selected and spectacular historical facts are presented here and they concern inventions, decorations, extreme cases and lack of ergonomic wear comfort (e.g. medieval armour). Thus, inventions like, for instance, armours dating back to pre-conquest Mexico—e.g. quilted cotton jackets soaked in brine as a support for overlapping plates of hardwood or coconut fibres armour against shark's teeth—will not be mentioned here, although these inventions are very interesting from the textile engineer's point of view.

Ballistics is the science of mechanics which concerns the movement and effects of different projectiles, especially bullets, on different objects. Therefore, protection against any projectile can be termed a ballistic protection. As there are different types of ballistic-resistant materials, this chapter will be totally devoted to PBP, which can be defined as a unique, close-to-body human protection. A common mistake or a mental shortcut concerning the term 'bulletproof materials' is the concept of bulletproof vests. In reality, PBP materials are not bulletproof, but they are bullet resistant. This means that ballistic-protective equipment utilized to protect against any ballistic danger is not completely immune to different dangers, missiles, projectiles, etc.; for example, retarding and/or mitigating the speed of dangerous objects with potentially penetrating abilities until the ballistic protection can stop it.

In addition to a concise history of PBP and its existence and evolution over the centuries, this chapter will present characteristics of the most contemporary PBPs, their composition and designs, including soft body armours, some hard body armours, helmets with concise information concerning protection from stabbing and sharp edges penetration. Additionally, different types of contemporary PBP will be presented, e.g. law enforcement, military personal protection, including sappers and K9 protection for dogs. The role of ballistics, types of guns and most of all PBP requirements, standardization and classification for contemporary PBP is also discussed in this chapter. The key market players are mentioned several times when needed. Some additional and concise facts are mentioned to reach the knowledge of the reader, e.g. future trends, especially for so-called liquid armour.

2. History of body protection

2.1. Early stage

The history of PBP is a history of conflicts between humans mixed with developments of weapons and materials utilized for protection. The first weapon was a human fist, a kick, a stone or a stick. Later, different bludgeon-like objects were developed for attack and self-protection. Due to its historical connotations, the club weapon is strongly associated with primitive cultures of the caveman era or Neanderthal era; however, this weapon is still in use but it is now called a baseball bat [1]. Thus, the first sorts of personal protections against blunt trauma, slashing, bites or stabbing can be dated back to beginning of humankind. These were made of different kinds of animal skins and furs [2–7]. The subject literature often mentions the skin of oxen as a source of the precious leather, which was utilized as armour by the Mongol army as early as the thirteenth century BC, a detailed description of which is given in an historic review [8]. Another important animal skin came from rhinoceros, which was utilized in China in the eleventh century BC [9], and from wild buffalo; these materials protected the torsos of the warriors. It seems that aspects of the ergonomics of personal protections were found to be important at that time. In order to increase the mobility of the warriors, some armies utilized textiles made of silk or linen. Linen was a very popular material, so there is nothing unusual in the fact that it was utilized as a component or a main raw material for linothorax. This element was a type of upper body armour used by the ancient Greeks. Different historical sources mention flax materials and associate weavers as producers of a base for an armour for the Greek army. In the early fourth century BC, the Athenian army underwent some military reforms, including the replacement of heavier bronze or chain mail body armour with linen corselets. As mentioned above, the key factor influencing changes was increasing the mobility of the soldiers by giving them armour that was 'light but protected the body equally well', in addition to reducing the size of the shield carried [3, 10, 11].

Some contemporarily reconstructed 10 mm composites were composed of 19 layers of low mass surface linen woven fabrics, while others achieved the same thickness using only 10 layers of much thicker linen. When it came to flexibility and resistance to stress, the performances of the 10-mm thick 19-layer and 10-layer pieces demonstrated very similar mechanical deformation parameters [3].

2.2. Fancy metal armour

Very often different armies around the globe utilized cloth garments with some metal (bronze) plates or scales attached to them. This form of protection was popular among the Assyrian (900–600 BC) and later Greek armies. There were further Gallic (eighth and ninth centuries), English (eighth century) and Frankish (ninth century) developments of this style.

The armour worn by a soldier presented in **Figure 1** is a perfect example of scales armour combining small, usually iron plates. Another historical source [12] suggested a simplified differentiation between metal elements attached to the system. According to this study, the

Figure 1. English soldier, eighth century; source: Grafton [6]. Arms and Armour: A Pictorial Archive from Nineteenth-Century Sources (Dover Pictorial Archive), Dover Publications, New York, USA, 1995.

external small plate armours are divided into three categories: lamellar, scale and an undefined category. When utilized to protect a torso, scale armour overlaps downwards and is predominantly mounted on a continuous substrate, usually of textile or leather. The second type is lamellar armour, where lamels overlap upwards and are connected continuously to the substrate, but are connected with each other by different cords.

Another very important innovation in armour was a mail armour (or chain mail). The exact invention moment and its place is unknown, but it was widely utilized in Europe in the Middle Ages quickly spread and was further developed and transformed by different nations, e.g. France, England and Germany. As mentioned before in the text of this chapter, the roles that were assigned to armour in the Middle Ages were protection and decoration but also presentation of social status and intimidation of the enemy. The examples of such types of armours are presented in **Figure 2(a, b** and **c)**. The chain mail is noticeable on the arm of the silhouette (b) in **Figure 2**.

Together with the development of firearms, suitable protective armour was developed to absorb the impact of firearms. As a consequence, body armours (including helmets) were made of different types of metals and their mass became very high. In fact, the period from the fifteenth to sixteenth century abounds in the greatest amount of very sophisticated, decorative armour for different level soldiers and noblemen in Europe.

A simplified version of protection is Polish Hussar top armour and a helmet, as presented in **Figure 3**. It was made of steel, iron, brass and leather. It is less decorative and the elements of tiles protecting the arms and shoulders suggest that requirements for increased mobility of the wearer were taken into consideration when designing this armour.

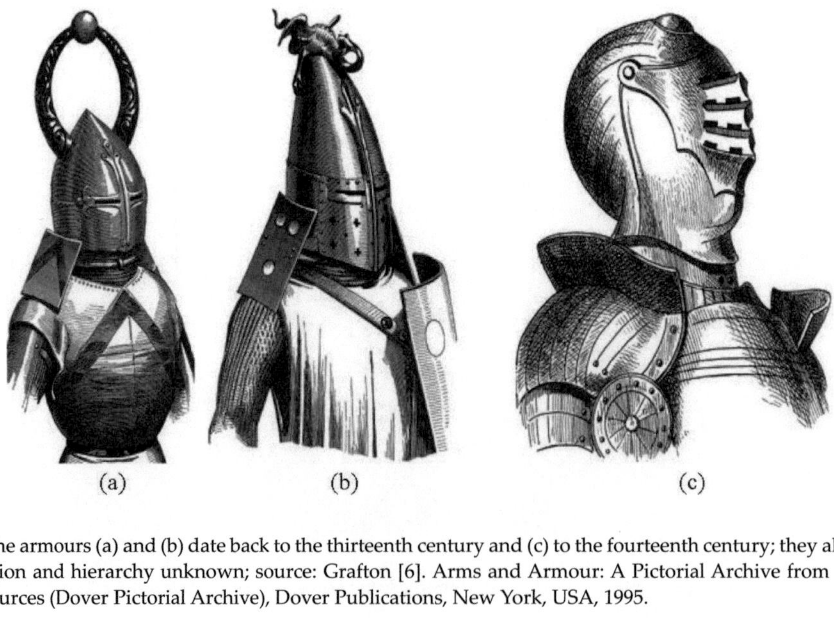

Figure 2. The armours (a) and (b) date back to the thirteenth century and (c) to the fourteenth century; they all come from France, region and hierarchy unknown; source: Grafton [6]. Arms and Armour: A Pictorial Archive from Nineteenth-Century Sources (Dover Pictorial Archive), Dover Publications, New York, USA, 1995.

Figure 3. Half armour for a Polish Hussar, dating back to 1683–1700. Made of steel, iron, brass and leather, presented at an exposition in The Art Institute of Chicago, Chicago, Illinois, USA; source: own photo, archives of the author.

Independent of the fact that further development of armour was going in the direction of increasing mobility and visibility, the armours of the nobles were still very decorative, thick, extremely expensive and also very heavy.

2.3. Japanese input

Japanese armour was also highly variated; the scale of its decorativeness depended on the position occupied by the warrior. The full armour set was worn only by the emperor, by a shogun (a military chef in the twelfth century; later the noble/the most skilled soldier), by a samurai who was a higher class swordsman or by the soldiers (but this armour was very heavy and uncomfortable). Up to the thirteenth century, there were two distinct types of armour: the *yoroi*, a heavy, box-like and very ornate and expensive style for mounted samurai and the *do-maru*, a simple suit of armour for the regular foot soldier, which was wrapped around the body of the soldier. Both were constructed from small metal plates connected together using leather elements (stripes). The sets of plates were fixed with silk cords to make a light but resilient armour plate. With time, mainly during the fourteenth century, samurai decided to adopt and adapt the *do-maru* as it was much lighter and more convenient for fighting on the battlefield. The more demanding battle conditions of the fifteenth and sixteenth centuries for samurai forced armour producers to reduce the mass and improve overall comfort of wear [7] (**Figure 4**).

It is very interesting to discover that Japanese shin guards called *suneate* were something even more than the inspiration for contemporary sport shin guards. In the Momoyama period of 1568–1600, armour craftsmanship reached a summit of excellence in providing the samurai with a suit of armour that would allow him to perform movements as freely as possible while providing him with the maximum protection against arrow, blade and ball. In this period of time, shin guards consisted of a number of vertical metal plates joined by chain mail on a cloth backing, with a leather patch on the inside of each to prevent rubbing against the stirrup

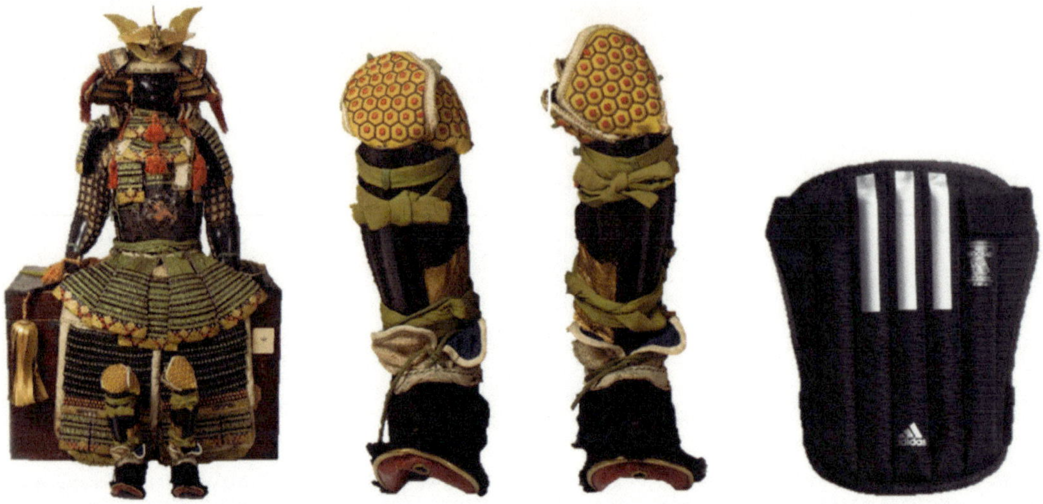

(a) (b) (c)

Figure 4. (a) Samurai armour, Edo period, eighteenth century: Kabuto bachi signed 'Masuda Myōchin Minbu Ki no Munesada Saku' and dated February 1757. This samurai armour has the kamon of the Inaba family. Kamon is a Japanese symbol used to decorate, identify and distinguish an individual or a family (source of (a) and (b): Giuseppe Piva Japanese Art, Milan, Italy; for more information see http://www.giuseppepiva.com/en/services) [13]; (b) shin protectors of a samurai armour from (a); (c) a contemporary football shin guard by Adidas; source: http://www.soccershoo.com/product/adidas-11-anatomic-lite-shinpads-x-large/[14].

when riding a horse [7]. A contemporary shin guard has a very similar construction, having a five-stick construction in the protection area to absorb blows. This construction allows the distribution of the impact force along the tibia and to the lateral muscle [15].

2.4. Armour improvements and the dusk of knighthood

There has been no special consistency in the use of armour over the centuries; however, it is believed that the use of different armours can be attributed to economic and technical reasons (e.g. external conditions, access to resources, lifestyle, fashions and a constant need not only for improvement in wear comfort but also for improvement in protective abilities of the armours due to the increasingly lethal potential of weapons). Different forms of cuirasses and plates, with reinforced or not reinforced coats, remained in use throughout the fourteenth and fifteenth centuries. In parallel and later, breastplates and backplates became popular, which covered the chest from the lower neck to just below the rib cage. They were made of steel, which was a relatively new development in the fourteenth century. The thickness of the steel shield varied between 1 and 2.5 mm, depending on the need for protection [16]. During the seventeenth century, the development of heavier, thicker plates to stop projectile-like missiles from guns offered an armour that was plain but functioned sufficiently well [16].

Overall, one may conclude that knights were wearing better and better protective equipment and any introduced improvements increased wear comfort and ergonomic requirements. However, armour was expensive and many young, prospective knights would need to either gain money from tournaments or capture armour in battle. Around the seventeenth century, the presence of knights became less important as the whole concept of army organization and troops had changed. Mastering horsemanship and swordplay remained important in the eighteenth and nineteenth centuries, until the mechanized warfare of World War I made it redundant [16].

2.5. World War I

At the very beginning of World War I, very few armies were equipped with any form of body protection. Due to a huge number of casualties, the French Army introduced a modernized protective helmet. Starting in 1915, body armour was utilized on the battlefield, but only in a limited capacity. The equipment consisted of steep plates giving protection from shrapnel [4]. This single-shot French weapon—the same as that mounted on the FT-17 Renault tank—could fire a small explosive shell which was able to pierce ¾ inch of armour plate at 2500 yards (2286 m) and it proved to be especially effective at suppressing snipers and machine guns [17]. The German Army introduced silicon–nickel breastplates in May 1917 for its soldiers. Later, the British Army offered the highest variety of protection to their soldiers and the most interesting ones were the 'Best Body Shield' and 'Portobank' armoured waistcoats. An average level of protection and a low mass burden was provided by Berkeley's Flexible Armour Guard, the Franco-British Cuirass and Wilkinson's Safety Jacket (mass of about 3 lbs). The British Munitions Inventions Board conducted some experiments with textile materials, e.g. kapok, flax, cotton, sisal, hemp and silk. The idea behind the work was to slow and trap projectiles in the textile materials. The studies focused on silk, which turned out to be the most resistant of the tested fibres. The armour made of silk gave similar protection as a shield made of manganese steel.

2.6. World War II—Flak jacket development

One of the greatest body protections was the Medical Research Council (MRC) Body Armour, which consisted of three separate 1-mm thick steel plates, weighing a total of 1.1 kg to protect the heart, main vessels and lungs (first version); lungs, liver, parts of the spine (second version) and abdomen (third version). This armour was made to be comfortable and it could withstand a 0.38 calibre pistol bullet at 4.5 m and 0.303 calibre bullet at 640 m. MRC was introduced by the British Army in 1941. It turned out that this armour was far from ideal. Not only was it difficult to perform rapid movements while wearing it, but it caused wounds on the body. Two other types of body armour invented by the British Army and consequently tested by the American Army—the so-called the Armourette and the Wisbrod Armoured Vest—were characterized as increasing the load on the soldier, decreasing overall mobility and efficiency; therefore, the advantages of wearing this armour were at a minimum level [4].

After an analysis of the wounds of the soldiers in the US Air Force, it turned out that about 70% were due to low velocity missiles, namely 'flak' fragments. The term flak refers to anti-aircraft guns operated during World War II by the German **Flugabwehrkanone**. The flak-protective vest, produced firstly in the United Kingdom and then in the United States, was known as the Flyer's Vest M1 and its modification was known as M2 (only the front of this version was armoured) [4]. Both versions weighed around 3.6 kg. They were standardized in October 1943 in the United States and they initially used cotton covering fabric, but changed to ballistic nylon by DuPont over the next few years. Many other modifications were made, standardized and propagated among soldiers, e.g. M6–M8 [4].

2.7. Further developments

Until 1952, numerous different body armours were designed and redesigned, including the concept of adding 12 layers of nylon fabric to increase the flexibility of the body protection (so-called M-1951 body armour). During the Vietnam War, M-1955 was issued and propagated among soldiers. M69, which contained a ballistic nylon filling covered by waterproof vinyl plastic, was issued at a later point in the conflict. Twelve layers of ballistic nylon fabric protected the front and 10 layers protected the back of the soldiers. At this point in time, zippers of Velcro were already in use to close vests; the average mass was 4 kg and the cost was 35 USD [4].

In 1962, studies concerning the hardening of armour were conducted and further development concerning an enamel layer that could make the armour layer harder and more resistant continued under the title Hard Face Composite armour in the Goodyear Aerospace Corporation, Ohio, US. The performance of armour containing a ceramic front and glass-reinforced plastic (GRP) at the back was proven against small round projectiles. At present, the term glass fibre reinforced plastic (GFRP) is commonly utilized in composite related terminology, commonly referred to as 'fibreglass'; these reinforcements are made of high tensile strength glass fibres [18]. After 1965, body armour classified and standardized in the United States utilized three types of ceramics as protection, namely aluminium oxide (Al_2O_3; in the US Army use), silicon carbide (SiC) and boron carbide (B_4C; Navy, Air Force and Marines). These ceramic plates were introduced into the textile covers of the vests as monolithic plates [4, 18].

2.8. Post-Kevlar® invention era

One of the most important inventions of the twentieth century, apart from computers or life-saving heart-lung machines, is a synthetic fibre from the group of aromatic polyamides [19]. Nowadays, it is known by the name Kevlar®. The polymer out of which the fibre was made was invented in 1964 by the chemist Mrs. Stephanie Kwolek who worked in a research lab at DuPont. DuPont was looking for chemical combinations that would make stronger fibres for fabrics. In the course of lab experimentation, Kwolek heated a newly mixed combination of substances [20]. The mixture presented unexpected features. The new era of personal body protection had begun.

The initial tests performed on the fabrics made of Kevlar® proved its ability to stop a wide range of projectiles. The next stage was to elaborate a new set of bullet-resistant vests for law enforcement. In the course of these experiments, five plain weave fabrics woven from Kevlar® 29 yarns of 1000 denier (about 111 tex) were tested and turned out to be the most bullet resistant. Soon, it was discovered that this amount of layers could stop some of the projectiles, but could not prevent non-lethal injuries, which are nowadays called blunt trauma. A great number of modifications to the protective set of fabrics, as well as the design of the protection itself, were introduced. One of the most meaningful was the personal armour system for ground troops (PASGT) weighing 4.5 kg (medium size). PASGT was made of Kevlar® and this abbreviation refers to both vests and helmets[1] made of Kevlar®. These were utilized by all military services from the mid-1980s to around the middle of the last decade [4, 19, 21].

3. Existing high-performance fibres for ballistic protection

1. **Arom**atic polyam**ides**, also known as **Aramids**, are chemical, man-made, synthetic polymers utilized for production of flame retardant and ballistic protection fibres. Aramids belong to the polyamide (PA) group, together with aliphatic polyamides (e.g. Polyamide 6 or 6.6 (PA 6 and 6.6)) and Polyphthalamides, also known as polyamides with semi-aromatic chains, e.g. Polyamide 6T (PA 6T). A very specific aramid known as Kevlar® was commercialized by the Du Pont Company in 1972. The discovered substance was characterized as having a super-rigid molecular chain and a fibre made of it had an ultra-high modulus [19]. Kevlar® aramid fibre is based on poly(*p*-phenylene terephthalamide; PPD-T), one of the *para*-oriented aromatic polyamides that was obtained by S. Kwolek. PPD-T can be prepared in the frame of a *classical synthesis* based on a low temperature polycondensation of p-phenylene diamine (PPD) and terephthaloyl chloride (TCI) in a dialkyl amide solvent and other methods, e.g. direct polycondensation reaction [19, 22].

   Apart from well-known types of Kevlar®—Kevlar®, Kevlar® 29, Kevlar® 49, Kevlar® 68, Kevlar® 100, Kevlar® 119, Kevlar® 129 and Kevlar® 149—relatively novel varieties include Kevlar® AP (15% higher tensile strength than K-29) [23] and KM2 Plus fibre [24] with enhanced ballistic resistance for armour applications [25]. **Table 1** presents a comparison of some high-performance fibres for ballistic purposes. Another type of aramid-based para-aramid fibre is Twaron® (a brand name of Teijin Aramid). It is a heat-resistant and strong synthetic fibre which was developed in the early 1970s by the Dutch company Akzo [26].

2. **Poly-p-phenylenebezobisoxazole (PBO)** is one of the polybenzoxazoles containing an aromatic heterocyclic ring. In 1994, Dow Chemicals, together with Toyobo, developed a new spinning technology that brought about the development of this substance. The process of polymerization of PBO takes place when polyphosphoric acid from 4,6-diamino-1,3-benzenediol dihydrochloride mixes with terephthalic acid. The fibre of PBO is made of highly oriented molecular chains, but its crystallite size is small compared to that of p-phenylene terephthalamide (PPTA), the component of para-aramid. PBO fibres consist of fully extended chains, like other high-modulus and high-strength fibres, but it also consists of highly oriented chains. Presently, there are two commercially available PBO-type fibres produced by Toyobo Co. Ltd. They are known in the market as ZYLON® AS (as spun) and ZYLON® HM (high modulus). Although nowadays the producer of these fibres does not clearly indicate their potential ballistic application on the website of the company, PBO-based fibre has about 1.6 times higher tensile strength than Kevlar, and these fibres are mentioned in the literature as ballistic fibres [27–34].

[1]Helmets are the topic of a separate sub-chapter; see **Helmets as elements of Personal Protection**.

3. **Ultra-high-molecular-weight polyethylene** (**UHMWPE, UHMW**) is a subset of the thermoplastic polyethylene. Also known as high-modulus polyethylene (HMPE) or high-performance polyethylene (HPPE), it has extremely long chains. UHMWPE is a type of polyolefin. It is made up of extremely long chains of polyethylene, which all align in the same direction. It derives its strength largely from the length of each individual molecule (chain). UHMWPE is synthesized from monomers of ethylene, which are bonded together to form the base polyethylene product. This type of fibre is produced from the polymer via the process of gel spinning. It is about 40% stronger than most aramid fibres. Due to these fibres' properties, they are predominantly utilized in ballistic protection. During the gel-spinning process, the fibre-forming polymer is dissolved in a solvent and spun through a spinneret. The polymer itself is in a gel-like state, which means it is only partially liquid. After spinning, the filaments have a highly oriented structure and the liquid crystals are aligned along the axis of the fibre. Currently, there are two major types of these fibres that are commercially available. They are produced by Honeywell and DSM under the names Spectra® [35] and Dyneema®, respectively [36]. Different series of these fibres have been established as their properties were modified during the production process, e.g. Spectra® 900, 1000 and 2000; Dyneema® SK25, SK60, SK65, SK71, SK75 and SK75. Both Spectra® and Dyneema® are utilized in ballistic protection products as they are characterized as having high-energy absorption and can dissipate the hit wave more easily compared with other existing ballistic fibres. These fibres are utilized for both soft and hard ballistic-protective products [2, 37].

4. **M5 Fibre (polyhydroquinone-diimidazopyridine) -** The synthesis of **poly[2, 6-diimidazo[4, 5-b4′,5′-e]pyridinylene-1,4-(2,5-dihydroxy)phenylene]polyhydroquinone-diimidazopyridine**) and the utilization of a conventional air gap wet-spinning of this solution in methanesulfonic acid led to the creation of a fibre known as M5 fibre which is a high-performance fibre originally developed by Akzo Nobel and currently produced by Magellan Systems International (Magellan) [28]. The crystal structure of this fibre features typical covalent bonding in the main chain direction, but it also features a hydrogen-bonded network in the lateral dimensions. The problems related to processing the fibre—and especially with obtaining an optimal crystal orientation and, as a consequence, optimal ultimate mechanical properties (e.g. average fibre strength was 4 GPa)—have led to modifications in this fibre's production process. These are expected to correspondingly increase ballistic impact performance [38].

5. **Endumax®** belongs to the group of fibres based on polyethylene. According to the specification given by Tejin, the producer of this fibre, Endumax®, can be produced in wide or thin tapes. Thus, one has a need of only 25% of the quantity of matrix material conventionally used to produce a UD or composite (compared with conventional thin multifilament yarns). Therefore, one makes some savings not only in terms of weight but also in terms of the amount of chemicals needed. Due to the material's high stiffness and shape at low matrix content, the pressure needed to produce a shaped anti-ballistic plate is approximately 60 bar—significantly lower than for other **UHMWPE** materials, which may require up to 200 bar pressure. When applied into shaped plates, this polymer retains its original form and protection performance levels, even if they have been exposed to temperatures and/or moisture levels above the normal working range (e.g. during storage) [39].

Commercial name of the fibre	Producer	Chemical substance	Density [g/cm³]	Tensile modulus [GPa]	Tensile strength [GPa]	Elongation at break [%]	Product forms and their applications	References
Kevlar® 29	DuPont[2]	poly-para-phenylene terephthalamide (PPTA)	1.44	71	2.9	3.6	Ballistic applications, ropes, cables, protective apparel such as cut-resistant gloves, helmets, vehicular armouring.	[2, 20, 28, 32, 40–43]
Kevlar® 49			1.45	78	3.4	3.3	High-modulus type used primarily in fibre optic cable, textile processing, plastic reinforcement, ropes, cables and composites for marine sporting goods and aerospace applications.	
Kevlar® KM2/KM2 Plus			1.44	82	3.9	3.3	Kevlar® KM2® and KM2® Plus technology help provide protection from select fragmentation and small arms threats. KM2® Plus represents the highest grade protective fibre for military use. Kevlar® KM2 is meant for helmets and vests for military and high-performing spall liners. Kevlar® KM2 Plus is a high tenacity, high toughness and finer denier fibre used in vests and helmets for both military and law enforcement officers	[2, 25, 28, 32, 40, 43]
Twaron®	Teijin	PPTA and poly-para-phenylenediamine (PPD)	1.44	70	3.2	3.3	High ballistic protection, lower weight, greater comfort and longer lifetimes, protects against penetration of bullets, and fragments as well as stabbing	[2, 20, 43–45]
Spectra 900	Honeywell	Ultra-high-molecular-weight polyethylene (UHMWPE)	0.97	2.4-3.5	2.4-3.5	4.0	Stronger than steel and 40% stronger than aramid fibre. Capable of withstanding high-load strain-rate velocities. Spectra® fibre, one of the world's strongest manmade fibres, is commonly used to produce bullet-resistant Spectra Shield® body and vehicle armour and helmets	[2, 20, 35, 37]
Dyneema® SK75/ SK78 but Dyneema® different grades: HB212, HB210, HB80, HB56, etc. is meant for Hard Body Protection	DSM		0.97	109-132	3.3-3.9	3.0-4.0	Dyneema® Force Multiplier Technology can reduce the weight of armour by up to 20%. The result is exceptionally protective performance without compromising comfort, agility, or function. Comfortable protection against handguns, shrapnel and knives. Dyneema® is used in ballistic helmets, vests and shields. It is utilized in protective equipment to safeguard soldiers, law enforcement officers, commercial pilots and high-profile civilians. Grades of hard ballistic protections (inserts, helmets and shields) are HB212, HB210, HB80, HB56, HB50, HB26 and HB2, respectively, from the highest to the lowest performance and aerial density [g/m²] from 0.136 ± 0.005 up to 0.261 ± 0.005	[2, 20, 36]

Commercial name of the fibre	Producer	Chemical substance	Density [g/cm³]	Tensile modulus [GPa]	Tensile strength [GPa]	Elongation at break [%]	Product forms and their applications	References
M5	Magellan Systems International LLC; with DuPont starting from 2005	Polyhydroquinone-diimidazopyridine	1.70	271	3.9	1.4	Fibres are lighter and give more effective protection from different threats, including bullets, fragments, IEDs and mines, than other existing fibres. Potential future applications of the fibre include fragmentation vests and helmets, composites for use in-conjunction with ceramic materials for small arms protection and structural composites for vehicles and aircraft	[2, 20, 28, 29]
Zylon ® AS	Toyobo	Poly-phenylene benzobisoxazole (PBO)	1.54	5.8	180	3.5	Although Zylon is a high-performance fibre and was utilized in the past for ballistic purposes, the website of the producer of this fibre does not provide information about potential ballistic applications for this fibre	[2, 20, 27]
Zylon ® HM			1.56	5.8	270	2.5		
Endumax®	Teijin	Ultra-high-molecular weight-polyethylene (UHMWPE)	0.97	170	2.8	1.7	Endumax lightweight plates enable ballistic protection gear to meet requirements in terms of protection, flexibility and low weight. Protective unidirectionals (UDs) and composites (like insert plates), protective panels and helmets made from Endumax, have a very high degree of braking energy, resulting in a particularly high-stopping power for bullets and fragments. Endumax can be applied in both soft and hard body armour	[39, 46]

[1]Only selected fibers are presented in **Table 1**. If these are classified as soft body armour, it means that they are utilized predominantly in this application, but it does not exclude them from being applied elsewhere.

[2]DuPont, Honeywell and other companies offer a large number of different high-performance and high-modulus types of fibres and fabrics, e.g. Kevlar 68, Kevlar 100, Kevlar 119, Kevlar 129, Kevlar® XP™ [41], Kevlar Protera® Fabric Arc Flash Performance [42], S-900 family, S-1000 family. This technology, made from a high-toughness resin combined with Kevlar® KM2® Plus fibre, helps manufacturers provide a 20% decrease in helmet weight while maintaining the same performance. Kevlar® XP™ for hard armour provides a minimal back face deformation. Although the materials have exceptional mechanical properties, only very few of these materials are designated by their producers for ballistic protections.

Table 1. Comparison of typical high-performance fibres predesignated for ballistic purposes in soft body armour (fabrics, panels, composites)[1].

3.1. Ballistic structures

The high-performance yarns are collected together into a form of woven fabrics, knit structures and non-woven materials. These arrangements/structures allow the dissipation of bullet impact energy reasonably quickly. The essence of bullet-resistant materials is that when grouped together into panels (e.g. a specific/determined number of identical materials placed one onto another) and put into vests, they constitute bullet-resistant vests.

3.2. Action mechanism of the bullet-resistant vest

The ballistic performance of woven structures in the ballistic packet of the bullet-resistant armour depends strictly on the mechanical properties of yarns and fibres (type of high-strength fibre, type of weave, linear density of the yarn and weave density). In the case of woven fabrics utilized as a protective packet, the fabrics are woven densely in the form of plain weave. It was observed that the density of the yarns packed to form the weave for ballistic fabrics is the best from 0.6 up to 0.95 [2, 28, 43]. Below 0.6, fabrics are simply too loose to meet ballistic requirements and above 0.95 the yarns are tightly packed and can be damaged during the fabric production process. When hitting the surface of protection made of woven fabrics, projectiles cause a deformation of these fabrics that starts to spread the yarns apart. This takes place especially when the calibre of the bullet is small. In such a case, it can penetrate the fabric without or with very limited yarn damage. In case of larger calibres, the yarns tend to lock up the bullet in one of the inner layers, while the initial layers and yarns within them are damaged due to the bullets passing through them. The bullets are slowed down by the initial layers and the impact energy of the bullets reduces and is dissipated. In order to dissipate impact energy quickly and to offer the maximal protection, Unidirectional (UD) shields are used for both soft and hard panels, e.g., Dyneema®UD by DSM and Spectra®Shields by Honeywell. In Dyneema®UD, all the yarns (groups of filaments) are positioned parallel to each other, in the same plane, rather than being woven together. In UD configuration, the fibres of Dyneema®UD allow energy transmission from the place where the bullet strikes by energy distribution along the fibres much faster than in conventional woven fabrics. This is due to the fact that the absorption power of the yarn in woven fabrics is lost at the cross points of warps and wefts. It has been proven that instead of supporting the impact energy dissipation process, the cross points (or crossover points) rather hamper this process [2, 35, 36, 43]. Spectra Shield™ is not a woven material, but a thin, flexible ballistic composite made from layers of unidirectional fibres held in place by flexible resins. These Spectra fibres of a single layer are arranged in a way which does not allow them to cross each other. The fibres of the second layer are placed and held in a different direction compared with the fibres of the first layer, e.g. the fibres of the first layer are kept under 0° and the fibres of the second layer are kept under 90°, but all in the same plane. Then, both layers are sealed between two thin sheets of polyethylene film. A similar solution is applied in Twaron Unidirectional Laminate UD41—or those combining UD41 with other Twaron materials—and offers several advantages for engineering these modern ballistic-protective vests. They provide enhanced protection against bullets and fragments, as well as more comfort and excellent performance–weight ratios. Twaron UD41 is a unidirectional laminate suitable for soft body armour. Consisting of four plies of unidirectional Twaron fibre lines (plied in a 0°/90°/0°/90° configuration, as presented in **Figure 5**), it makes full

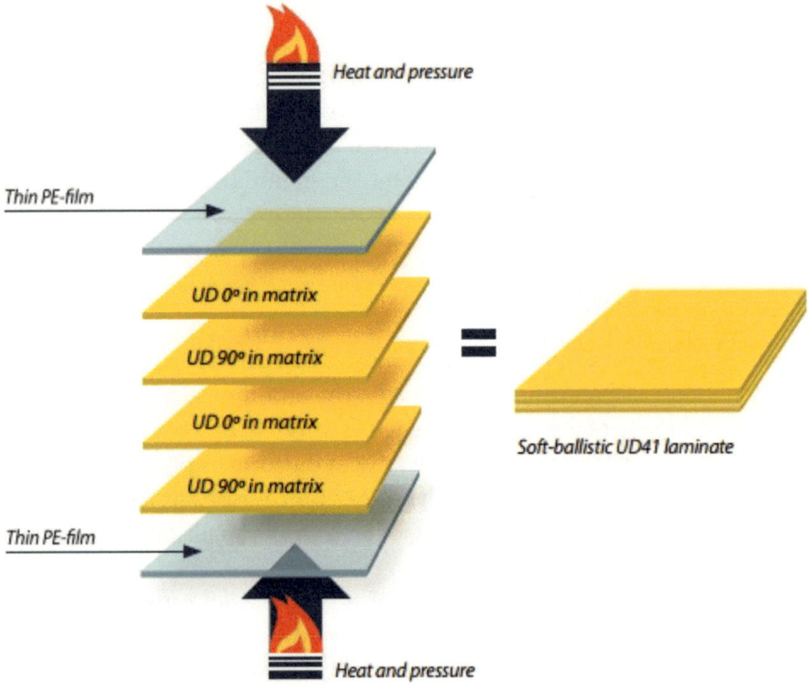

Figure 5. Schematic presentation of a single ballistic laminate UD41 by Tejin; PE stands for polyethylene; UD stands for unidirectional; source: Ballistics material handbook—Twaron by Teijin [http://www.teijinaramid.com/wp-content/uploads/2016/05/Teijin-Aramid-Ballistics-Material-Handbook.pdf].

use of Twaron's high-fibre tenacity and avoids the crimping of typical woven material. So-called smart UD technology aligns the parallel Twaron fibres in each layer, and each layer is constructed in a resin matrix. The top and bottom UD plies are then laminated to ensure maximum abrasion resistance [47].

Some examples of fabric and composite materials utilized for ballistic protection in soft armours are presented in **Figures 6–14**.

Figure 6. Plain weave fabric made of Kevlar 49® yarns by DuPont; source: author's own photo archives. Note: Plain weave single fabric; 85 warps/10 cm and 85 wefts/10 cm. Both warps and wefts are made of Kevlar. Total area density: 300 g/m^2. Thickness: 0.35 mm.

Figure 7. Ballistic panel Gold Shield® GV-2018 with Kevlar® by Honeywell; source: author's own photo archives. Note: The sample is composed of two panels and each of the panels consists of two layers. The first layer consists of sets of the parallel yarns positioned vertically (V) and the second layer contains the yarns positioned horizontally (H). Thus, the scheme of the sample construction is: (V + H) + (V + H). Total area density: 510 g/m². Thickness: 0.78 mm.

Figure 8. Ballistic panels Gold Shield® GV-2018 with Kevlar® by Honeywell. The sample is delaminated/stratified, showing two separate layers of a single panel with yarns positioned perpendicularly; source: author's own photo archives.

Figure 9. Ballistic panel Gold Shield® GN-2119 with Kevlar® by Honeywell; source: author's own photo archives. Note: Sample constitutes a single panel, made of two layers. The first layer consists of sets of the parallel yarns positioned vertically (V) and the second layer contains the yarns positioned horizontally (H). Thus, the scheme of the sample construction is: V + H. Total area density: 107 g/m². Thickness: 0.1 mm.

Figure 10. Ballistic panel Gold Shield® GN-2119 with Kevlar® by Honeywell. The sample is delaminated/stratified, showing two separate panels with yarns positions perpendicularly; source: author's own photo archives.

Figure 11. Ballistic panel Spectra Shield ® SR-1226 by Honeywell; source: author's own photo archives. Note: Sample is composed of two panels and each of the panels consists of two layers. The first layer consists of sets of the parallel yarns positioned vertically (V) and the second layer contains the parallel yarns positioned horizontally (H). Thus, the scheme of the sample construction is: (V + H) + (V + H). Total area density: 253 g/m^2. Thickness: 0.7 mm.

Figure 12. Ballistic panel Spectra Shield ® SR-1226 by Honeywell. The sample is delaminated/stratified, showing two panels with two layers each, and the yarns in each panel are positioned perpendicularly; source: author's own photo archives.

Figure 13. Ballistic panel Spectra Shield ® II SA-4144 panel by Honeywell; source: author's own photo archives. Note: From a distance, Spectra Shield® II SA—4144 is difficult to distinguish from Spectra Shield® SR-1226. The organoleptic assessment of the samples allows an easy verification. Spectra Shield® SR-1226 is thicker, stiffer and has a more waxy tactile sensation.

Figure 14. Spectra Shield ® II SA-4144 panel by Honeywell. The sample is delaminated/stratified, showing two separate panels with two layers of yarns in each of them, positioned perpendicularly; source: author's own photo archives.

4. Contemporary personal ballistic protection (PBP)

Ballistic protection, especially soft ballistic protection armour, has undergone a significant material and design revolution in recent years. Armour panels, meaning the essential elements providing protection in ballistic vests, consist of a ballistic panel, which is usually a set of woven or non-woven structures. They are protected by a cover from environmental influences. Of course, neither the cover for the panel nor the carrier—another element of the ballistic vest—is intended to provide ballistic protection. The carrier refers to the textile elements, usually made of nylon, which are visible from the outside when the person is wearing the ballistic vest. The principal purpose of the carrier is to support and secure the panels to the wearer's body. This subchapter predominantly discusses so-called soft body armour and hard body armour—both of these types of protection are mentioned in the classification presented below. However, the functions, designs and the materials utilized in these two groups are different.

4.1. Soft body armour

Soft body armour consists of flexible panels of ballistic materials. This type of armour is designed to protect from assaults with pistols and revolvers—generally, handguns. Due to its lower weight when compared with hard body armour, it is rather intended to be utilized for extended daily wear, for several hours, e.g. armour worn by law enforcement officers, correctional officers and guards. If it is worn under a uniform, it is called concealable armour. The soft armour panels are typically constructed of multiple layers of ballistic-resistant materials, e.g. Kevlar fabrics, Spectra or Dyneema UD non-woven materials as presented in **Figures 6–14**. The number of layers in the panel influences the panel's overall performance, which means the ability to resist the energy of projectiles. Normally, each layer is supposed to absorb and dispatch a certain amount of energy, which is less and less when transferred to the next layer closer to the body. When a projectile strikes the panel, the yarns and the fibres catch the bullet due to the mutual interlacing of the yarns (in a woven structure) or superimposing yarns from different layers of the panel (unidirectional panels). These fibres have the ability to absorb and dissipate (disperse) the energy of impact, which is passed on from the bullet to the panel and to be specific to each of the panel's layers gradually. This process causes the bullet to deform or 'mushroom'. This ability of the panel to absorb and disperse the energy of the bullet is the key to its ability to reduce blunt force injury to the body resulting from bullets that do not perforate the panels. As the fibres in a panel jam a bullet, the energy of the bullet pushes the panel into the body of the wearer, potentially resulting in injury to the torso. The cone made in layers of a panel is schematically presented in **Figure 15**. This type of non-penetrating injury can cause severe contusions (bruises) and can cause damage to the internal structures of the body (musculature, bones, ligaments, organs and vascular system) that may even result in death [43, 46].

4.2. Hard body armour

Hard armour, on the other hand, consists of rigid panels or plates. Hard armour is designed to offer greater protection against higher threats than soft armour could bear on its own. Hard

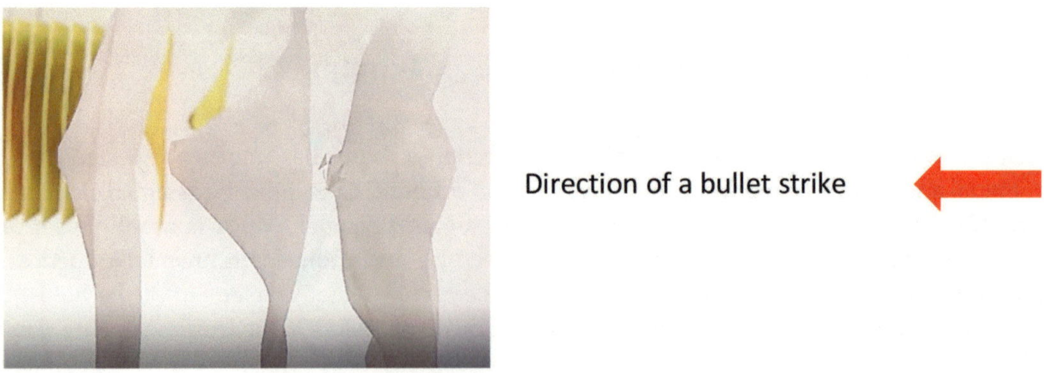

Figure 15. Characteristic cone shapes in the layers of a soft protective panel elements after the bullet struck and perforated them, heading in the direction of the next layer. The cones become smaller due to the smaller amount of bullet kinetic energy transferred to each consecutive layer; prepared on the basis of a Tensylon™ commercial released by DuPont.

armour plates are used in tactical armour. Tactical armour is typically a combination of a hard armour plate and soft armour panels, making it thicker and heavier than soft armour alone. The side of the panel that faces away from the body is referred to as the strike face, because it is the side that is intended to be hit by the bullet. The other side of the panel that is worn against the body is referred to as the wear face or body side.

Hard armour plates may be constructed from ceramics, compressed laminate sheets, metallic plates or composites that incorporate more than one material, e.g. CeraShield™ and Cercom® by Coorstek or Tensylon [43, 48–50] as presented in **Figure 16**. Tensylon™ is currently being used in-theatre as a spall liner in mine-resistant ambush protected (MRAP) vehicles, e.g. Tensylon™ HSBD 30A is a bi-directional laminate. The hard armour plates act in one of the following ways: they can capture and deform the bullet or they can break up the bullet. In both instances, the armour then absorbs and distributes the force of the impact. Although some hard armour plates are designed to be used by themselves in a carrier, in the majority of cases, they are designed to be used in-conjunction (IC) with a soft armour panel. Many hard armour plates are designed to be used with a specific soft armour panel to achieve a desired level of ballistic protection. They are introduced into the ballistic set by adding pockets to the front and rear of a soft armour's carrier. The hard armour plates are inserted into these pockets over a portion of the underlying soft armour panel. The hard armour plate component of the IC armour is clearly marked to identify the corresponding model of soft armour panel with which it is to be utilized. The most commonly used ceramics that can be used as stand-alone monolithic plates for armour purposes are aluminium oxide (Al_2O_3), silicon carbide (SiC) and boron carbide (B_4C). Al_2O_3 is usually the most economical alternative, but the final protection solutions using Al_2O_3 are heavier, since Al_2O_3 has the highest density and the lowest ballistic efficiency of the three ceramic types [49, 51]. B_4C is the hardest ceramic; but at high-impact pressures, an amorphization process weakens the ceramic. This is problematic when the threat is an armour-piercing projectile at high velocity. Ceramics with a small grain size usually perform better than ceramics with larger grain sizes.

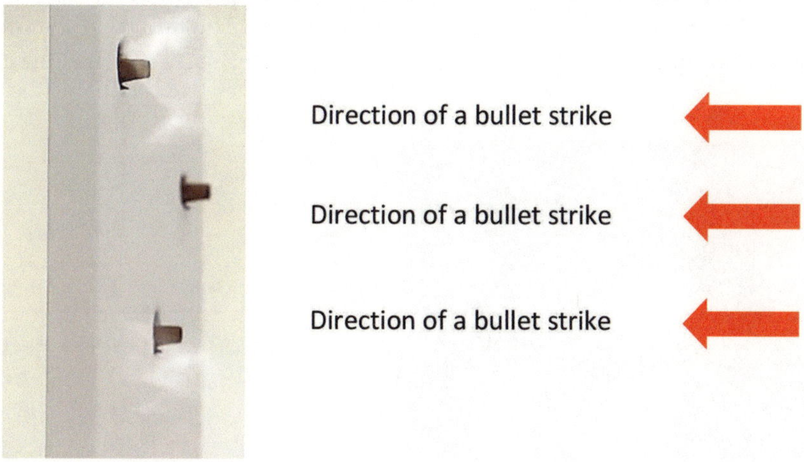

Figure 16. The characteristic indentions in the hard body armour and mushroomed projectiles after they struck the panel and perforated it; prepared on the basis of a Tensylon™ commercial released by DuPont.

4.3. Combination armours

Combination armours are specially designed to provide protection against both firearms and edged or stabbing weapons. It means that in these armours, the protective panels are composed of layers of materials that are stab resistant as well as layers of materials that are ballistic resistant. These types of armours are also called *dual threat* or multiple threat armours. The National Institute of Justice (NIJ) in the United States provides a list of those combination armours that have been tested and found to be compliant with both NIJ Standard-0101.06 and NIJ Standard-0115.00 for both ballistic and stab resistance [52]. There is an example of a multi-threat protection by DSM shown in **Figure 17**.

4.4. Anti-blunt trauma plates

These plates are made of layers of ballistic-resistant fabrics, metals, laminate sheets or other materials. They are referred to as trauma packs or plates because they are intended to minimize blunt force trauma injury to the torso resulting from a bullet striking an armour. Sometimes these plates are inserted into the vest carriers to provide some additional ballistic protection, although this is not their prime purpose. They can be easily distinguished from traditional hard armour plates due to their size, typically anti-blunt trauma plates smaller and thinner. These anti-blunt trauma plates are usually placed in the centre of the chest in pockets in front of (or, less commonly, behind) the front soft armour panel. As is the case of armour panels, the orientation of trauma plates and packs matters. They are marked using the same convention as is used for armour panels. Some armours incorporate multiple trauma packs.

4.5. In-conjunction armour (IC)

Usually, hard body armour plates are meant to be used with a specific type of soft body armour panel to accomplish a specific and required level of ballistic protection. In such case, these hard body armours are not designed to be used alone. These hard body armour plates are inserted into the pockets in the front and at the back of the soft body armour carrier.

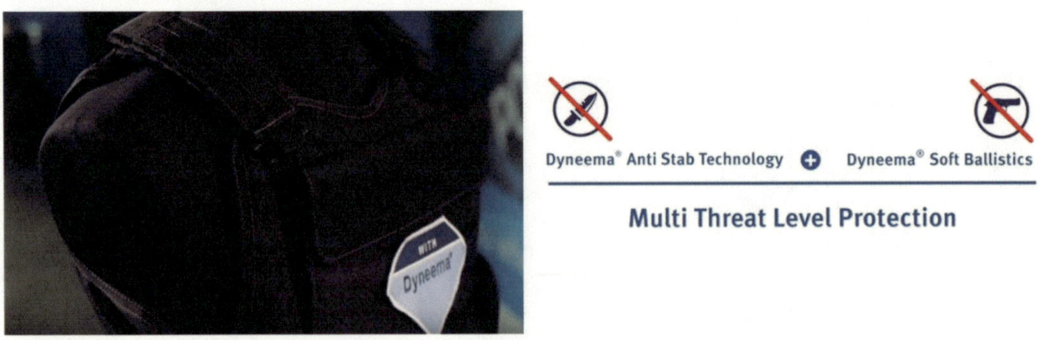

Figure 17. Dyneema® Multi-threat level protection with pictograms describing its properties by DSM; source: http://www.dsm.com/products/dyneema/en_GB/applications/personal-armor/ballistic-anti-stab-vests.html [53].

5. Contemporary classification of PBP

The classification of the materials for impact protection is quite complex due to many existing materials and existing protection elements. In **Figure 18**, a simplified graphical version of this classification is presented. It takes into consideration three major approaches: textile personal protection versus non-textile personal protection, ballistic personal protection versus non-ballistic personal protection and hard body armour versus soft body armour.

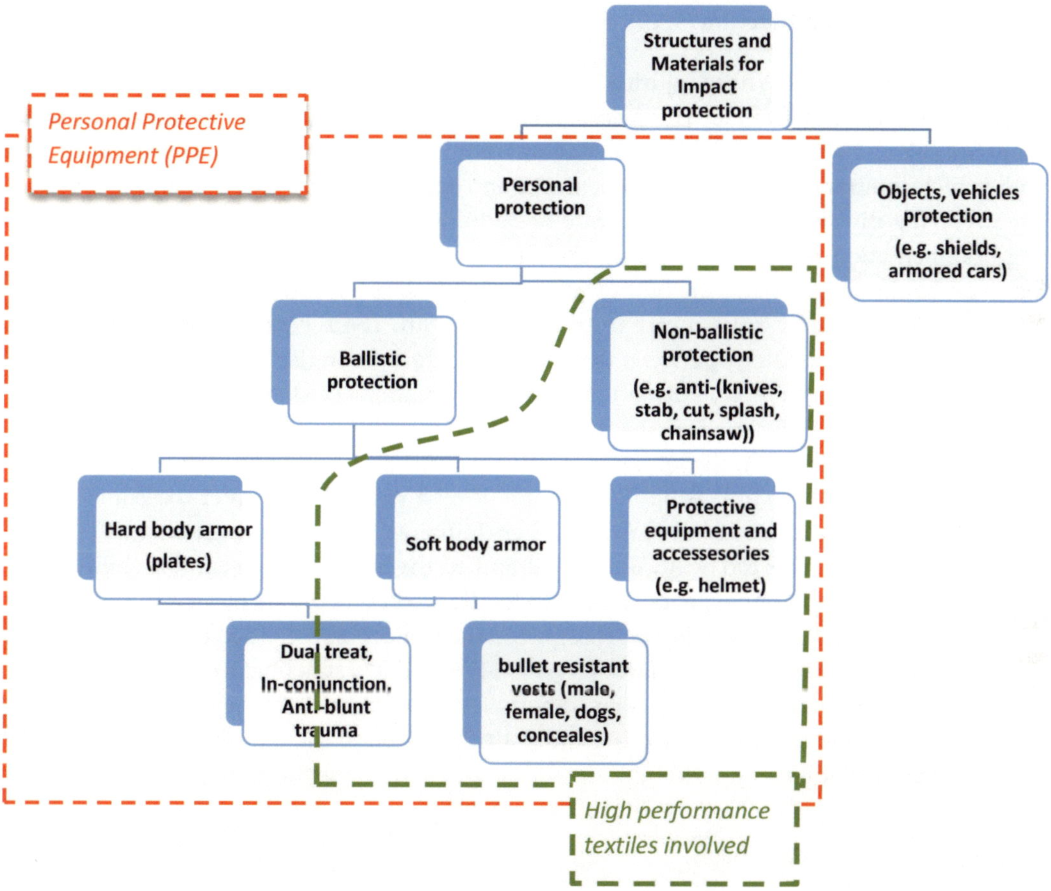

Figure 18. Contemporary classification of PBP elements.

6. Helmets as elements of personal protection

Similar to protective vests or other elements of body armour, helmets underwent significant changes and developments in terms of their design and applied materials. Helmets began as head protection made from woven fabric with some elements of leather and ceramics. Later on, helmets were made from metal and had three main functions—protection, deterrence and decoration. Today, the contemporary advanced combat helmet (ACH) has only one main function—protection. One of the key advances, which has influenced ballistic helmets, was the development of aramid fibres in the 1960s, which led to Kevlar®-based helmets. The Department of Defense (DoD) in the United States and other relevant national and international institutions have continued to invest in research to improve helmet performance, through better design and materials, as well as better manufacturing processes. **Tables 2–6** and **Figures 19–23** present an overview of contemporary ballistic helmets, which are available on the market.

6.1. History of head protection

A variety of threats lead to head injuries in the battlefield. Since World War II, the predominant threats have been from fragmentation and ballistic threats from explosions, artillery and small arms fire; blunt trauma caused by translation from blast, falls, vehicle crashes and impact with vehicle interiors and from parachute drops and exposure to primary blasts. Injuries, usually involving division of tissue or rupture of body tissue coming from an explosive source (e.g. fragmentation from bombs, mines and artillery), dominate all injuries, including bullets. Non-battle causes, including blunt traumatic injuries, produced nearly 50% of the hospitalizations for traumatic brain injury in Iraq/Afghanistan. There is no biomechanical link in the current test methodology between the back face deformation (BFD) assessment and head injuries from behind-helmet deformation. Different variations of steel helmets were used by forces in the United Kingdom and the British Commonwealth during World War I and later. Since 1945, the escalation in the lethality of ballistic threats, resulting in higher fatalities and injuries, is observed. The bullets and shrapnel in World War II had greater mass and higher velocities. As was the case in World War I, soldiers initially resisted wearing helmets. They felt that the 3.5 lb helmet was too heavy and that it limited the hearing, vision and mobility of the wearer. However, the troops quickly accepted the trade-off when they observed the lethality of the munitions on the battlefield and recognized the protection provided by the helmet. The personnel armor system for ground troops (PASGT) was the first helmet to utilize Kevlar. The name and the abbreviation PASGT refer to both vests and helmets made of Kevlar and they were used by all military services from the mid-1980s to around the middle of the last decade. These helmets are still being used by some services but will be replaced in the future [21].

Timeline	Year 1980
Helmet types	Personnel armour system for ground troops (PASGT)
Material composition	Aramid fibre (Kevlar®)
Protection level	Originally, only Level II and contemporarily produced PASGT helmets met Level IIIA ballistic penetration protection according to the contemporary standard of NIJ 0106.01 as presented in product information folders of producers of these helmets
Technical parameters	Presently provide a full-coverage-style helmet, four-point adjustable retention system
Provider(s)/Price	3M company; https://www.3m.com/~625 USD

Table 2. An overview of personnel armour system for ground troops (PASGT) helmet.

Timeline	Year 2001—MICH & Year 2003—ACH
Helmet types	Modular integrated communications helmet (MICH) Advanced combat helmet (ACH)
Material composition	MICH is a composite helmet made of Kevlar intended for use by special forces. ACH was elaborated on the basis of MICH. ACH uses ballistic fibre such as Kevlar or Twaron
Protection level	MICH: Level IIIa High Cut MICH Helmet
Technical parameters	Mass of MICH without accessories is about 1.36 kg
Provider(s)/Price:	MICH/cost starts at 449.99 USD; source: http://infidelbodyarmor.com/helmets-c-10/

Table 3. An overview of modular integrated communications helmet (MICH) and advanced combat helmet (ACH).

The US Special Operations Command designed and developed the modular integrated communications helmet (MICH) as a replacement for PASGT (**Figure 19**). MICH incorporated several changes, including improved Kevlar aramid-fibre reinforcement, leading to better protection. These helmets also allowed better fit and integration of communication headsets. MICH was adopted by the US Army in 2001–2002 as its basic helmet and renamed the advanced combat helmet. The Marine Corps decided to use a design profile that was similar to the PASGT and called it the light weight helmet (LWH). There were also developments in helmet retention systems. The MICH, ACH [54] and LWH helmets switched to a multi-pad and four-point retention system (**Figure 20**) that had better impact protection while providing increased comfort. The next major advance in helmet technology resulted from a combination of advances in materials and manufacturing processes. As soon as UHMWPE was developed and later on adopted by key players producing ballistic shields, the producers of ballistic helmets adopted it as well.

The future assault shell technology (FAST) helmet is significant for its early use of UHMWPE material and its novel design (**Figure 21**). The next solution, enhanced combat helmet (ECH), delivers much better protection against fragments compared with ACH, due to a shift to unidirectional UHMWPE fibre in a thermoplastic matrix. The shift was also enabled by a new generation of preforms and manufacturing methods appropriate for UHMWPE.

Timeline	2010
Helmets types	FAST
Material composition	Ballistic shell made of a hybrid composite of carbon, unidirectional polyethylene and woven aramid
Protection level	Level IIIa
Technical parameters	Thickness of the shell: 6.43 mm
Provider(s)/Price	Ops-Core: 1,662 USD; source: http://www.ops-core.com/ fast-mt-super-high-cut-helmet/

Table 4. An overview of future assault shell technology (FAST) helmet.

Timeline	2012/2013
Helmet types	ECH
Material composition	Dyneema HB80 UD composite
Protection level	Level IIIa
Technical parameters	Head protection when parachuting for military paratroopers, head protection from bumping objects for military ground forces and head protection against handgun rounds and ballistic fragments. The ECH's profile is very similar to the ACH but is thicker. The helmet's shell is made of an ultra-high-molecular-weight polyethylene material. It protects 35% better against small-arms fire and fragmentation than the ACH.
Provider(s)/Price	3M company in collaboration with Ceradyne; 3M™ Defense Protection Systems/starts from 500 USD

Table 5. An overview of enhanced combat helmet (ECH) helmet.

The US government launched the helmet electronics and display system-upgradeable protection (HEADS-UP) programme in 2009. It leverages multiple efforts—in the areas of ballistic materials (transparent and non-transparent), high-resolution miniature displays and sensors—to design a modular-integrated headgear system that takes into account the relevant ergonomic considerations. HEADS-UP has focused on developing a technical data package of design options and trade-offs to build a modular, integrated headgear system.

The major threats that have caused head injuries in recent conflicts can be classified into three groups: ballistic, blunt and blast. Fragmenting weapons, including artillery, mines, mortars and other sources of explosions, are the principal source of wounding on the modern battlefield. These weapons, including improvised explosive devices (IEDs), have a multitude of fills/wounding mechanisms.

There is limited information on the effect of primary blast on the head. Traumatic brain injury (TBI) associated with blast exposure in operation enduring freedom (OEF)/operation Iraqi freedom (OIF) is estimated at up to 20% of deployed service personnel. The current helmet is not designed with considerations for primary blast, but there is substantial experimental evidence that the ACH helmet is protective against primary blast for most direct exposures.

Timeline	2013/2014
Helmet types	HEADS-UP
Material composition	UHMWPE
Protection level	Level IIIa
Technical parameters	Improved ballistic materials, non-ballistic impact linear materials and designs, transparent as well as heads-up display technologies, and better eye, face and hearing protection. The helmet displays integrated electronics in the helmet, a HEADS-UP display powered by an android phone, and a pneumatic linear system that meets the 14 feet/second impact requirement to reduce traumatic brain injuries.
Provider(s)/Price	Gentex Corporation/price unknown

Table 6. An overview of helmet electronics and display system-upgradeable protection (HEADS-UP) helmet.

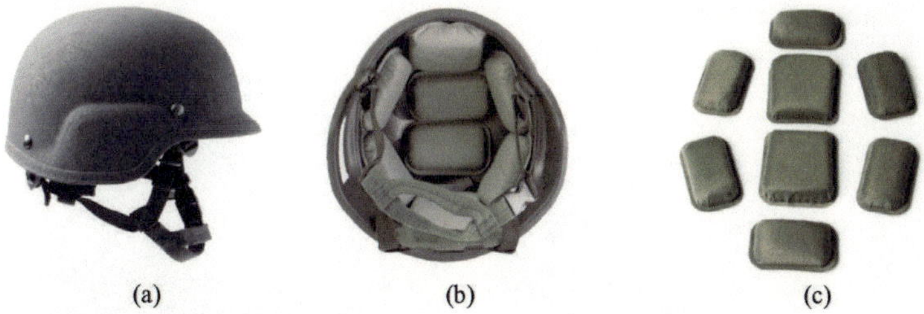

Figure 19. PASGT ballistic helmet; (a) side view on skull; (b) interior of the helmet; (c) pads of the helmet; source: CopQuest, Inc.; images available at https://www.copquest.com/protech-pasgt-tactical-helmet_89-1015.htm.

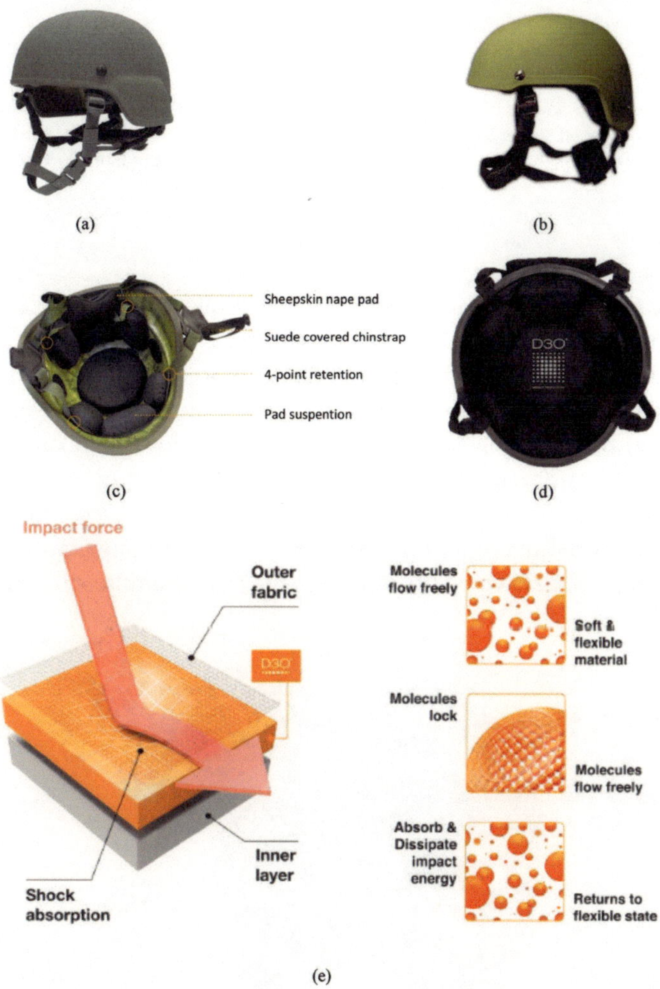

Figure 20. MICH and ACH ballistic helmets. (a) Side view of a MICH helmet; source: http://infidelbodyarmor.com/helmets-c-10/. (b) Side view of an ACH helmet; source: http://www.gentexcorp.com. (c) Interior of an ACH helmet; source: http://www.gentexcorp.com. (d) The 3M™ Ultra-Light Weight (ULW) Ballistic and Bump Helmet is manufactured by Ceradyne, Inc., a 3M™ Company source: https://www.d3o.com/defence/3m/. (e) Schematic presentation of impact protecting materials utilized in the pads of the helmets—e.g., (d). D3O technologies are based on non-Newtonian principles—in its raw form, the material's molecules flow freely, allowing it to be soft and flexible, but on impact, lock together to dissipate impact energy and reduce transmitted force; source: https://www.d3o.com/.

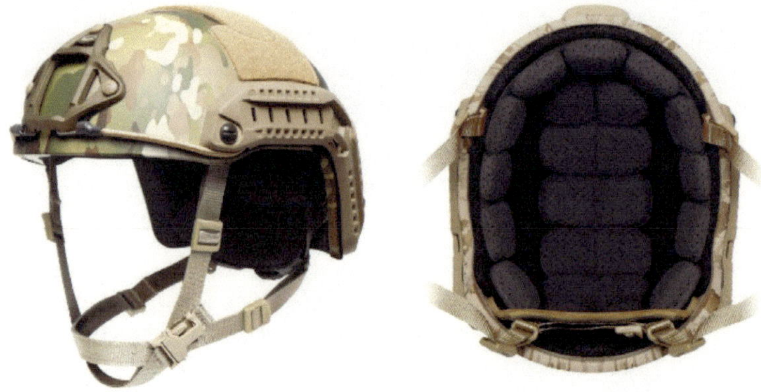

(a) (b)

Figure 21. (a) A side view of a FAST Maritime (MT) Super High Cut Helmet. Ear cut geometry is 16 mm higher than the FAST Ballistic High Cut (XP) shell shape, allowing for clearance of larger headset style communications devices. (b) An interior of the helmet; source: http://www.ops-core.com/system-platforms/ballistic-helmets/fast.

6.2. Primary blast

For severe TBI from blast exposure, there may be clear neurological changes, including reduced levels of mentation, unconsciousness and other dysfunctions. For milder exposures, possible consequences include neurological deficits, depression, anxiety, memory difficulty and impaired concentration. Epidemiological data, experimental results and computational models suggest that the ACH helmet does not exacerbate blast exposure. Modern ballistic wounding is generally differentiated between rifle and handgun rounds by velocity. For example, high-velocity tumbling rounds such as typical 5.56 mm projectiles (800 m/s or above muzzle velocity) have qualitatively different wounding behaviour than .22 calibre handgun ammunition (~330 m/s muzzle velocity), although they have similar diameters [21].

Figure 22. A side view of enhanced combat helmet (ECH); source: https://en.wikipedia.org/wiki/Enhanced_Combat_Helmet.

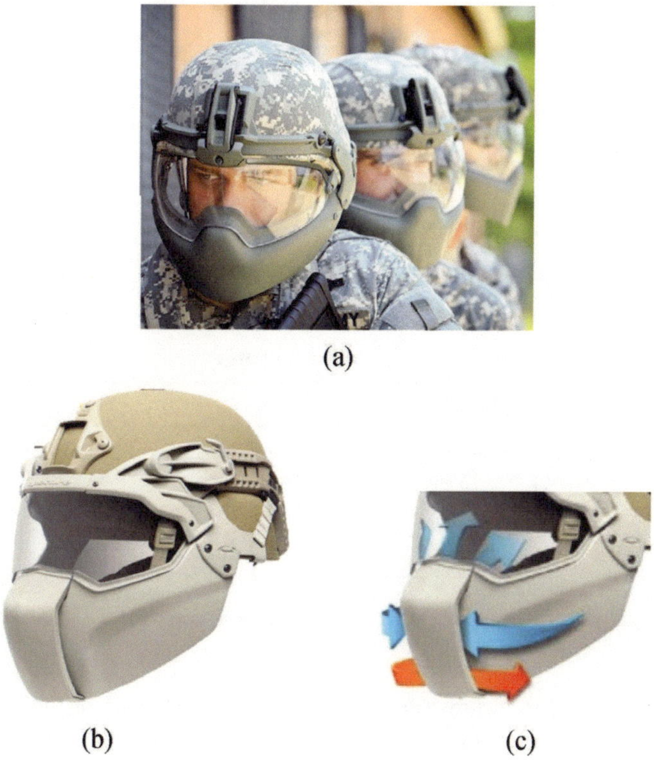

Figure 23. (a) HEADS-UP worn by soldiers; (b) a side view of HEADS-UP; (c) schematic presentation of streams of air inhalation and exhalation in the HEADS-UP; source: http://www.gentexcorp.com/assets/base/Brochures/GroundCatalog.pdf.

7. Testing methodology for PBP

7.1. Panels

The most recognizable is the work performed by NIJ in the United States, which elaborated and has improved the methodologies for testing different sorts of protections against ballistic and non-ballistic threats. This standard entitled Ballistic Resistance of Body Armour, NIJ Standard-0101.06 also classifies different personal body armours into six types (IIA, II, IIIA, III, IV + a special type) by level of ballistic performance, which are presented in **Table 7**. A special test class is defined to allow armour to be validated against threats that may not be covered by the first five standard classes presented in **Table 7**. A typical range configuration (a place where the ballistic tests take place) is presented in **Figure 24**. If the test is to be performed using a handgun rounds, the armour panel should be fixed at the distance 5.0 ± 1.0 m from the muzzle of the test barrel, and for rifle rounds, the armour panel should be fixed 15 ± 1.0 m from the muzzle of the test barrel. In order to minimize the possibility of excessive yaw at impact, or for other range configuration reasons, the distance may be adjusted for each threat. However, the distance should not be less than 4 m for any tested rounds [55].

NIJ Standard-0101.06*

Protection level	Ammunition type	Projectile mass	Projectile velocity
IIA	9 mm FMJ RN 0.40 S&W FMJ	8.0 g ≅ 124 gr 11.7 g ≅ 180 gr	373 m/s ± 9.1 m/s 352 m/s ± 9.1 m/s
II	9 mm FMJ 357 Magnum JSP	8.0 g ≅ 124 gr 10.2 g ≅ 158 gr	398 m/s ± 9.1 m/s 436 m/s ± 9.1 m/s
IIIA	0.357 SIG FMJ FN 0.44 Magnum SJHP	8.1 g ≅ 125 gr 15.6 g ≅ 240 gr	448 m/s ± 9.1 m/s 436 m/s ± 9.1 m/s
III	7.62 mm FMJ	9.6 g ≅ 147 gr	847 m/s ± 9.1 m/s
IV	0.30 AP	10.8 g ≅ 166 gr	878 m/s ± 9.1 m/s
A special type defined by the Standard-0101.06 separately			

*The data concern only new and unworn armour, although the standards foresee also the tests on conditioned armour. In such case, the allowed parameters, namely the velocities of the projectiles are lower.

FMJ RN–Full Metal Jacketed Round Nose; JSP–Jacketed Soft Point; SIG–Schweizerische Industrie Gesellschaft–Swiss Industrial Company, a producer of the type of the gun; FN–Flat Nose; SJHP–Semi Jacketed Hollow Point; AP–Armor Piercing.

Table 7. An overview of the personal body armour levels of protection according to the NIJ Standard-0101.06 [55].

Calibre of the gun is the approximate internal diameter of the barrel, or the diameter of the projectile it fires, e.g. 40 calibre of a firearm refers to the barrel diameter of 0.40 of an inch. The metric system is also popular to describe the diameter of a bullet or a barrel. In such case, the diameter is given in millimetres, e.g. 9 mm.

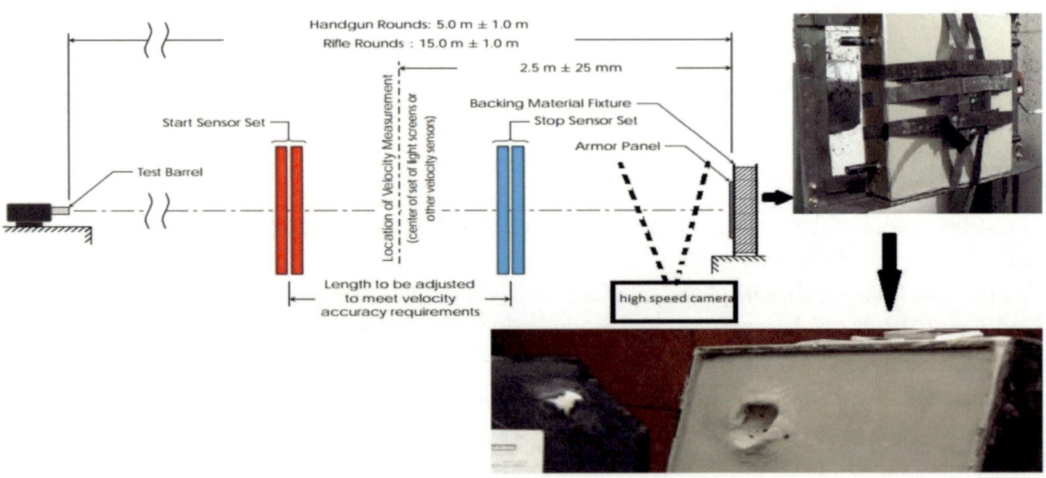

Figure 24. A graphical presentation of a set up for testing the armour panels at the laboratory testing range and images presenting backface deformation (backface signature depth) of a backing material and its response to ballistic impact on the plasticine (or clay) informing about the potential trance in a human body (blunt trauma) that that the fired projectile may cause. Although in an original graphical presentation of a set up presented in the Standard-0101.06 of NIJ, there is no high-speed camera, in many cases, this type of the camera is placed to observe and record the collision of projectiles with a protective panels; source: adapted from graphic [53]; images: top—Oregon Ballistic Laboratory test; bottom—modified on the base of IIIA Soft Body Armor Penetration in Ballistics Clay—AR500 Armor®.

Grain [gr] is a unit of mass. 1 g is approximately 15.43236 grains.

Other present or past standards being partially equivalents of NIJ Standard-0101.06 are:

- MIL-STD-662F, Military Standard: V50 Ballistic Test for Armor [56]

The purpose of this standard is to provide some general guidelines for procedures, equipment, physical conditions and terminology for determining the ballistic resistance of metallic, non-metallic and composite armour against small calibre arms projectiles. The ballistic test procedure described in this standard determines the V50 ballistic limit of armour. This test method standard is intended for usage in ballistic acceptance testing of armour and for the research and development of newly created armour materials.

- *UK Standard—UK/SC/5449*; Ballistic Test Method for Personal Armours and Lightweight Materials [57].

- *NATO Standard*—STANAG 2920 [58].

Ballistic test method for personal armour materials and combat clothing—STANAG 2920 is used to measure materials ability to stop fragments and shrapnel. The measuring technique was originally developed for body armour but now see general use in all situations where fragments are the primary concern. For instance, STANAG 2920 is used to measure Add-on-Armour systems for armoured vehicles.

Tests according to STANAG 2920 are conducted by shooting Fragment Simulating Projectiles (FSPs) onto the test specimen with different velocities while measuring the velocity of each FSP. By altering the velocities, after a number of shots, an estimate of the ballistic limit can be obtained, which is the speed up to which the material defeats the fragment.

7.2. Helmets

Based on the US testing protocols, one may say that the test protocol involves several shots with a specific calibre gun at the specific speed and the specific location on the helmet to prove the resistance (total or not) of the helmet. Overall, tests are aiming at assessment of three parameters, namely resistance to penetrate (RTP), the backface deformation (BFD) and probability V50, that the helmet is equally likely to stop or not stop an object striking at a specific velocity. The original army first article testing (FAT) procedure protocol consisted of 20.9 mm shots (four helmets and shots at five specified locations on a helmet). A manufacturer's helmet design was deemed to pass FAT for penetration if there were zero penetrations out of the 20 shots. In 2012, with Director of Operational Test and Evaluation's (DOT&E's) approved a new two-stage protocol. It involves performing a 0-out-of-22 test in the first part, and if the helmet design passes the first part of the test, then a second 17-out-of-218 plan is executed, for a total of 240 shots and a combined acceptable number of penetrations of 17 [21].

The helmets are tested in the laboratories having a very similar set up for testing as in case of testing the protective panels, which is presented in **Figure 24**. However, instead of the panel, one utilizes a model of the head with a helmet on it.

For combat helmets, the current testing methods and measures have no connection to research on head and brain injury. The lack of connection between injury and current test methods and measures is a significant concern. During such test, the helmet, needs to be tested, is fastened to a headform packed with modelling clay, and a rifle-like device is used to fire various projectiles into the helmet. The clay is used as a recording medium for: (1) assessing penetration should the projectile or portions thereof pass through the helmet into the clay as presented in **Figure 25** and (2) measuring the deformation of the helmet, where a trace is left in the clay surface as a result of the ballistic impact pushing the helmet into the clay.

One of the critical issues with the clay (Roma Plastilina #1) is that the clay is time and temperature sensitive in that, its properties can change significantly over a 45-minute period as it cools. These effects are likely to affect BFD measurements [21].

7.3. Resistance to penetration (RTP)

RTP is measured by shooting a given ballistic projectile at a set of helmets and counting the number of complete penetrations. Most ballistic impacts penetrate the helmet to some degree, so the DOT&E FAT distinguish between complete and partial penetrations. A complete penetration in RTP testing is defined as: *Complete perforation* of the shell by the projectile or fragment of the projectile as evidenced by the presence of that projectile, projectile fragment or spall in the clay, or by a hole which passes through the shell. A partial penetration is defined as 'any fair impact that is not a complete penetration'. The intuitive notion is that a projectile that penetrates the shell is able to cause more serious head injuries than a projectile that does not, but there is no other linkage between what is measured and head injury [21].

7.4. Backface deformation (BFD)

After mounting the headform in the test fixture and mounting the helmet on the headform, the helmet is removed from the headform, and the clay surface is scanned with, for instance, a Faro® Quantum Laser Scan Arm laser or in other way. The helmet is then reattached to

Figure 25. A ballistic test set up for testing a MICH ballistic helmet against 9 mm calibre; source: www.kick-az.com.

the headform, and the shot taken. The helmet is again removed from the headform and inspected for penetration and perforation. The clay is rescanned with the FARO laser to calculate BFD or the clay indention is measured other way. It is unclear how well BFD from ballistic impact characterizes the effect of blunt force trauma, which is one of the main types of brain injury that the helmet is intended to protect against. The choice of the helmet BFD threshold values—25.4 mm for front and back shots and 16 mm for side and crown shots does not have a scientific basis [21].

7.5. V50 test

It refers to estimating the bullet speed at which there is a 50% chance of penetration. This test uses a witness plate mounted inside the headform rather than packing the headform with clay as is done with RTP/BFD testing. Because of this difference, the DOT&E FAT protocol defines a V50 complete penetration as a shot where impacting projectile or any fragment perforates the witness plate resulting in a crack or hole, which permits light passage. A break in the witness plate by the helmet deformation is not scored as a complete penetration. The definition of what constitutes a penetration, and how such penetrations are measured, differs between RTP and V50 tests. V50 specifies a 'hole which permits light passage' whereas RTP does not [21].

8. Stab resistance of personal body armour (PBA)

This type of protection is applied to reduce the risk of stabbing by knives or spikes, etc. especially in places like correctional facilities. There is a standard entitled stab resistance of personal body armour by NIJ, 0115.00 [52], which establishes minimum performance requirements and methods of test for the stab resistance of personal body armour intended to protect the torso against slash and stab threats. The threat posed by a knife depends, among other things, on its sharpness, pointedness, style, handle and blade design, attacking angle, the physical condition of the attacker and the skill of the attacker. Because these parameters can vary widely from one situation to the other, armours that will defeat a standard test blade may not defeat other knife designs under similar conditions or the same knife design if other attacking parameters are changed [52]. The threats analysed in this standard came from hand-delivered impacts with sharp-edged and/or pointed instruments which points or tips lie near the centreline of the clenched fist holding the weapon. PBAs covered by this standard are classified into one of two distinct protection classes depending on the type of threat. Within each threat protection class, the armour is further classified into one of three protection levels. The levels of protection indicate the stab energy the vest is expected to bear. The first protection class is intended to deal with threats that might be expected on 'the street' from high quality, commercially machined edged knife blades. This class is referred to as the 'Edged Blade' class. The second protection class is intended to deal with threats that might be expected in correctional facilities and these are weapon types constructed by inmates. These are lower quality knife blades and spike style improvised from other materials. This class is referred to as the

'Spike' class. The three levels of protection presented in this standard were derived from the frequency distribution of the energy that can be delivered by a male population using several stabbing techniques. The lowest energy level corresponds to the 85th percentile (corresponds to the E1, level 1), the next energy level corresponds to the 90th percentile (E1, level 2) and the highest energy level corresponds to the 96th percentile (E1, level 3). For any given protection level, the test protocol requires the knife blade or spike to impact the armour test sample at two distinct energy levels. At the given condition, a maximum blade or spike penetration of 7 mm is allowed. The penetration limit was determined through research indicating that internal injuries to organs would be extremely unlikely at 7 mm of penetration. The test protocol then requires an overtest condition where the knife blade or spike kinetic energy is increased by 50% (e.g. 24 + 50% × 24 = 36, E2, level 1). At this higher energy condition, called 'E2', a maximum blade or spike penetration of 20 mm is allowed. This overtest is required to ensure that there is an adequate margin of safety in the armour design [52] (**Table 8**).

During stab-resistant drop tests, the front and back panels of two complete body armours are tested for resistance to stab penetration using a different testing set up than the one for ballistic tests. The exemplary test set up is presented in **Figure 26**. The stabbing element is dropped on the fixed protective panel to observe and measure whether there is a perforation.

Apart from an angle indent into the panel, the tests take into consideration the stab energy level and the weapon type itself.

Protection level	'E1' Strike Energy [J]	'E2' Overtest Strike Energy [J]
1	24 ± 0.50	36 ± 0.60
2	33 ± 0.60	50 ± 0.70
3	43 ± 0.60	65 ± 0.80

Table 8. Stab-resistant protection level strike energies [52].

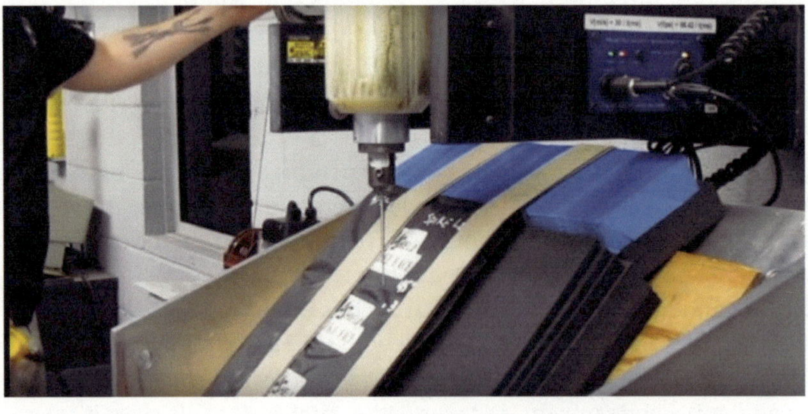

Figure 26. Stab-Resistant Panel NIJ 0115.00 Level II Spikes Test; source: http://www.aashield.com/.

9. Other protective elements

9.1. Female armour

The bust area of a female body armour is shaped by a unique process, which eliminates cutting or cut depending on the producer of the front ballistic panel. It enhances wear comfort and mobility without sacrificing ballistic performance. The example of a female ballistic personal protection is in **Figure 27**.

9.2. Improvised explosive devices (IED) and sappers

The risks that many active soldiers present at the current world scenes may face are very often related to IED, which are homemade bombs unique in their construction, size and potential lethal effect. Therefore, IED remains in the contrast with commercially available weapons, which is controlled. Since IED is unique, it is impossible to be fully prepared and protected against its effects. Persons located in the explosion radius get affected by the penetration of the elements being hidden in the IED when it is constructed, e.g. spikes, nails, etc. or in other way because they are in the blast region of IED. Blast is defined as a detonation of liquid or solid explosive material results in the generation of gaseous products in the pressure range of 150,000 atmospheres or 1.5 billion Pascals (1.5 GPa) and temperature of 3000 Kelvin. In many cases, if there is any suspicion about the bomb and there is a time to make an attempt to disarm it, a person called sapper may be sent to the scene. Clearing the scene is only one of the duties of a sapper, among demolitions, bridge-building, field defences as well as building, and reparations. Sappers are called pioneers, combat engineers or field engineers. The personal protection for

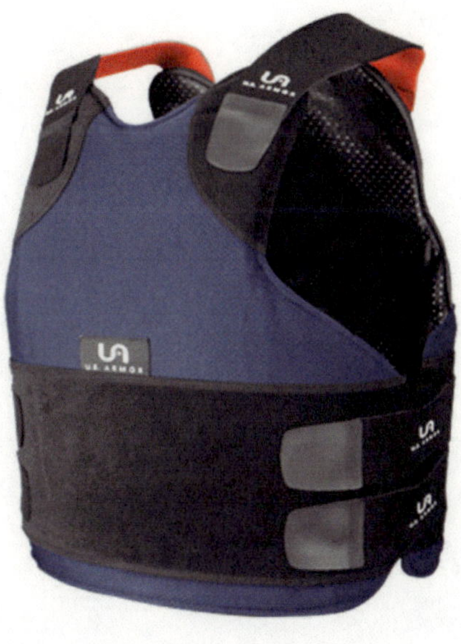

Figure 27. (a) Female ballistic-resistant vest called Enforcer XLT; source: http://www.usarmor.com/products/concealable/enforcer-xlt.

sappers called explosive ordnance disposal (EOD) suit or a blast suit is far more complexed that in case of law enforcement, correctional officers or regular soldiers as it needs to cover the whole body of the sapper. In order to enhance the protection, some of the parts of the bomb suit overlap. An example of the EOD personal protective equipment can be Med-Eng EOD 9 or Med-Eng EOD 10, mass about 33.4 kg. These are new generation bomb suits and helmets composed of jacket, an integrated groin protector, trousers and boot covers for integrated blast protection against IEDs and bombs. In order to improve the protection, some studies on specific elements of the bomb suits are performed, e.g. development of pelvic textile protection for soldiers [59].

9.3. Concealable body armour

The idea, which stands behind producing this type of protection, is to hide the fact of wearing it on the body. Therefore, the carrier of the ballistic panel is having a form of a suit vest. It is meant for politicians and business people. Usually, the available level of protection in case of the concealable body armour is IIIa according to NIJ.

9.4. Protection for police dogs

A police dog, often called K9 (**Figure 28**), is a specially trained type of dog to assist law enforcements and also military service in searching for drugs and explosives, searching for lost people, looking for crime scene evidence and protecting police officers who handle them. These dogs are equipped in protective vests, which are usually dual threat protections, it means that they protect dogs from firearms and stabbing. These vests are cut from the same

Figure 28. Detroit police K9 dog wearing bullet-resistant vest, 9 August 2016; source: Robin Buckson/Detroit News available at http://ktar.com/story/1220231/detroit-police-dogs-to-sport-bullet-resistant-vests/.

Kevlar® bullet-resistant cloth as their human partners, covering all the vital organs. The vests are designed so that the dogs can wear without reducing their mobility and efficiency. The dogs' protective vests, like protective ballistic vests for humans, undergo field tests.

9.5. Three-dimensional (3D) structures for ballistics

One of the interesting and significant innovations in soft body armour is a completely different structure of the protective fabric from these, which were presented in this chapter so far, namely 3D structures for ballistics. The idea of this invention is to produce a single 3D fabric capable of replacing many 2D fabrics or non-wovens currently and traditionally utilized. It requires producing a very thick 3D fabric, which would be able to compensate and even surpass the capabilities of existing 2D panels for soft body armour to dissipate the kinetic energy of the threats [2]. The advantage of these structures over the 2D structures is the fact that they can be produced using both conventional and 3D weaving machines. These are yarns in *z*-direction that hold together warps and wefts in these structures. Another advantage of this solution is the fact that yarns do not crimp and due to the structure they create, they allow a high longitudinal wave propagation after the contact with a projectile. One of the recognized producers of 3D structures for ballistics, 3Tex, developed production process and structures to create 3D fabrics for body and vehicle armour (3WEAVE® based on S2-Glass) [2, 60].

As it was mentioned in the text of this chapter, plain weave utilized in case of woven structures made of Kevlar yarn minimizes warps and wefts slippage; however, it also provides the highest crimp of the yarns, which as a consequence limits the propagation of the wave energy along the yarns. In 3D orthogonal structures, one observes a high yarn coverage and a low yarns crimp. There are some comments in the literature [2] saying that this type of element for soft body armour may replace currently popular multi-layer system of body protection; however, the author believes that orthogonal 3D structure for ballistics will be rather utilized in the future as an alternative but will not replace the traditional 2D structures due to cost of weaving machines for 3D structures, the fact of easiness to produce UD structures and a ballistic panel for soft body armour made of Kevlar woven fabrics, and finally, the fact that 2D structures are working perfectly well at this point.

10. Future trends

There are two major directions in which the development of ballistic protections can head. The first direction mentioned here is non-Newtonian fluids as an element supporting the existing ballistic panels in stopping the projectiles and as shock absorber element. The second direction is work on auxetic materials, which are materials having a negative Poisson's ratio.

10.1. Non-Newtonian fluids for liquid body armour

For a better understanding of this topic, it is suggested to start with an explanation what the Newtonian liquids are. The term Newtonian liquid was given after Sir Isaac Newton (1642–1726) who characterized the flow behaviour of fluids with a simple linear relation between shear stress [mPa] and shear rate [1/s]. This relationship is now known to the world as Newton's Law of Viscosity where:

$$\tau = \eta\gamma, \tag{1}$$

where τ is a shear stress in Pascal [Pa], η is the viscosity of the fluid, in (N × s)/m^2 = [Pa × s], which is Pascal-second, and γ is a shear rate. γ is the rate at which a progressive shearing is applied to the substance, measured in reciprocal seconds [s^{-1}].

Liquid has a definite volume, but not a definite form [61]. A common property of liquid is that they can only transmit a pressure to solid or liquid surfaces bounding the liquid. Tangential forces on such surfaces will first occur when there is a relative motion between the liquid and the solid or liquid surface [60–62]. Such forces are frictional forces on the surface of bodies moving through air or water. When we study the flow of water in the bath or in a river, we can see that the flow velocity is greatest in the middle of the water or river and is reduced to zero at the edges of the bathtub or a riverbank. The phenomenon is explained by the notion of tangential forces, between the water layers that try to slow down the flow. The volume of flowing liquid is nearly constant. As a consequence, liquids are considered to be incompressible. A fluid is a material that deforms continuously when it is subjected to anisotropic states of stress. Usually, highly viscous fluids do not obey this linear law and therefore they belong to the *non-Newtonian fluids*. For non-Newtonian fluids in simple shear flow, a viscosity function is the following:

$$\eta(\gamma') = \frac{\tau}{\gamma'}, \tag{2}$$

where: η (γ') is called the apparent viscosity [Pa × s]; τ is shear stress [Pa].

In order to calculate the shear stress τ of the non-Newtonian fluids, one may apply the following formula:

$$\tau = \eta(\gamma')\gamma' \tag{3}$$

Shear thickening fluids (STFs) are characterized by an increase in viscosity when the shear rate increases achieves a critical value. The incorporation of STFs to Kevlar® fabrics is being investigated to improve ballistic protection capabilities as well as to enhance stab resistance. Shear thickening is defined in the British Standard Rheological Nomenclature as the increase of viscosity with increase in shear rate [60–63].

There are two main types of the non-Newtonian media: (1) fluids with the maximal (zero-shear-rate) Newtonian viscosity and (2) yielding viscoplastic materials. Numerous intermediate and superimposing situations can also exist [61–64].

10.2. Non-Newtonian substances

A popular non-Newtonian substance is a combination of polyethylene glycol (liquid phase) in combination with a silica powder (solid phase). The mixture contains billions of silica nanoparticles, more than 100 times finer than a human hair evenly distributed in the glycol.

In the liquid state, the particles have a weak molecular surface charge so they do not clamp together. However, when the object affects the liquid, it changes radically the state as its kinetic energy forces the particles to stick together in the lattice, which is a strong chemical bond called hydro-cluster. The liquid becomes as hard as ceramic for a very short period of time (parts of seconds). As soon as the kinetic energy is spent, the bond (lattice) is released and the solution becomes liquid again. The features of the mixture of polyethylene glycol and a silica powder are utilized to enhance abilities of ballistic protections, e.g. fabrics made of Kevlar are soaked in this mixture.

10.3. Auxetic materials

These are solid materials that have negative Poisson's ratio [65–67]. It means that when they are submitted to the stretch in the longitudinal direction, they expend latterly (get bulky or thicker) and when they are compressed, they are getting narrower in the direction perpendicular to the direction of the compressive force. This characteristic is due to the complex microstructure of these solid materials [64, 65]. Based on these structures present in nature, e.g. cristobalite or same selected human tissue, some man-made forms of auxetic structures are created, e.g. metallic, ceramic and foams. One of the typical and known examples of an auxetic foam [65, 67, 68], having high crack resistance [68]. Due to the extraordinary microstructure of these materials, the idea of imitating it and shaping textiles along the lines of this unique microstructure appeared. The liquid crystalline polymer (LCP) was developed with laterally attached rods in a main chain of this polymer [69]. The orientation of the laterally attached rods is parallel to the polymer chain axis. Under tensile stress, full extension of the polymer main chain forces is leading to an expansion in the direction normal to the chain axis and hence to auxetic behaviour. The auxetic fibres have great potential to be used in fibre-reinforced composites. Recently invented auxetic yarn [70] has the ability to response to external force and to the moisture present in the surrounding by using moisture-activated shrinking filament. The invented fibre is a combination of two components, one component is a moisture-sensitive shrinking filament with relatively high modulus of elasticity such as modified cellulosic fibres, e.g. cotton or rayon. The other component is an elastic material of lower modulus of elasticity. When the fibre is in wet state, the moisture-sensitive shrinking component shrinks and a pulling force is applied along to the elastic component causing it to deform and form helices. There were several different attempts to prepare fabrics based on the auxetic materials structure. Some studies have revealed that an auxetic effect can successfully be induced by using rotating units such as squares [71], rectangles [72]. An auxetic woven textile structure utilizing a double helix yarn in a composite material was produced [73]. The yarn was a reinforcement in the composite. Auxetic materials can also be used for vibration damping and shock absorbency in case of bomb blast curtains, which can open a large number of pores under tension allowing the shock wave through but leaving the curtains intact to catch glass and other debris. A commercially available product—Zetix™ helical-auxetic fibre technology is a perfect example an anti-ballistic application of auxetic structures into protective textiles. Zetix™ is used in a variety of products, including body armour and seat belts. XTEGRA [74] is another producer of auxetic textiles structures, which provides the fabrics for armour, blast panels and thermal protection systems. This type of protective materials utilizes a very modest amount of expensive high-performance fibres and a great amount of their cheap replacements.

Author details

Izabela Luiza Ciesielska-Wróbel

Address all correspondence to: iza.ciesielska@poczta.fm

Department of Materials, Textiles and Chemical Engineering, Centre for Textile Science and Engineering, Ghent University, Zwijnaarde, Gent, Belgium

References

[1] Tony Thomas. A Crime Report "Man Attacks Stranger with Baseball Bat Outside Walmart" Presented on. Available from: http://www.wsbtv.com/news/local/gwinnett-county/police-man-attacks-stranger-with-baseball-bat-over-parking-spot-at-walmart/462827295 [Accessed: 30 January 2017]

[2] Chen X, Chaudhry I. Ballistic protection. In: Scott RA, ed. Textiles for Protection. 1st ed. Cambridge, UK: Woodhead Publishing; 2005

[3] Aldrete GS, Bartell S, Aldrete A. Reconstructing Ancient Linen Body Armor (Kindle Locations 184-187). Baltimore, Maryland, USA: Johns Hopkins University Press; 2013

[4] Dunstan S. Flak Jackets: Twentieth Century Military Body Armour (Men at Arms Series, 157). London, UK Osprey Publishing; 1984

[5] Dean B. Handbook of Arms and Armor (Kindle Locations 2180-2187). New York, USA: The Metropolitan Museum of Art; 1915

[6] Grafton CB. Arms and Armor: A Pictorial Archive from Nineteenth-Century Sources (Dover Pictorial Archive) (Kindle Locations 42-43). New York, USA: Dover Publications; 1995

[7] Turnbull S. Samurai Armies 1550-1615 (Men-at-Arms) (Kindle Locations 483-492). Botley, Oxford, UK: Osprey Publishing; 1979

[8] May T. The Mongol Empire. A Historical Encyclopedia, ABC-CLIO ™. CA, USA: Santa Barbara; 2017

[9] Peter Mansoor. Armor – Protective Clothing in Encyclopedia Britannica. Available from: https://www.britannica.com/topic/armour-protective-clothing [Accessed: 30 January 2017]

[10] Everson T. Warfare in Ancient Greece: Arms and Armour from the Heroes of Homer to Alexander the Great. Stroud, UK: Sutton Publishing; 2004

[11] Griffith GT. Peltasts, and the origins of the Macedonian phalanx. In: Dell HJ, editor. Ancient Macedonian Studies in Honor of Charles F. Edson. Thessaloniki: Institute for Balkan Studies; 1981. pp. 161-167

[12] Dawson T. Armour Never Wearies': Scale and Lamellar Armour in the West, from the Bronze Age to the 19th Century. Stroud, Gloucestershire, UK: History Press; 2013

[13] Giuseppe Piva. Samurai Armor Collection from Giuseppe Piva Japanese Art, Milan, Italy. Available from: http://www.giuseppepiva.com/en/services [Accessed: 31 January 2017]

[14] A football shin guard by Adidas. Available from: http://www.soccershoo.com/product/adidas-11-anatomic-lite-shinpads-x-large/ [Accessed: 31 January 2017]

[15] Ankrah S, Mills NJ. Performance of Football Shin Guards for Direct Study Impacts. Sports Engineering. 2003. pp. 207-219

[16] Walker PF. History of Armour 1100-1700. Wiltshire, UK: Crowood Press; 2013

[17] Henry M. US Marine Corps in World War I 1917-18 (Men-at-Arms) (Kindle Locations 595-598). Botley, Oxford UK: Midland House, West Way; 1999

[18] Wanberg J. Composite Materials Fabrication Handbook (Kindle Locations 2811-2813). Stillwater, MN, USA: Wolfgang Publications Inc; 2009

[19] Rowell R. The 12 Most Amazing American Inventions (Kindle Location 5). North Mankato, MN, USA: 12-Story Library, Peterson Publishing Company; 2015

[20] Yang HH. Kevlar ® Aramid Fiber. Chichester, West Susses, England: John Wiley & Sons; 1993

[21] Review of Department of Defense Test Protocols for Combat Helmets Committee on Review of Test Protocols Used by the DoD to Test Combat Helmets, Board on Army Science and Technology, Division on Engineering and Physical Sciences, National Research Council. The National Academies Press, Washington DC, USA, 2014.

[22] Higashi F, Nakano Y, Goto M, Kakinoki H. High-molecular-weight polyterephthalamide by direct polycondensation reaction in the presence of poly(ethylene oxide). Journal of Polymer Science: Polymer Chemistry Edition. 1980;**18**(3):1099-1104

[23] Information about Kevlar AP Fiber Provided by the Producer. Available from: http://www.dupont.com/products-and-services/fabrics-fibers-nonwovens/fibers/brands/kevlar/products/kevlar-belts.html [Accessed: 1 February 2017]

[24] Information about Kevlar XP Fiber Provided by the Producer. Available from: http://www.dupont.com/products-and-services/fabrics-fibers-nonwovens/fibers/brands/kevlar/products/kevlar-xp-soft-armor.html [Accessed: 1 February 2017]

[25] Information about Kevlar KM2 Fiber Provided by the Producer. Available from: http://www.dupont.com/products-and-services/fabrics-fibers-nonwovens/fibers/brands/kevlar/products/kevlar-km2-plus.html [Accessed: 1 February 2017]

[26] Jain A, Vijayan K. Thermal aging of Twaron fibers. High Performance Polymers. 2003;**15**: 105-129

[27] Information about Zylon Fibers. Available from: http://www.toyobo-global.com/seihin/kc/pbo/zylon_youto.html [Accessed: 9 February 2017]

[28] Chapman R. Developments in Protective Textiles. Leatherhead, UK: Pira International Ltd; 2009

[29] Bunsell AR, eds. Handbook of Tensile Properties of Textile and Technical Fibres. 1st ed. Cambridge, UK: Woodhead Publishing; 2009. ISBN: 9781845693879

[30] Eichhorn SJ, Hearle WS, Jaffe M. Handbook of Textile Fibre Structure: Fundamentals and Manufactured Polymer Fibres. Vol. 1. 1st ed. Cambridge, UK: Woodhead Publishing; 2009

[31] Alagirusamy R, Das A. Technical Textile Yarns, Industrial and Medical Applications. 1st ed. Cambridge, UK; 2009

[32] Wilusz E. Military Textiles. 1st ed. Cambridge, UK: Woodhead Publishing; 2008

[33] Horrocks AR, Anand SC. Handbook of Technical Textiles. 1st ed. Cambridge, UK: Woodhead Publishing; 2000

[34] Lawrence C. High Performance Textiles and Their Applications. 1st ed. Cambridge, UK: Woodhead Publishing; 2014

[35] Information about Spectra Fibers Provided by the Producer. Available from: https://www.honeywell-spectra.com/products/fibers [Accessed: 16 March 2017]

[36] Information about Dyneema® Based Products. Available from: http://www.dsm.com/products/dyneema/en_GB/applications/personal-armor.html [Accessed: 16 March 2017]

[37] Hearle WS. High-Performance Fibers. 1st ed. Cambridge, UK: Woodhead Publishing; 2001

[38] Cunniff PM, Auerbach MA. High Performance"M5" Fiber for Ballistic/Structural Composites [Internet]. 2002. Available from: http://web.mit.edu/course/3/3.91/www/slides/cunniff.pdf

[39] Information about Endumax. Available from: http://www.teijinendumax.com/wp-content/uploads/2015/04/brochure-endumax-4-20151.pdf [Accessed: 6 March 2017]

[40] Information about Kevlar Fibers. Available from: http://www.dupont.com/products-and-services/fabrics-fibers nonwovens/fibers/brands/kevlar/products/dupont-kevlar-fiber.html [Accessed: 16 March 2017]

[41] Information about Kevlar XP. Available from: http://www.dupont.com/content/dam/dupont/products-and-services/fabrics-fibers-and-nonwovens/fibers/documents/DSP_Kevlar%20XPS300_datasheet_K23806.pdf [Accessed: 16 March 2017]

[42] Information about Kevlar Protera. Available from: http://www.dupont.com/products-and-services/personal-protective-equipment/thermal-protective/brands/protera.html [Available: 16 March 2017]

[43] Sparks E. Advances in Military Textiles and Personal Equipment. 1st ed. Oxford, UK: Woodhead Publishing; 2012

[44] Information about Twaron. Available from: http://www.teijinaramid.com/brands/twaron [Accessed: 16 March 2017]

[45] Information about Teijin Products. Available from: http://www.teijinendumax.com/wp-content/uploads/2015/04/brochure-endumax-4-20151.pdf [Accessed: 16 March 2017]

[46] Holder E, Mason K, Sabol WJ. Selection and Application Guide to Ballistic-Resistant Body Armor For Law Enforcement. Corrections and Public Safety, NIJ Selection and Application Guide-0101.06, US Department of Justice, Office of Justice Programs: Washington DC, US; 2014

[47] Information about Products of Teijin Including UD41. Available from: http://www.teijinaramid.com/wp-content/uploads/2016/05/Teijin-Aramid-Ballistics-Material-Handbook.pdf

[48] Information about Products of Coorstek Company. Available from: https://www.coorstek.com/english/solutions/industries/defense-security/armor-components/body-armor-components [Accessed: 16 March 2017]

[49] Information about Ceramic Products of Coorstek Company. Available from: https://www.coorstek.com/media/2981/01091-ceramic-armor-overview.pdf [Accessed: March 16, 2017]

[50] Information about Tensylon ™. Available from: http://www.dupont.com/content/dam/dupont/products-and-services/personal-protective-equipment/vehicle-armor/documents/DPP_Tensylon30A_datasheet_K25929_2.pdf [Accessed: 6 March 2017]

[51] Rahbek D, Johnsen BB. Dynamic Behaviour of Ceramic Armour Systems, Norwegian Defence Research Establishment (FFI) [Internet]. 2015. Available from: https://www.ffi.no/no/Rapporter/15-01485.pdf [Accessed: 21 February 2017]

[52] Rice K, Higgins K. Stab Resistance of Personal Body Armor Standard–0115.00. U.S. Department of Justice, Office of Justice Programs, National Institute of Justice, Washington DC, USA, 2000. https://www.ncjrs.gov/pdffiles1/nij/183652.pdf

[53] Information about Products of DSM Company. Available from: http://www.dsm.com/products/dyneema/en_GB/applications/personal-armor/ballistic-anti-stab-vests.html [Accessed: 16 March 2017]

[54] McEntire J, Whitley P. Blunt Impact Performance Characteristics of the Advanced Combat helmet and the Paratrooper and Infantry Personnel Armor System for Ground Troops Helmet. US Army Research Lab Report No. 2005-12

[55] Rice K, Riley M, Foster A. Ballistic Resistance of Body Armor. Standard–0101.06, National Institute of Justice, U.S. Department of Justice, Office of Justice Programs, Washington DC, USA, 2008. https://www.ncjrs.gov/pdffiles1/nij/223054.pdf

[56] US Military Standard MIL-STD-662F: V50 Ballistic Test for Armor, 18 December 1997.

[57] UK Standard - UK/SC/5449; Ballistic Test Method for Personal Armors and Lightweight Materials.

[58] Ballistic Test Method for Personal Armor Materials and Combat Clothing NATO Standard - STANAG 2920.

[59] Freier KS. Development of Pelvic Protection Textiles for Soldiers: Personal Protective Equipment in a System Approach to Counter IED Threats. Saarbrucken, Germany: Akademikerverlag, OmniScriptum GmbH & Co. KG; 2015

[60] Information about Products of 3-D Products for Ballistic Protection by 3Tex Inc. Available from: http://www.3tex.com/node/7.html [Accessed: 4 April 2017]

[61] Irgens F. Rheology and Non-Newtonian Fluids. Switzerland: Springer Internet Publishing; 2014

[62] Malkin AY. Non-Newtonian viscosity in steady-state shear flows. Journal of Non-Newtonian Fluid Mechanics. 2013;**192:**48-65

[63] Galindo-Rosales FJ, Rubio-Hernández FJ, Sevilla A. An apparent viscosity function for shear thickening fluids. Journal of Non-Newtonian Fluid Mechanics. 2011;**166**:321-325

[64] Srivastava A, Majumdar A, Butola BS. Improving the impact resistance performance of Kevlar fabrics using silica based shear thickening fluid. Materials Science and Engineering: A. 2011;**529**:224-229

[65] Lim TC. Auxetic Materials and Structures (Engineering Materials). Springer Singapore Edition; 2015

[66] Hu H, Zulifqar A. Auxetic textile materials - A review. Journal of Textile Engineering & Fashion Technology. 2016;**1**(1):1-20

[67] Stręk AM. Production and study of polyether auxetic foam. Mechanics and Control. 2010;**29**(2):78-87

[68] Chinta SK, Veena KV, Karoshi VR. Auxetic fibers. International Journal of Engineering Research & Technology. 2012;**1**(9):1-10

[69] Liu P, He C, Griffin A. Liquid crystalline polymers as potential auxetic materials: Influence of transverse rods on the polymer mesophase. Abstracts of Papers of the American Chemical Society1998;**216**:U108

[70] Lee W, Lee S, Koh C, Heo J. United States Patent US 7858055 B2 Moisture sensitive auxetic. Material, 2010

[71] Grima JN, Evans KE. Self expanding molecular networks. Chemical Communications. 2000;**16**(16):1531-1532

[72] Grima JN, Gatt R, Alderson AE, Evans K. On the auxetic properties of Rotating rectangles' with different connectivity. Journal of the Physical Society of Japan. 2005;**74**(10):2866-2867

[73] Miller W, Hook P, Smith CW, Wang X, Evans KE. The manufacture and characterisation of a novel, low modulus, negative Poisson's ratio composite. Composites Science and Technology. 2009;**69**(5):651-655

[74] Information about Auxetic Fabrics. Available from: https://www.advancedfabrictechnology.com [Accessed: 5 April 2017]

Chapter 9

Efficiency of Medical Workers' Uniforms with Antimicrobial Activity

Urška Rozman, Daniela Zavec Pavlinić, Emil Pal,
Vida Gönc and Sonja Šostar Turk

Additional information is available at the end of the chapter

http://dx.doi.org/10.5772/intechopen.68830

Abstract

Antimicrobial finishing of textiles protects users from pathogenic microorganisms, which can cause medical and hygienic problem. The use of such textiles particularly increases in healthcare facilities, where reduction and transmission of pathogenic bacteria are important factors for preventing nosocomial infections. In the present study, the efficiency of fabric with silane quaternary ammonium compounds (Si-QAC) applied as active agents was evaluated. A test was performed according to ATCC 100-1999 Test Method after 0-, 24- and 48-hour incubation times. The treated textiles were effective against *Enterococcus faecalis, Staphylococcus aureus* and *Klebsiella pneumoniae,* but were not effective for Gram-negative *Escherichia coli* and *Pseudomonas aeruginosa.* Testing was also performed in hospital environment at infectious department where working clothes made of treated fabric were compared to normal working clothes. Antimicrobial textiles were not effective in a hospital environment, where average microbial count on medical workers' uniforms without antimicrobial protection was 1.4×10^9 cfu/mL, and 1.3×10^9 cfu/mL for uniforms made of antimicrobial material. Our conclusion is that quantities of application rates for Si-QAC should be higher or should be improved with applying another antimicrobial coating to obtain complex with dual activity.

Keywords: antimicrobial textiles, silane quaternary ammonium compounds, antimicrobial finished textiles test methods, medical workers' uniforms, healthcare-associated infections

1. Introduction

Hospital textiles, together with moisture and heat, create the right conditions for growth, proliferation and long-term survival of many microorganisms, which can serve as a vector of cross-transmission of healthcare-associated infections (HCAI) [1, 2]. Medical workers' uniforms can easily become contaminated and there are several reports of uniforms contamination by pathogenic microorganisms [3–5]. HCAI do not only represent complications in the treatment of patients [6, 7], but also cause economic damage with annual financial losses in Europe estimated at 7 billion EUR. Therefore, the medical institutions are increasingly using fabrics with antimicrobial activity, for reducing the transmission of HCAI. Various antimicrobial textile materials are developed using a variety of active agents including triclosan, metals and their salts, phenols, quaternary ammonium compounds (QAC), and organometallics [8]. The aim of our study was to test the efficiency of fabric with silane quaternary ammonium compounds (Si-QAC) applied as active agents in laboratory environment as well as in hospital environment.

2. Properties of antimicrobial textiles

The finishing process in textile manufacturing industry gives special, enhanced antimicrobial characteristics to the fabric, that can be biocidal (i.e. include agents that kill microorganisms) or biostatic (i.e. inhibit the microorganisms' growth). Various antimicrobial textile materials are developed using a variety of active agents which include synthetic antimicrobial agents such as triclosan, metal and their salts, phenols, quaternary ammonium compounds (QAC), and organometallics. When evaluating antimicrobial treated textiles the principle of antimicrobial agent fixation (**Figure 1**) to the fabric is crucial, since it is closely related to the mechanism of antimicrobial activity [9, 10]. Leaching antimicrobials that are not chemically bound to the textile are gradually and persistently released from the textile into their surroundings where they demonstrate their antimicrobial activity. On the other hand, the chemically bound antimicrobials are bound to the surface of the textile fibres and act as a barrier for the microorganisms with which they come into contact (principle of bio-barrier formation) [11, 12]. Some of the most common antimicrobial textile finishes, their principles of functioning with positive and negative features are presented in **Table 1**.

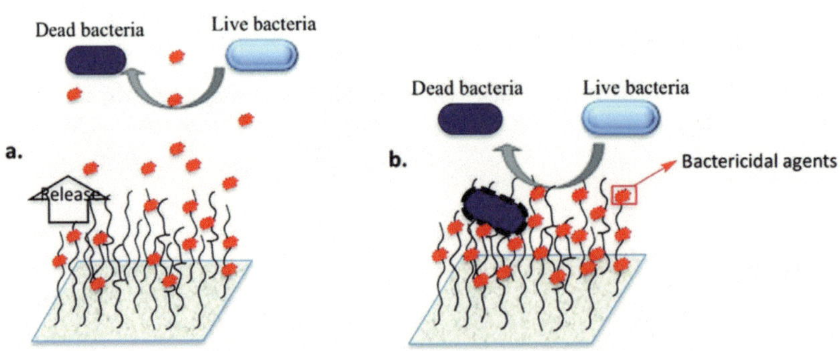

Figure 1. Principle of antimicrobial agent fixation and mode of action (a) leaching antimicrobial agent released from the textile, (b) contact killing coating bonded to the textile [11].

http://dx.doi.org/10.5772/intechopen.68830

Antimicrobial agent	Principle of functioning	Positive/negative features
N-halamines [13, 14]	Electrophilic substitution of Cl in the N-Cl bond with H in the presence of water and results in the transfer of Cl+ ions that can bind to acceptor regions on microorganisms. This hinders enzymatic and metabolic processes, leading to the destruction of the microorganisms	+ biocides that are active for a broad spectrum of bacteria, fungi and viruses – as an N—H bond, which does not have antimicrobial properties, is formed in the substitution reaction, further exposure of the agent to dilute sodium hypochlorite is needed for regeneration of its antimicrobial activity
Triclosan	By using electro chemical mode of action active substance penetrate and disrupt cell wall, leading to leaking of metabolites and disabling cell functions [15]	+ it is not water solubel and does not leach out [16]
Chitosan [17–19]	Positively charged amino groups can bind to the negatively charged bacterial surface, resulting in the disruption of the cell membrane and an increase in its permeability. Such antimicrobial function is very similar to that determined for QAS. Chitosan can also interact with the DNA of microorganisms to prevent protein synthesis	+ nontoxicity, biocompatibility and biodegradability – weak adhesion to cellulose fibres, resulting in a gradual leaching from the fibre surface with repetitive washing
Metals (Ag, Co, Zn), metal compounds and nanoparticles of metals [20, 21]	Oxidative stress resulting in damage to the lipids, binding and inactivation of intracellular proteins [22], inhibiting active enzyme centres in microorganisms [16] and losses replication ability of microorganisms DNA	+ nano-sized inorganic particles possess high surface area/volume ratio and display unique physical and chemical properties Ag: low toxicity to animals' cell [23]
Bioactive Plant-based Antimicrobial Agents [24, 25]	mostly secondary metabolites such as alkaloids, steroids, tannins, and phenol compounds resembling endogenous metabolites, ligands, hormones, signal transduction molecules or neurotransmitters	+: enable the production of safe, nontoxic, skin and environment friendly bioactive textile products
QAC [26–30]	Interactions between the cationic ammonium group of the QAC and the negatively charged cell membrane of the microbe; these interactions consequently result in the formation of a surfactant–microbe complex. This in turn causes the interruption of all essential functions of the cell membrane: the denaturation of proteins and interruption of protein activity, causing disruption of the cell structure [22]	+: active against a broad spectrum of microorganisms such as Gram-positive and Gram-negative bacteria, fungi and certain types of viruses –: QASs have an inherent weakness: leaching from the textile. To fix a QAS on textile fibres, sol–gel technology has also been used in antimicrobial textiles. This enables the formation of a nanocomposite polymer network with an organic–inorganic hybrid structure (e.g Si-QAC)

Table 1. Principles of functioning with positive and negative features of selected antimicrobial textile finishes.

Today, there are different available antimicrobial agents on the market to give textiles antimicrobial characteristics. Each antimicrobial agent has different modes of action and is chosen according the end use of a textile product [31].

2.1. Methods for application of antimicrobial agents

Antimicrobial agent can be applied by different methods. Active agent can be added before the polymer extrusion in the spinning process producing fibres or films. In case of antimicrobial fibres, the antimicrobial agent is integrated into the fibre, while antimicrobial films act as a barrier to microorganisms, but can make fabric impermeable to airflow leading to heath stress [32, 33].

Another procedure to obtain textile with antimicrobial protection is adding the active agent in the phase of textile finishing, where active agent can be applied on fibres or on the flat surfaces. The application can be obtain during processes of exhaustion, pad-dry-curing, coating, spraying and foam technique [16, 32].

2.2. Functionalization of textiles by Si-QAC

Quaternary ammonium compounds (QAC) are a chemical class of cationic surface active agents [12]. Once the microorganisms are exposed to the QAC, the agent absorb and penetrate into the cell wall where it reacts with the cytoplasmic membrane (lipid or protein) leading to membrane disorganization. Low-weight intracellular material is leaking out, which leads to degradation of proteins and nucleic acids and finally autolytic enzymes cause lysis of the cell wall [34, 35]. Although being chemically bonded to the textile fibres and their concentration does not change with time, they cannot ensure the permanent antimicrobial activity of agents because the settling of dead microorganisms on the bio-barrier can greatly reduce or even eliminate their effectiveness [36]. There is proven antimicrobial activity of QAC on natural cellulosic fibres, polyester/cotton blends and secondary acetates [16, 22].

The active ingredient dimethyltetradecyl[3-(trimethoxysilyl)propyl]ammoniumchlorid of tested antimicrobial agent consist of a Silan group, positively charged ammonium group and long non-polar chain (**Figure 2**). The antimicrobial agent is applied using Foulard process. In the clamping frame, the Silan group is bound to the fibre using heath and a condensation reaction, where QAC agent is covalently bound to the fibre (**Figure 3**) [37]. The permanent connection with the neighbouring molecules is formed and the active agent evenly surrounds the fibre.

The SEM imaging of fabric samples (**Figure 3**) shows that covalent bounding of active agent did not cause any noticeable degradation of the fibres and did not affect fibre morphology, indicating that important functional and aesthetic qualities of the fabric were retained. There are no obvious film layers and particles attachments on the treated textiles surfaces.

2.3. Durability of antimicrobial treatment using Si-QAC protection

Ideal antimicrobial agents should be effective against a broad spectrum of agents, have low toxicity, be compatible with other finishes, be easy to apply and also be durable and

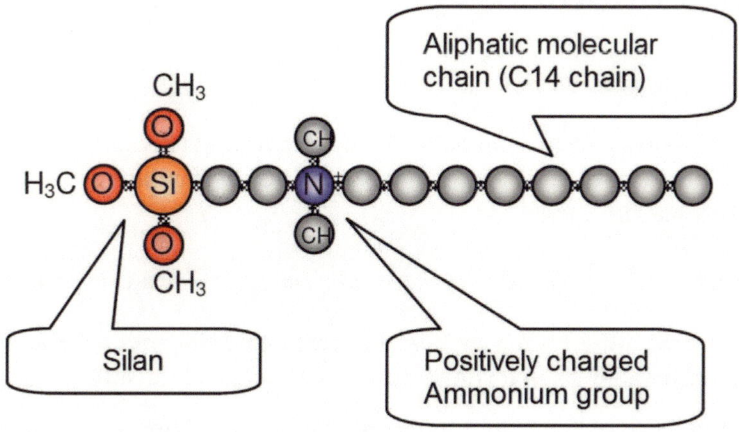

Figure 2. Schematic chemical structure of tested antimicrobial agent.

persistent in washing process. Therefore for many textile applications the wash durability is a key parameter for assessing the performance of antimicrobial treatments [31]. Durability is fundamental to ensuring the antimicrobial performance throughout the life cycle of a textile product. Washing and wear parameters, which can of course lead to a loss of the antimicrobial efficiency, are the key factor for assessing the antimicrobial performance. It is expected that a durable antimicrobial textile finish should survive at least 50 machine washes in line with industrial practices [38]. In our survey Si-QAC treated textiles were washed according to RAL-GZ 992 procedure in 25 and 50 cycles. Antimicrobial protection was decreasing with the number of washing cycles and was totally lost after 50 washing cycles.

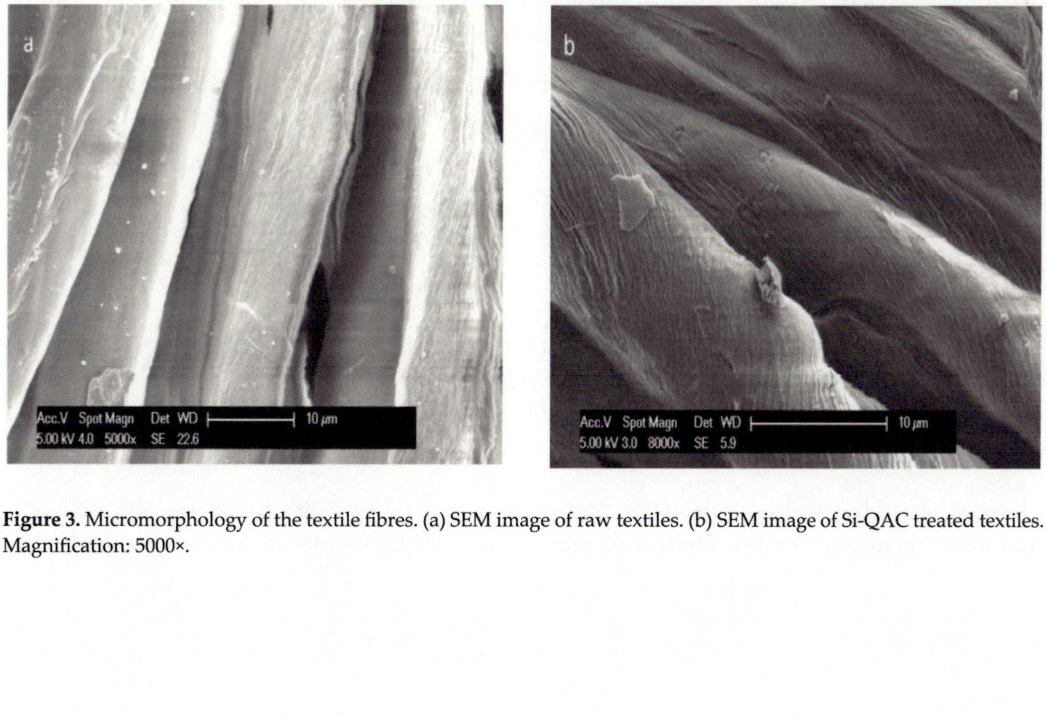

Figure 3. Micromorphology of the textile fibres. (a) SEM image of raw textiles. (b) SEM image of Si-QAC treated textiles. Magnification: 5000×.

3. Testing and characterization

3.1. Standards and test methods

Proof of activity against bacteria can be performed according to accepted standards and test methods where tested material (e.g. fabric) are compared to standard or an identical material without antimicrobial finishing.

3.1.1. JIS L 1902:2008 (testing for antibacterial activity and efficacy on textile products)

This standard specifies the testing method for evaluating the antibacterial activity against bacteria on antibacterial finished textile products. Using one of the following methods:

- Qualitative test (halo method) to evaluate antibacterial activity of textile products by the existence of halos, and applicable to those of which antibacterial finishes chemicals can diffuse into the plate agar culture medium.
- Quantitative test to evaluate antibacterial activity of textile products by bacteriostatic activity value, bactericidal activity value, bacterial decrease value. The absorption method shall be applied for high humidity conditions and printing method for low humidity conditions. Bacteria are shaken out after the incubation of inoculated test pieces.

The measurement of bacterial concentration shall be carried out by viable plate counting or luminescence method for determining ATP (Adenosine Tri-phosphate) concentration [39].

3.1.2. DIN EN ISO 20743

This International Standard specifies quantitative test methods to determine the antibacterial activity of all antibacterial textile products including nonwovens, with different types of used of antibacterial agents or the application methods. The user can select out of three inoculation methods for antibacterial activity determination:

- Absorption method where the test bacterial suspension is inoculated directly onto tested surfaces;
- Transfer method where the test bacteria are placed on an agar plate and transferred onto tested surfaces; and
- Printing method where the test bacteria are placed on a filter and printed onto tested surfaces.

For measuring the enumeration of bacteria colony, the viable plate counting method and the ATP luminescence method are also specified [40].

3.1.3. AATCC 147 (parallel streak method)

The objective of the method is detecting bacteriostatic activity on textile materials. The method is useful for obtaining a rough estimate of activity. Increasing degrees of sensitivity is indicated where growth of the inoculum organism is decreasing from one to the other end of each streak and from one to the next streak. Tested surfaces are placed in intimate contact with the inoculated (with test bacterium) agar surface and incubated. Antibacterial activity of the tested material is indicated in clear area of interrupted growth underneath and along the side of the test material [41].

3.1.4. AATCC Test Method 100-1999 (antibacterial finishes on textile materials: assessment of)

This method provides a quantitative procedure for the evaluation of the antibacterial activity degree. Test and control swatches are inoculated with test organisms and incubated. By shaking in known amounts of neutralizing solution, the bacteria are eluted from the swatches. The number of eluted bacteria is determined and percentage reduction by the treated specimen is calculated [42].

3.2. Study in laboratory environment

Textiles pieces (50/50 PES/COT white fabric, 195 g/m^2) with and without Si-QAC dimethyltetradecyl[3-(trimethoxysilyl)propyl]ammoniumchlorid were sterilized, dried, artificially contaminated (**Figure 4**) and tested for antimicrobial activity according to ATCC 100-1999 Test Method after 0, 24 and 48 hour incubation times. Ready-made cultures of *Escherichia coli* (DSM 1562),

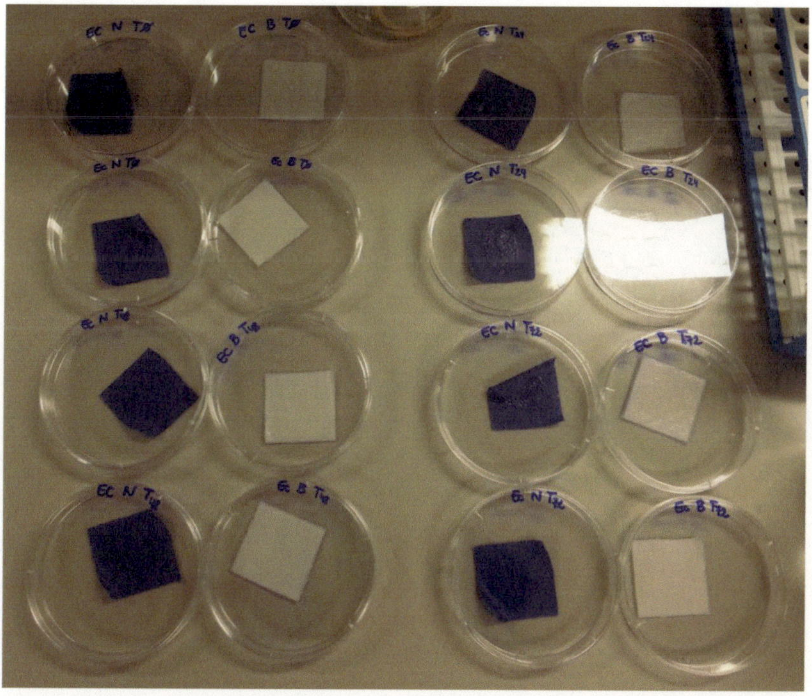

Figure 4. Artificially contaminated tested textile pieces with *E. coli* suspension (first and third column textile pieces with Si-QAC, second and fourth column textile pieces without Si-QAC).

Enterococcus faecalis (ATTC 29212), *Klebsiella pneumoniae* (ATCC 13883), *Staphylococcus aureus* (ATCC 25923) and *Pseudomonas aeruginosa* (ATCC 27853) were taken from the freezer and grown in nutrient broth (Tryptiyc Soy broth) for 4 days in an incubator at 37°C. Efficiency rate after 24 or 48 hour incubation was calculated using serial dilution, viable plate counting and colony forming units per millilitre (cfu/mL) calculation. Appropriate selective growth medium (e.g. VRBD agar, Kanamycin Esculin Azide agar, *Klebsiella pneumoniae* selective agar, Baird Parker agar, Cetrimide agar) were used.

The performance of fabrics with antimicrobial activity varies and depended on tested microbial species and it is to some extent related to the bacteria Gram categorization (e.g. Gram negative and Gram positive). When comparing textile with antimicrobial activity to nontreated textile the differences in reduction rate were 3.22×10^2 cfu/mL for *E. faecalis*, 1.87×10^4 cfu/mL for *S. aureus* two representatives of Gram-positive bacteria and 6.81×10^3 cfu/mL for Gram-negative *K. pneumoniae* (**Figure 5**), but treated textiles was not effective for Gram-negative *E. coli* and *P. aeruginosa* (**Figure 6**).

QACs are membrane active agents with a target site predominantly at the cytoplasmic membrane [43]. Since the cell wall structure varies between G+ and G− bacteria, the differences in sensitivity to QAC are expected. The cell wall of G+ bacteria represents much less effective barrier for the entry of antiseptics and disinfectants, which may explain the sensitivity of these organisms to many antibacterial agents including QACs [44–46]. Gram-negative bacteria are generally more resistant to antiseptics since their outer membrane acts as a barrier that limits

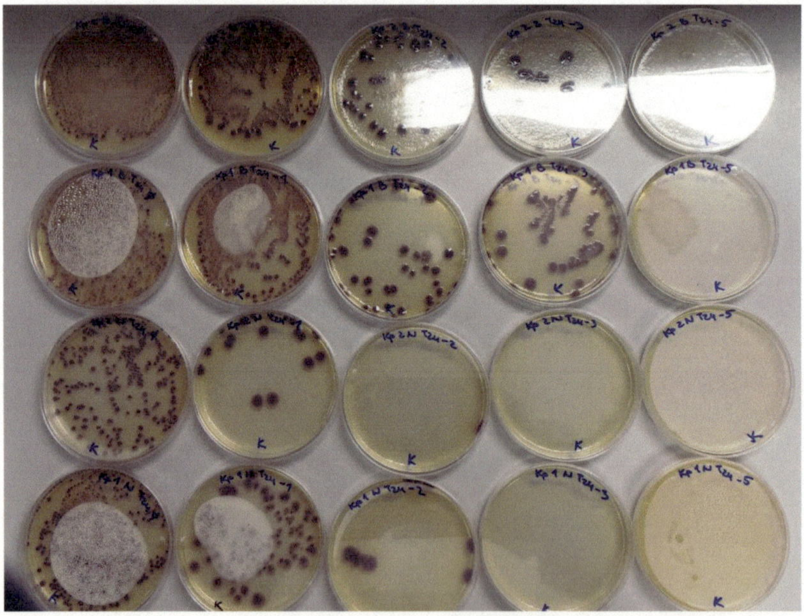

Figure 5. Colonies of *K. pneumoniae* after plating serial dilutions of suspension obtained from artificially contaminated textile pieces that were incubated for 24 hours. First and second row: samples without Si-QAC, third and fourth row: samples with Si-QAC.

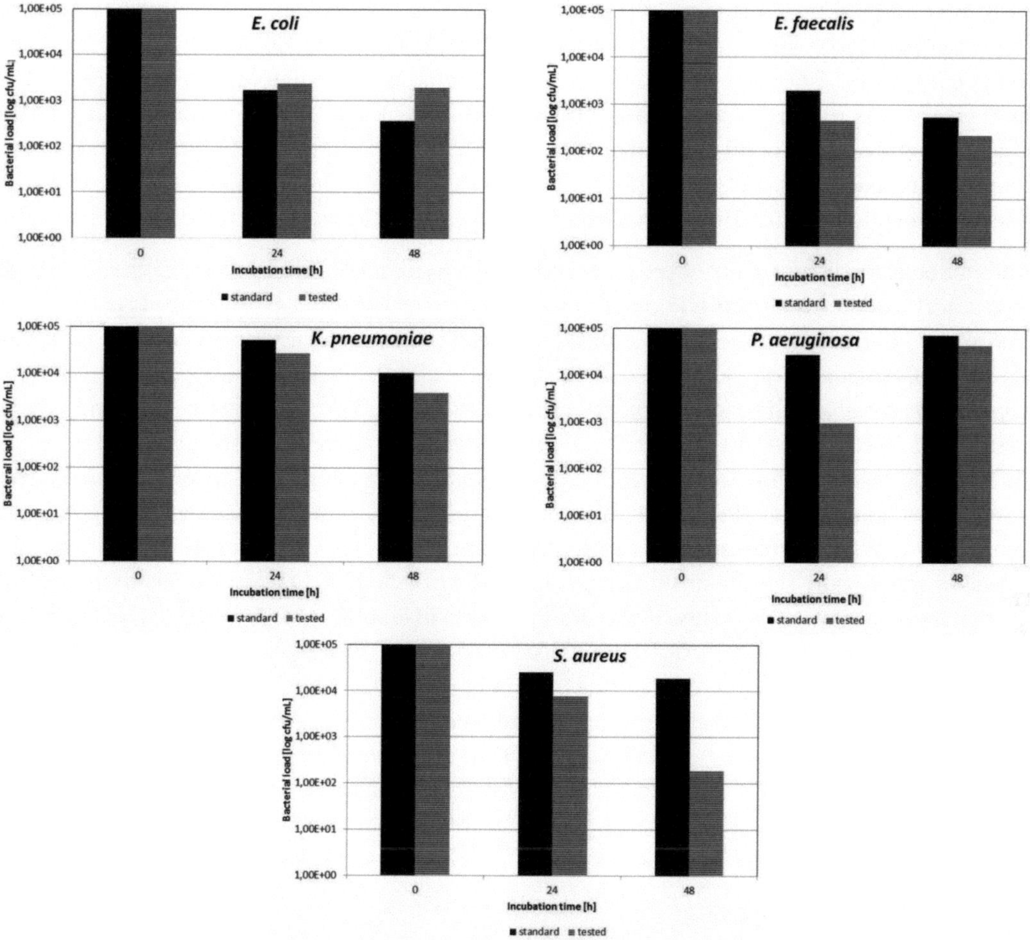

Figure 6. Bacteria per mL of eluting solution after 0-, 24- and 48-hour incubation time for 1 × 10^5 cfu/mL inoculum on standard and tested textile pieces.

the entry of many chemically unrelated types of antibacterial agents [47–52]. The formation of biofilm (in case of *P. aeruginosa*) can also account for the reduced sensitivity of bacteria [53].

3.3. Study in hospital environment

Medical workers' uniforms were made of antimicrobial material and worn for one day at infectious department by 10 nurses. Normal working clothes were used for negative control. Testing was repeated in two cycles. At the time of testing patient contaminates with MRSA (methicillin-resistant *Staphylococcus aureus*), CRE (carbapenem-resistant Enterobacteriaceae) and bacteria producing ESBL (extended-spectrum beta-lactamases) were hospitalized on the department. After eight-hour shift, uniforms were transferred to laboratory (each uniform in separate sterile bag) and tested for presence of microorganisms. Method of sterile elution was performed as follows: 2 L of sterile elution solution (0.9% NaCl + 0.2% Tween 80) (JIS L 1902,

2008) was added to uniform in a sterile bag and shaken for 30 min at 300 rpm on a shaking machine (Heidloph vibramax 100), after 10, 20 and 30 min uniform was gowned per hand in the closed bag. 100 mL of eluate was sterile filtrated through 0.8 μm 0.45 μm membrane filter which was further transferred to 100 mL 0.9% NaCl and shaken vigorously for 1 min. Efficiency rate after 24 or 48 hour incubation was calculated using serial dilution, viable plate counting and colony forming units per millilitre (cfu/mL) calculation. MALDI-TOF analysis of selected morphologically different colonies was used for microorganism identification.

Antimicrobial textiles were not effective in a hospital environment, where average microbial count on medical workers' uniforms without antimicrobial protection was 1.4×10^9 cfu/mL, and 1.3×10^9 cfu/mL for uniforms made of antimicrobial material. Species of *Acinetobacter lwoffii* (G−), Bacillus spp. (G+), *Pseudomonas aeruginosa* (G−), *Pseudomonas alcaliphila, Pseudomonas stutzeri, Pseudomonas manteili, Pseudomonas putida, Staphylococcus epidermidis* (G+), *Staphylococcus lugduensis, Staphylococcus warneri* and *Stenotrophomonas maltophilia* (G−) were identified. Our results are consistent with others who had tested uniforms with antimicrobial protection in the real life hospital environment. Burden et al. found no evidence that either antimicrobial scrub product decreased bacterial contamination of HCWs' uniforms or skin after an 8-hour workday [54] or experimental uniforms provided results only slightly in agreement with in vitro data [55]. Despite the fact that the effectiveness of an antimicrobial material was demonstrated in laboratory test, the effectiveness in hospital environment was not present. Bacterial load on medical workers' uniforms was much higher than the concentrations of microorganisms used in laboratory test method, which could also have an effect on the antimicrobial efficiency, since the adsorption of dead microorganisms can affect antimicrobial function of chemical bound agent [12]. Therefore, our conclusion is that quantities of application rates of Si-QAC should be higher. Another approach to improve the effectiveness of antimicrobial protection is the tailoring of different antimicrobial coatings to obtain complex with dual activity (e.g. applying silver chloride and Si-QAC composite) [36, 56–61].

4. Application potentials and environmental concerns

By implementing medical workers' uniforms with antimicrobial activity the risk for cross-transmission is reduced, while the fabric itself is protected against potential growth of newly microorganisms. Within the decrease of cross-infections, the protection of users (medical workers, patients, workers in food industry, etc…) is better; moreover, the comfort and well-being at workplace are improved.

4.1. Medical workers' uniforms, beddings, and patients' wear

Work clothing of health workers is uniform, has the agreed cut and colour, and is always clean and daily fresh [62]. Its purpose is to protect health workers from direct contact with the patient's secretions and consequently the transfer of microorganisms from patient to medical personnel and later to other patients [63]. The protective and preventing antimicrobial products can be used in different end application and they are ideal to be used in healthcare sectors. Workers' uniforms and bed sheets, which build microenvironment around human body, made of antimicrobial treated fabrics can for example protect humans against nosocomial infections in the hospital environments. Antimicrobial treatment of textiles aimed for hospital environment is usually marketed with the function of preventing the growth of microorganisms [54]. Antimicrobials for textiles need to fulfil many different criteria including efficacy against microorganisms, suitability for textile processing, durability and a favourable safety and environmental profile [31]. In the field of medical textiles, Ag (silver) and Si-QAC agents are widely used [31].

4.2. Preventing cross-contamination and HCAI

The origin of infection can be an infected person (e.g. patients, health workers or visitors) the environment. Microorganisms can be part of the patient's normal flora that can cause the nosocomial infection during the process of diagnosis, treatment and care of immunocompromised patients. Microorganisms are able to survive on environmental surfaces up to several weeks [64] providing a significant biotransfer/cross-contamination/cross-infection potential [65] that should not be overlooked. Microorganisms can also survive in the patient's abiotic environment, such as contaminated equipment for care, diagnosis and treatment (textiles), food, water, and disinfectants and on surfaces [66], which shows that one of the possible sources of nosocomial pathogens can be inappropriately disinfected textiles [67]. The environment can play a marked role in the nosocomial transmission of microorganisms [68], where garments of healthcare workers are an important aspect of the environment that can easily become contaminated and are therefore recognized as a possible vehicle of nosocomial infections. Hospital textiles could be a source of HCAIs, contributing to the transmission of pathogens both through indirect contact, via hospital staff, endogenously, and by means of aerosols [68, 69]. The characteristics of the textile in question, together with humidity and heat, can create the right conditions for the proliferation of numerous microorganisms [2, 3]. Pathogenic bacteria such as MRSA [3], *P. aeruginosa* and *K. pneumoniae* [4] and *C. difficile* [5] were also detected on uniforms of physicians and nurses. Surveys show that hospital textiles can be the source of nosocomial infections with streptococci [70], enterococci [71], *Bacillus cereus* [72], staphylococci [73] and coliform bacteria [74]. In addition, there is continuously increasing trends of antimicrobial resistance among HCAI pathogens leading to raising healthcare costs, prolonged hospital stays, treatment failures, and sometimes death [75].

4.3. Impact on the environment and human health

Although synthetic antimicrobial agents effectively inhibit the growth of microbes, most of them are toxic, can cause adverse effects on human health, and have environmental issues [8]. Recommended application rate of a silane quaternary ammonium compounds are ca. 10,000 mg/kg [76]. Si-QAC consumption in textiles was 1128 tonnes in 2004, most of it being a particular type of Si-QAC [77]. Exposure pathways and potential health effects need to be considered in order to evaluate the safety of antimicrobial compounds for humans. Si-QAC is readily biodegradable (after 6 days, 70% of the substance is degraded) [78] and is also expected to rapidly hydrolyze [79]. The type of active material (free or bound), the concentration in the product, the routes of exposure and the frequency of use all influence the extent to which humans may be exposed to antimicrobials in a textile product [80]. Although Si-QAC is a corrosive chemical, the USEPA does not expect any severe effects of Si-QAC use on human health [79], there is however evidence of skin sensitization for Si-QAC [78]. For any substance that addresses microorganisms there is a need to consider the potential for development of resistance. Although durable antimicrobials are increasingly popular for the consumer market, questions regarding their use will continue to rise regarding their effect on the evolution of resistant microbes [81]. Due to relatively large quantities of recommended application rates, the bacterial resistance should be considered. Most biocides used on commercial textiles can induce bacterial resistance to these substances, which can lead to increased resistance to certain antibiotics in clinical use [82], but any possible clinical significance of this remains to be tested [22]. Although the number of studies elucidating the association between antimicrobials resistance and resistance to clinical isolates has been limited, recent laboratory studies have confirmed the potential for such a link (e.g. transfer of triclosan resistance) in *E. coli* and *Salmonella enterica*. Thus, widespread use of antimicrobials may represent a potential public health risk in regard to development of concomitant resistance to clinically important antimicrobials [83].

5. Conclusion

Working cloths made of textiles with proven antimicrobial protection could be an ideal solution for areas with frequent infections such as healthcare facilities. Ideal antimicrobial agents should be effective against a broad spectrum of agents, have low toxicity, be compatible with other finishes, be easy to apply and be durable and persistent in washing process. Efficiency of antimicrobial textiles could be improved with higher application rates of Si-QAC or additional binding of another antimicrobial agent, but in this connection, the effect on antimicrobial resistance evolution should be considered.

Author details

Urška Rozman[1]*, Daniela Zavec Pavlinić[2], Emil Pal[3], Vida Gönc[1] and Sonja Šostar Turk[1]

*Address all correspondence to: urska.rozman@um.si

1 Faculty of Health Sciences, University of Maribor, Maribor, Slovenia

2 Titera d.o.o., Murska Sobota, Slovenia

3 General Hospital Murska Sobota, Murska Sobota, Slovenia

References

[1] Malnick S, Bardenstein R, Huszar M, Gabbay J, Borkow G. Pyjamas and sheets as a potential source of nosocomial pathogens. Journal of Hospital Infection [Internet]. 2008;**70**(1):89-92. Available from: http://linkinghub.elsevier.com/retrieve/pii/S0195670108002351 [Accessed: November 8, 2016]

[2] O'Hanlon SJ, Enright MC. A novel bactericidal fabric coating with potent in vitro activity against meticillin-resistant Staphylococcus aureus (MRSA). The International Journal of Antimicrobial Agents [Internet]. 2009;**33**(5):427-431. Available from: http://linkinghub.elsevier.com/retrieve/pii/S092485790800527X [Accessed: November 8, 2016]

[3] Boyce JM, Potter-Bynoe G, Chenevert C, King T. Environmental contamination due to methicillin-resistant *Staphylococcus aureus*: Possible infection control implications. Infection Control and Hospital Epidemiology [Internet]. 1997;**18**(9):622-627. Available from: http://www.ncbi.nlm.nih.gov/pubmed/9309433 [Accessed: November 8, 2016]

[4] Wiener-Well Y, Galuty M, Rudensky B, Schlesinger Y, Attias D, Yinnon AM. Nursing and physician attire as possible source of nosocomial infections. American Journal of Infection Control [Internet]. 2011;**39**(7):555-559. Available from: http://linkinghub.elsevier.com/retrieve/pii/S0196655311001179 [Accessed: November 8, 2016]

[5] Perry C, Marshall R, Jones E. Bacterial contamination of uniforms. Journal of Hospital Infection [Internet]. 2001;**48**(3):238-241. Available from: http://linkinghub.elsevier.com/retrieve/pii/S0195670101909624 [Accessed: November 8, 2016]

[6] Haley RW, Culver DH, White JW, Morgan WM, Emori TG, Munn VP, et al. The efficacy of infection surveillance and control programs in preventing nosocomial infections in US hospitals. The American Journal of Epidemiology [Internet]. 1985;**121**(2):182-205. Available from: http://www.ncbi.nlm.nih.gov/pubmed/4014115 [Accessed: November 8, 2016]

[7] Harbarth S, Sax H, Gastmeier P. The preventable proportion of nosocomial infections: An overview of published reports. Journal of Hospital Infection [Internet]. 2003;**54**(4):258-266; quiz 321. Available from: http://www.ncbi.nlm.nih.gov/pubmed/12919755 [Accessed: November 8, 2016]

[8] Murugesh Babu K, Ravindra KB. Bioactive antimicrobial agents for finishing of textiles for health care products. Journal of the Textile Institute [Internet]. 2015;**106**(7):706-717. Available from: http://www.tandfonline.com/doi/full/10.1080/00405000.2014.936670 [Accessed: November 8, 2016]

[9] Yuan Gao Y, Cranston R. Recent advances in antimicrobial treatments of textiles. The Textile Research Journal [Internet]. 2008;**78**(1):60-72. Available from: http://journals.sagepub.com/doi/10.1177/0040517507082332 [Accessed: January 31, 2017]

[10] Abidi N, Kiekens P. Chemical functionalisation of cotton fabric to impart multifunctional properties. Tekstilec. 2016;**59**(2):156-161

[11] Schindler W, Hauser P. Chemical finishing of textiles [Internet]. Woodhead Publishing Limited; 2004 [cited 2017 Jan 27]. 212 p. Available from: https://books.google.com/books?hl=sl&lr=&id=S42kAgAAQBAJ&oi=fnd&pg=PP1&dq=Chemical+Finishing+of+Textiles&ots=u-U5JjrZ8B&sig=gZTCkhB2NzLLGezYG27b5M3vDpc

[12] Simoncic B, Tomsic B. Structures of novel antimicrobial agents for textiles—a review. The Textile Research Journal [Internet]. 2010;**80**(16):1721-1737. Available from: http://journals.sagepub.com/doi/10.1177/0040517510363193 [Accessed: January 27, 2017]

[13] Barnes K, Liang J, Wu R, Worley SD, Lee J, Broughton RM, et al. Synthesis and antimicrobial applications of 5,5′-ethylenebis[5-methyl-3-(3-triethoxysilylpropyl)hydantoin]. Biomaterials [Internet]. 2006;**27**(27):4825-4830. Available from: http://www.ncbi.nlm.nih.gov/pubmed/16757023 [Accessed: January 27, 2017]

[14] Liang J, Chen Y, Barnes K, Wu R, Worley SD, Huang T-S. N-halamine/quat siloxane copolymers for use in biocidal coatings. Biomaterials. 2006;**27**(11):2495-2501

[15] Levy CW, Roujeinikova A, Sedelnikova S, Baker PJ, Stuitje AR, Slabas AR, et al. Molecular basis of triclosan activity. Nature [Internet]. 1999;**398**(6726):383-384. Available from: http://www.ncbi.nlm.nih.gov/pubmed/10201369 [Accessed: March 17, 2017]

[16] Bischof Vukušić S, Gaan S. Antimicrobial protective textiles. In: Bischor Vukšić S, Vujasinović E, editors. Functional protective textiles [Internet]. Zagreb: University of Zagreb, Faculty of Textile Technology; 2012 [cited 2017 Mar 17]. p. 101-44. Available from: http://www.worldcat.org/title/functional-protective-textiles/oclc/814043279

[17] Oktem T. Surface treatment of cotton fabrics with chitosan. Coloration Technology [Internet]. 2003;**119**(4):241-246. Available from: http://doi.wiley.com/10.1111/j.1478-4408.2003.tb00179.x [Accessed: January 27, 2017]

[18] Huang K-S, Wu W-J, Chen J-B, Lian H-S. Application of low-molecular-weight chitosan in durable press finishing. Carbohydrate Polymers. 2008;**73**(2):254-260

[19] Lim S-H, Hudson SM. Synthesis and antimicrobial activity of a water-soluble chitosan derivative with a fiber-reactive group. Carbohydrate Research. 2004;**339**(2):313-319

[20] Kohen R, Nyska A. Oxidation of biological systems: Oxidative stress phenomena, antioxidants, redox reactions, and methods for their quantification. Toxicologic Pathology [Internet]. 2002;**30**(6):620-650. Available from: http://www.ncbi.nlm.nih.gov/pubmed/12512863 [Accessed: January 27, 2017]

[21] Shahidi S, Wiener J. Antibacterial Agents in Textile Industry. In: Varaprasad B, editor. Antimicrobial Agents [Internet]. Rijeka: InTech; 2012 [cited 2017 Feb 1]. p. 337-406. Available from: http://www.intechopen.com/books/antimicrobial-agents/antibacterial-agents-in-textile-industry

[22] McDonnell G, Russell AD. Antiseptics and disinfectants: Activity, action, and resistance. Clinical Microbiology Reviews [Internet]. 1999;**12**(1):147-179. Available from: http://www.ncbi.nlm.nih.gov/pubmed/9880479 [Accessed: January 27, 2017]

[23] Raffi M, Hussain F, Bhatti TM, Akhter JI, Hameed A, Hasan MM. Antibacterial characterization of silver nanoparticles against *E. coli* ATCC-15224. Journal of Materials Science and Technology [Internet]. 2008;**24**(2):192-196. Available from: http://www.omnis-mg.hr/Radovi/EC-16nm-pakistan.pdf [Accessed: March 17, 2017]

[24] Cowan MM. Plant products as antimicrobial agents. Clinical Microbiology Reviews [Internet]. 1999;**12**(4):564-582. Available from: http://www.ncbi.nlm.nih.gov/pubmed/10515903 [Accessed: February 1, 2017]

[25] Ciocan D, Ioan B. Plant products as antimicrobial agents. Ann "Alexandru Ioan Cuza"University SecII a [Internet]. 2007 [cited 2017 Feb 1];**8**(1):151-6. Available from: http://www.gbm.bio.uaic.ro/index.php/gbm/article/view/402

[26] Ascenzi JM. Handbook of disinfectants and antiseptics. New York: Marcel Dekker, Inc.; 1996. 161 p.

[27] Gilbert P, Moore LE. Cationic antiseptics: Diversity of action under a common epithet. Journal of Applied Microbiology [Internet]. 2005;**99**(4):703-715. Available from: http://doi.wiley.com/10.1111/j.1365-2672.2005.02664.x [Accessed: February 1, 2017]

[28] Ahlström B, Chelminska-Bertilsson M, Thompson RA, Edebo L. Long-chain alkanoylcholines, a new category of soft antimicrobial agents that are enzymatically degradable. Antimicrobial Agents and Chemotherapy [Internet]. 1995;**39**(1):50-55. Available from: http://www.ncbi.nlm.nih.gov/pubmed/7695328 [Accessed: February 1, 2017]

[29] Judeinstein P, Sanchez C, Schubert U, Hüsing N, Lorenz A, Sanchez C, et al. Hybrid organic–inorganic materials: A land of multidisciplinarity. Journal of Materials Chemistry [Internet]. 1996;**6**(4):511-525. Available from: http://xlink.rsc.org/?DOI=JM9960600511 [Accessed: February 1, 2017]

[30] Nalwa HS. Handbook of organic-inorganic hybrid materials and nanocomposites. California: American Scientific Publishers; 2003. 810 p.

[31] Windler L, Height M, Nowack B. Comparative evaluation of antimicrobials for textile applications. Environment International. 2013;**53**:62-73

[32] Mao J. Durable antimicrobial finish for cotton with new technology. AATCC Review. 2002;**2**(12):15-7

[33] Belkin NL. Effect of laundering on the barrier properties of reusable surgical gown fabrics. American Journal of Infection Control [Internet]. 1999;**27**(3):304-305; author reply 307-308. Available from: http://www.ncbi.nlm.nih.gov/pubmed/10358238 [Accessed: March 17, 2017]

[34] Salton MR. Lytic agents, cell permeability, and monolayer penetrability. The Journal of General Physiology [Internet]. 1968;**52**(1):227-252. Available from: http://www.ncbi.nlm.nih.gov/pubmed/19873623 [Accessed: January 31, 2017]

[35] Denyer SP. Mechanisms of action of antibacterial biocides. International Biodeterioration and Biodegradation. 1995;**36**:227-245

[36] Logar N, Danijela K, Tomšič B, Pavko Čuden A, Simončič B. Tailoring of a dual-active antibacterial coating for polylactic acid fibres. Tekstilec. 2016;**59**(4):289-297

[37] Achwal WB. Antimicrobial finishes and their modifications. Colourage. 2003;**1**:58-59

[38] Sun G, Worley S. Chemistry of durable and regenerable biocidal textiles. The Journal of Chemical Education. 2005;**82**(1):60-64

[39] Japanese Industrial Standard. JIS L 1902:2008 testing for antibacterial activity and efficacy on textile products. Japanese Standards Association; 2008. 35 p.

[40] ISO/TC 38 T. ISO 20743:2013(en), Textiles–determination of antibacterial activity of textile products [Internet]. 2013. Available from: https://www.iso.org/obp/ui/#iso:std:59586:en [Accessed: December 19, 2016]

[41] AATCC Test Method 147-2004. Antibacterial Activity of Textile materialss: Parallel streak method. AATCC Technical Manual. Research Triangle Park, NC USA;2012. pp. 257-8.

[42] AATCC Test Method 100-1999. Antibacterial Finishes on Textile Materials: Assessment of, AATCC Technical Manual. Research Triangle Park, NC USA; 2004. pp. 149-51.

[43] Hugo WB. Disinfection mechanisms. In: Russel AD, Hugo WB, Ayliffe GAJ, editors. Principles and practice of disinfection, preservation and sterilization. 3rd ed. Oxford, England: Blackwell Science, Oxford, England; 1999. p. 258-83.

[44] Russel AD. Modification of the bacterial cell envelope and enhancement of antibiotic susceptibility. In: Stuart-Harris CH, Harris DM, editors. The Control of Antibiotic-Resistant Bacteria. London, England: Academic Press, Ltd.; 1981. pp. 119-165

[45] Russell AD. Mechanisms of bacterial resistance to biocides. International Biodeterioration and Biodegradation. 1995;**36**:247-265

[46] Russell AD, Chopra I. Understanding antibacterial action and resistance. Hemel Hempstead UK: E. Horwood; 1990. 246 p.

[47] Hammond SM, Lambert PA, Rycroft AN. The Bacterial Cell Surface [Internet]. Dordrecht: Springer Netherlands; 1984. Available from: http://link.springer.com/10.1007/978-94-011-6553-2 [accessed: January 31, 2017]

[48] Hancock REW. Alterations in outer membrane permeability. Annual Review of Microbiology [Internet]. 1984;**38**(1):237-264. Available from: http://www.ncbi.nlm.nih.gov/pubmed/6093683 [accessed: January 31, 2017]

[49] Nikaido H. Prevention of drug access to bacterial targets: permeability barriers and active efflux. Science (80-). 1994;**264**(5157):382-8.

[50] Paulsen IT, Park JH, Choi PS, Saier MH. A family of gram-negative bacterial outer membrane factors that function in the export of proteins, carbohydrates, drugs and heavy metals from gram-negative bacteria. FEMS Microbiology Letters [Internet]. 1997;**156**(1):1-8. Available from: http://www.ncbi.nlm.nih.gov/pubmed/9368353 [accessed: January 31, 2017]

[51] Russell AD, Gould GW. Resistance of enterobacteriaceae to preservatives and disinfectants. The Journal of Applied Bacteriology [Internet]. 1988;**65**(s17):167S–195S. Available from: http://doi.wiley.com/10.1111/j.1365-2672.1988.tb04651.x [accessed: January 31, 2017]

[52] Vaara M. Agents that increase the permeability of the outer membrane. Microbiology Reviews [Internet]. 1992;**56**(3):395-411. Available from: http://www.ncbi.nlm.nih.gov/pubmed/1406489 [accessed: January 31, 2017]

[53] Brown MRW, Gilbert P. Sensitivity of biofilms to antimicrobial agents. The Journal of Applied Bacteriology [Internet]. 1993;**74**(S22):87S–97S. Available from: http://doi.wiley.com/10.1111/j.1365-2672.1993.tb04345.x [accessed: January 27, 2017]

[54] Burden M, Keniston A, Frank MG, Brown CA, Zoucha J, Cervantes L, et al. Bacterial contamination of healthcare workers' uniforms: A randomized controlled trial of antimicrobial scrubs. Journal of Hospital Medicine [Internet]. 2013;**8**(7):380-385. Available from: http://www.ncbi.nlm.nih.gov/pubmed/23757125 [accessed: February 1, 2017]

[55] Condò C, Messi P, Anacarso I, Sabia C, Iseppi R, Bondi M, et al. Antimicrobial activity of silver doped fabrics for the production of hospital uniforms. The New Microbiologica. 2015;**38**:551-558

[56] Zhi Li †, Daeyeon Lee ‡, Xiaoxia Sheng †, Robert E. Cohen,* A, Michael F. Rubner* †. Two-Level Antibacterial Coating with Both Release-Killing and Contact-Killing Capabilities. Langumuir. American Chemical Society; 2006;**22**(24):9820-3.

[57] Dwivedi P, Narvi SS, Tewari RP. Coating made from *Pseudotsuga menziesii* phytosynthesized silver nanoparticles is efficient against *Staphylococcus aureus* biofilm formation. Nano Life [Internet]. 2015;**5**(4):1540006. Available from: http://www.worldscientific.com/doi/abs/10.1142/S1793984415400061 [accessed: January 31, 2017]

[58] Mohamed RR, Sabaa MW. Synthesis and characterization of antimicrobial crosslinked carboxymethyl chitosan nanoparticles loaded with silver. International Journal of Biological Macromolecules. 2014;**69**:95-99

[59] Wang X, Du Y, Fan L, Liu H, Hu Y. Chitosan-metal complexes as antimicrobial agent: Synthesis, characterization and structure-activity study. Polymer Bulletin [Internet]. 2005;**55**(1-2):105-113. Available from: http://link.springer.com/10.1007/s00289-005-0414-1 [accessed: January 31, 2017]

[60] El-Shafei A, Abou-Okeil A. ZnO/carboxymethyl chitosan bionano-composite to impart antibacterial and UV protection for cotton fabric. Carbohydrate Polymers. 2011;**83**(2):920-925

[61] Kaur P, Thakur R, Barnela M, Chopra M, Manuja A, Chaudhury A. Synthesis, characterization and *in vitro* evaluation of cytotoxicity and antimicrobial activity of chitosan-metal nanocomposites. Journal of Chemical Technology and Biotechnology [Internet]. 2015;**90**(5):867-873. Available from: http://doi.wiley.com/10.1002/jctb.4383 [accessed: January 31, 2017]

[62] Ribič H, Kramar Z. Preprečevanje okužb, povezanih z zdravstvom: skripta za študijski program Zdravstvena nega (VS). Jesenice, Slovenia: Fakulteta za zdravstvo Jesenice; 2016. 115 p.

[63] Kotnik Kevorkijan B. Enostavno širjenje bolezni. In: Južnič Sotlar M, editor. Bolnišnične okužbe: klostridijska driska. Ljubljana, Slobenia: Studio Moderna storitve d.o.o.; 2013. pp. 10-13

[64] Wilks SA, Michels H, Keevil CW. The survival of Escherichia coli O157 on a range of metal surfaces. The International Journal of Food Microbiology [Internet]. 2005;**105**(3):445-454. Available from: http://www.ncbi.nlm.nih.gov/pubmed/16253366 [accessed: December 19, 2016]

[65] Verran J, Redfern J, Smith LA, Whitehead KA. A critical evaluation of sampling methods used for assessing microorganisms on surfaces. Food and Bioproducts Processing [Internet]. 2010;**88**(4):335-340. Available from: http://linkinghub.elsevier.com/retrieve/pii/S0960308510001021 [accessed: December 19, 2016]

[66] Gubina M, Ihan A. Medicinska bakteriologija z imunologijo in mikologijo. Ljubljana: Medicinski razgledi; 2002. 543 p.

[67] Fijan S, Koren S, Cencič A, Šostar-Turk S. Antimicrobial disinfection effect of a laundering procedure for hospital textiles against various indicator bacteria and fungi using different substrates for simulating human excrements. Diagnostic Microbiology and Infectious Disease [Internet]. 2007;**57**(3):251-257. Available from: http://www.ncbi.nlm.nih.gov/pubmed/17046191 [accessed: December 19, 2016]

[68] Neely AN. A survey of gram-negative bacteria survival on hospital fabrics and plastics. The Journal of Burn Care & Rehabilitation [Internet]. 2000;**21**(6):523-527. Available from: http://www.ncbi.nlm.nih.gov/pubmed/11194806 [accessed: October 7, 2016]

[69] Borkow G, Gabbay J. Biocidal textiles can help fight nosocomial infections. Medical Hypotheses [Internet]. 2008;**70**(5):990-994. Available from: http://www.ncbi.nlm.nih.gov/pubmed/17959322 [accessed: October 7, 2016]

[70] Brunton WAT. Infection and hospital laundry. Lancet. 1995;**345**(8964):1574-1575

[71] Wilcox MH, Jones BL. Enterococci and hospital laundry. Lancet [Internet]. 1995; **345**(8949):594. Available from: http://www.ncbi.nlm.nih.gov/pubmed/7776811 [accessed: December 19, 2016]

[72] Barrie D, Hoffman PN, Wilson JA, Kramer JM. Contamination of hospital linen by Bacillus cereus. Epidemiology and Infection [Internet]. 1994;**113**(2):297-306. Available from: http://www.ncbi.nlm.nih.gov/pubmed/7925667 [accessed: December 19, 2016]

[73] Gonzaga AJ, Mortimer EA, Wolinsky E, Rammelkamp CH, Wolinsky E, Mortimer EA, et al. Transmission of Staphylococci by fomites. The Journal of the American Medical Association [Internet]. 1964;**189**(10):711-715. Available from: http://jama.ama-assn.org/cgi/doi/10.1001/jama.1964.03070100005001 [accessed: December 19, 2016]

[74] Kirby WMM, Corpron DO, Tanner DC, Finland M, Kenney M, Schneierson SS, et al. Urinary tract infections caused by antibiotic-resistant coliform bacilli. The Journal of the American Medical Association [Internet]. 1956;**162**(1):1. Available from: http://jama.jamanetwork.com/article.aspx?doi=10.1001/jama.1956.02970180003001 [accessed: December 19, 2016]

[75] European Centre for Disease Prevention and Control. Antimicrobial resistance surveillance in Europe 2010. Surveillance Report; 2012. p. 202

[76] Agency USEP. National Pesticide Information Retrieval System (NPIRS) [Internet]. 1998. Available from: https://cfpub.epa.gov/si/si_public_record_Report.cfm?dirEntryID=2893 [accessed: January 27, 2017]

[77] BI. Biocide Information [Internet]. 2004. Available from: http://www.biocide-information.com/ [accessed: February 13, 2017]

[78] National Industrial Chemicals Notification and Assessment Scheme. Chemical in Sanitized T99-19. [Internet]. 2007. Available from: https://www.nicnas.gov.au/ [accessed: February 13, 2017]

[79] McMahon T, Morrow M, Isbell D, Lantz T, Campbell McFarland J, Shamim N, et al. Reregistration Eligibility Decision for Trimethoxysilyl Quaternary Ammonium Chloride Compounds. Washington: United States Environmental protection Agency; 2007. 1-94 p.

[80] Wijnhoven S, Peijnenburg W, Herberts C, Hagens W, Oomen A, Heugens E. Nanosilver—A review of available data and knowledge gaps in human and environmental risk assessment. Nanotoxicology. 2009;**3**:109-138

[81] Dhende V, Hardin I, Locklin J. Durable antimicrobial textiles: Types, finishes and applications. In: Annis P, editor. Understanding and Improving the Durability of Textiles. Oxford, Cambridge, Philadelphia: Woodhead Publishing Limited; 2012. pp. 145-173

[82] Russel A. Biocide usage and antibiotic resistance: The relevance of laboratory findings to clinical and environmental situations. Lancet Infectious Diseases. 2003;**3**:794-803

[83] Yazdankhah SP, Scheie AA, Høiby EA, Lunestad B-T, Heir E, Fotland TØ, et al. Triclosan and antimicrobial resistance in bacteria: An overview. Microbial Drug Resistance [Internet]. 2006;**12**(2):83-90. Available from: http://www.ncbi.nlm.nih.gov/pubmed/16922622 [accessed: January 27, 2017]

Chapter 10

Functionalized Polypropylene Filaments for Flammability

Nazan Avcioğlu Kalebek

Additional information is available at the end of the chapter

http://dx.doi.org/10.5772/intechopen.69461

Abstract

A large number of textiles and indoor decorations including carpets, curtains, wallpaper, and so on are all easy to burn. Textile materials that are readily combustible can serve as one of the ingredients in a fire and pose a serious threat to human life and property in fire accidents. The demands for flame-retardant textile fabrics have a steady growth for the past decades and have become an urgent exigency. In this research, two new eco-friendly boron chemicals, boric acid, and borax decahydrate, were used as flame-retardant. The flame-retarding functions for bulked continuous filaments polypropylene (BCF PP) yarn are investigated through the applications of boron chemicals. It is determined that applying boron chemicals improves fire retardance.

Keywords: bulked continuous filaments (BCF), polypropylene (PP), boric acid, borax decahydrate, flame-retardant, flammability

1. Introduction

Fibers are the basic elements of a textile. They can be classified as belonging to one of the following two classes: (a) natural and (b) synthetic. Manufacturing of synthetic fibers was started in 1930. Even now, synthetic fibers are widely used in many applications. Annual consumption of synthetic fibers is increasing day by day because of using a large number of industrial, technical, and engineering applications. Total world output of man-made textile materials has sharply increased, whereas natural textile materials have reduced.

Synthetic fibers are manufactured with organic compounds which are obtained from coal, oil, and petroleum. These mentioned raw materials are abandoned in nature in large quantities. Advantages of synthetic fibers are as follows:

- manufacturing of synthetic fibers is produced in any climate zones and at any time of the year,
- less capital investment and labor,
- cost of synthetic fibers is higher than natural fibers,
- synthetic fibers may be used for many applications,
- can be manufactured with desired requirements, and
- better physical and mechanical properties [1–6].

Polyolefin fibers have polymer chains with higher molar mass, saturated, aliphatic, and hydrocarbons. The Federal Trade Commission (FTC) in the USA approves two names "olefin fibers" and "polyolefin fibers" in order to describe manufactured fibers. Synthetic polymer is composed of approximately 85% by mass polyolefin units. PP is the major commercial man-made fiber. The commercial advantages are summarized as follows:

- low densities (lighter than water),
- high cover power comparison with other textile fibers,
- simplest production and mostly used production method,
- high performance/cost ratio,
- low melting point,
- comfortable to skin,
- good tensile properties,
- good abrasion resistance [7].

PP is the first synthetic stereo regular polymer to achieve industrial importance. BCF is widely used in machine-woven carpet manufacturing industry such as pile yarns, apparel, upholstery, floor coverings, hygiene medical, geotextiles, car industry, automotive textiles, various home textiles, wall covering, and so on. This is expected that this trend will continue in the near future [8–10].

BCF yarn production is a continuous process, consisting melt spinning, drawing, and texturing (**Figure 1**). After extrusion from the spinneret, the molten polymer starts solidifying. Finishing oil is applied to the solidified multifilament yarn, and the yarn is subsequently stretched by about three times in the drawing zones, so that it acquires enough orientation and consequently proper mechanical properties. Texturizing of the continuous filament yarn emerging from the drawing zone follows next. This process imparts volume and softness to the yarn. After leaving the texturizer, the plug rests on a rotating cooling drum, where the process of setting the rather three-dimensional coils formed as a result of the compression in the plug gets completed. Parameters such as air pressure, air temperature, and cooling air play an important role in specifying the final properties of textured yarns [11].

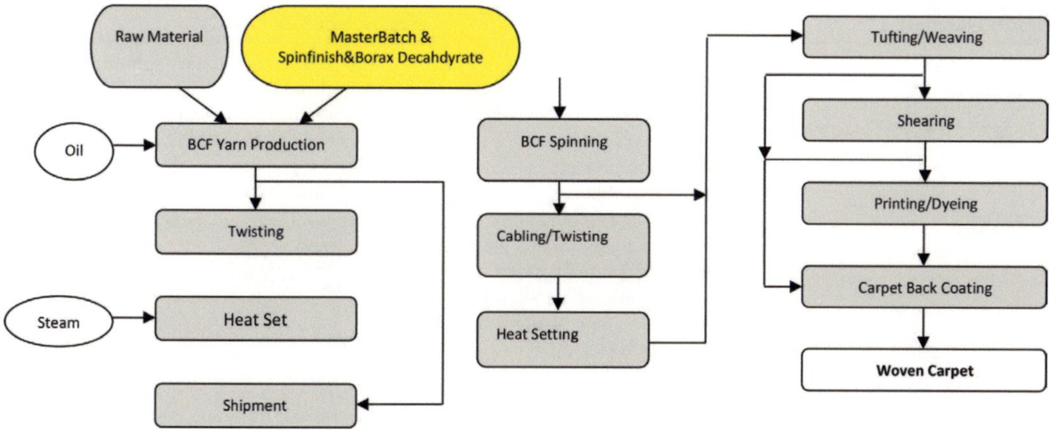

Figure 1. BCF carpet production process [11].

1.1. Flame-retardant

Flame-retardant (FR) fabrics are among few most profitable niche markets in the textile industry. Most of the textile materials are flammable, therefore applying flame-retardants to textile fabrics becomes necessary to assure humans safety under many circumstances. Most recently established federal regulations on the flammability of the fabric indicate that FR textiles will steadily increase in the near future. The increasing use of FR materials in industry has put lots of thrust on the scientific community to develop new polymer materials, FR chemicals, and fiber combinations to a wide range of end-use applications [12–14].

From 1980 to about 2000, very little new research was made about flame-retardant, which were shown chemical, physical, and performance properties and environmental problems (**Table 1**). The major development during these periods about flame-retardant began in 1980, and also amended in 1983 which required cigarette ignition resistance of outer cover fabrics. These were superseded in 1988 by the requirement that all UK domestic furnishing fabrics must have both cigarette and simulated match resistance when tested over unmodified polyurethane foam. These regulations promoted the development of flame-retardant back-coating which could be applied to any fabric thereby enabling many rich upholstered fabric designs to be rendered flame-retardant in spite of the fibers used [15].

Phosphorus compounds having flame-retardant properties in polypropylene may be both inorganic and organic as shown below. It is also common for them to be used in the presence of halogen- or nitrogen-containing compounds and especially those that generate intumescent char-forming characteristics. The range of phosphorus-containing flame-retardants is extremely wide since the element exists in several oxidation states. Phosphines, phosphine oxides, phosphonium compounds, phosphonates, elemental red phosphorus, phosphites, and phosphate are all used as flame-retardants in polymers [17].

Textile materials	Generic formulas	Comments
Cellulose (cotton, viscose) nondurable	Ammonium phosphates and mixtures with other salts, for example, ammonium bromide and sulphamate, ammonium polyphosphates, diammonium phosphate, dimethylphosphono (N-methylol) propionamide (Pyrovatex CP), guanidine phosphates, organophosphorus oxyanion salt, organic nitrogen-containing compound, organic N-and P-containing compounds, borax and combinations of the above	Available as proprietary formulations; semi-durability may be developed by postcuring to achieve-cellulose phosphorylation; addition of resins. dimethylphosphono (N-methylol) propionamide (Pyrovatex CP) bonds to cotton fibers covalently with high durability
Cotton-durable	Tetrakis (hydroxymethyl) phosphonium salt (THPX), tetrakis (hydroxymethyl) phosphonium salts-urea-ammonia (Proban CC), adduct condensates N-methylol and N,N′-dimethylol dialkyl phosphono-propionamides and derivatives	Typified by the ammonia-cured Proban® (Ciba/Huntsman) its variations and equivalents. Requires presence of cross-linking resins
Wool-durable to dry cleaning	Zirconium and titanium hexafluoride complexes tetrabromophthalic anhydride (TBPA)	Typified by Zirpro® may be used alone or with Zirpro® to reduce after flame times
Polyester-durable	Cyclic organophosphonate	Based on Antiblaze 19 (formerly mobil) and subsequently Antiblaze or Amgard CU (Rhodia) applied by thermofixation
Polyamide-durable	N- and S-containing polycondensates typically based on thiourea derivative-formaldehyde formulations	Applied in technical textiles
Acrclics	Back-coatings based on halogen-ATO formulations	The only commercial alternatives to using modacrylic fibers, for example, Kanecaron
Synthetic fibers-PP and PE	Phosphinic acid derivatives (Terevira CS), bisphenol-S oligomer (Toyobo GH) and melamine phosphate pentaerythritol	It can be applied on synthetic textiles as additives during polymerization and/or fiber extrusion
Synthetic fibers-PA, polyolefin, and nylon	Decabromodiphenyloxide, hexabromocyclododecane	It can be applied by coating. This shield acts as a barrier for heat, oxygen and mass transfer

Table 1. Examples of FR used in textiles [15, 16].

Only few of these chemicals are applicable to polyolefin especially polypropylene which are as follows:

- halogen phosphate,
- phosphine oxides and ammonium polyphosphates,
- phosphite,

- phosphonitrilic ester,
- phosphonopropionic acid amines,
- alkyl-aryl phosphonates,
- cyclic phosphonate esters,
- phosphonate isocyanurate,
- phosphoric acid,
- ammonium phosphate and tetrakis hydroxymethyl phosphonium hydroxide,
- Inorganic phosphoric acids such as ortho, meta, or pyrophosphoric acids,
- phosphonate ester, and
- organic phosphinate.

The reason for rapid expansion of PP as a fiber is easy processing, low production cost, and excellent properties [12]. However, PP is easily flammable and burns in air with a very hot and clean flame without giving any char residue. Its inherent flammability restricts its usage in many fields. For this reason, it is necessary to make PP flame-retardant, wherever required. Different methods can be used to obtain flame-retardant [13]. Additive is widely used for PP. In this approach, boron is used for flame-retardant.

The main purpose of present work is to investigate the combustion characteristics of flame-retardant borax decahydrate chemical compositions.

1.1.1. Boric acid

Boric acid (commercially known as Optibor) is a white triclinic crystal in water (5.46 wt%), alcohols, and glycerin [18]. It has the chemical formula H_3BO_3 (sometimes written as $B(OH)_3$), and exists in the form of colorless crystals or a white powder that dissolves in water. The three oxygen atoms form a trigonal planar geometry around the boron. The B—O bond length is 136 pm and the O—H is 97 pm. The molecular point group is C_3H. Crystalline boric acid consists of layers of $B(OH)_3$ molecules held together by hydrogen bonds of length 272 pm. The distance between two adjacent layers is 318 pm [19]. It is a weak acid and has a pH of 4 (saturated at room temperature). It is derived from boron. Upon heating in air to above 75°C, it loses part of its water.

Boric acid is found natively in its free state in some volcanic districts, for example, in the Italian region of Tuscany, the Lipari Islands, the US state of Nevada and in Turkey. In these volcanic settings, it issues, mixed with steam, from fissures in the ground. It is also found as a constituent of many naturally occurring minerals—borax, boracite, ulexite (boronatrocalcite), and colemanite. Boric acid and its salts are found in seawater. It is also found in plants, including almost all fruits.

Boric acid was first prepared by Wilhelm Homberg (1652–1715) from borax, by the action of mineral acids, and was given the name *sal sedativum Hombergi* ("sedative salt of Homberg").

However, borates, including boric acid, have been used since the time of the Greeks for cleaning, preserving food, and fire retardants in wood/cellulosic such as timber, plywood, particle board, wood fiber, paper products, textiles, and especially cotton products. Boric acid has also been used as fire retardant in epoxy intumescent coating, foam, and so on. When necessary, boric acid can be coated with silicone oil to alleviate its water solubility in water-based coating [18].

1.1.2. Borax decahydrate (disodium tetraborate decahydrate)

Boron is widely distributed on the earth, mainly in the form of boric acid or borate salts. World Health Organization (WHO) is reported that no any harmful effect is obtained [18]. It exhibits excellent buffering property. As a crystalline material, borax decahdyrate is stable under normal storage conditions. It will slowly lose water of crystallization if exposed to a warm and dry atmosphere. Conversely, exposure to a humid atmosphere can cause recrystallization at particle contact point, thus resulting in cracking. A mixture of ammonium chloride and borax was one of the treatments of cellulosic fabrics reported by Gay-Lussac in 1821. Due to its low dehydration, temperature, and waste solubility, sodium borates are only used as flame-retardants in cellulose insulation, wood timber, textiles, urethane foam, and coatings [18]. Borax decahydrate is environmentally friendly as flame-retardant for textile materials.

2. Material and methods

2.1. Material

For this research, 10 bobbins of BCF polypropylene yarns (1300, 1600, 1900, and 2500 dtex, 40 filaments) are produced on an industrial production line (**Table 2**). It is produced by changing only flow rate of pump. The natural color with trilobal filaments are produced by gravimetric feeding of 97% polypropylene granulates (MFI = 22) and 3.1% master batch to a Neugmac BCF machine (Neugmac S+) (**Figures 2–4**).

The temperature profile of the six heating zones of extruder is 180, 190, 220, 225, 227, and 229°C, respectively. Extruder pressure is 132 bar, extruder revolutaions is 54 rpm and extruder load is 23 kW. Pressure of spin pump before spinning is 45 bar and after spinning is 35 bar. Draw ratio in the drawing zone is 1:2.84. The temperature and pressure of hot air feeding, the

Melt flow index (MFI at 228°C)	21
Melting point (°C)	154
Polymer density (kg/L)	0.75
Heat conduction coefficient (J/ms °C)	0.18
Specific heat (kJ/kg °C)	2.34

Table 2. Properties of polypropylene granulates.

Figure 2. General view of Neugmac BCF machine.

Figure 3. Front side of machine.

Figure 4. Rear side of machine.

yarn to the texturizer are 138°C and 8.0 bar, respectively. Cooling air temperature, humidity, and velocity is 22°C, 72% rH, 0.9 m/s, respectively. Yarn is fed to the texturizer at a speed of 1100 m/min and is collected on cylindrical bobbins (1.5 kg) without intermingling. All yarns are twisted (110 turns/m) by a Volkman (Model: 0-5-0-C 8.02) two for one twister. Winding speed was 53.1 m/min. After twisting, the heat setting of the samples is carried out using the Rieter heat-setting machine (Model: Power Heat-Set GKK2500). Forty-five heat-set BCF polypropylene yarns were produced four levels of setting temperatures (100, 120, 130, and 140°C).

Boric acid and borax decahydrate were purchased from MERCK Chemicals (Gaziantep/TURKEY). Borax decahydrate (also called borax) is slightly soluble in cold water (4.71–12.83% by wt in 20°C), and highly soluble in hot water (30% at 60°C). Boric acid is soluble in water (5.25–15.9% by wt in 23°C). These are added to BCF PP yarns during the spin finish operations at the same time of master batch because of decomposition temperature. The amount of applied boric acid and borax decahydrate was prepared according to chemical properties, which were written on the information chart of chemicals. The ratio is chosen 9.65 for borax decahydrate and 8.37 for boric acid. First, borax decahydrate (9.65 g at 20°C) and boric acid (8.37 g at 23°C) is resolved in 500 ml warm water at 25°C and 500 ml spin finish oil at 20°C separately (**Figure 5**). Two different solutions were prepared, one liquid solution contains boric

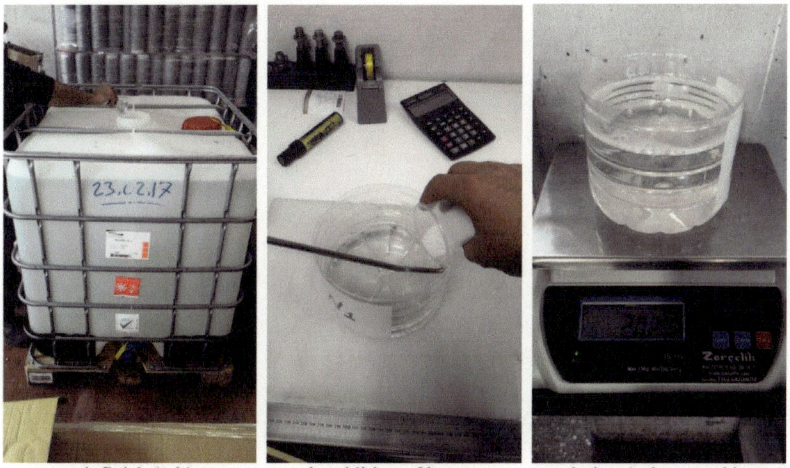

Figure 5. Preparation of solution. (a) spinfinish (1 L), (b) addition of boron and (c) solution (solvent and boron).

acid, water, and spin finish, one liquid solution contains borax decahydrate, water, and spin finish. Technical properties of boric acid and borax decahydrate are given in **Tables 3** and **4**.

2.2. Method

Flammability test express as percent volume that supports flaming combustion in flowing mixture of oxygen and nitrogen (**Figure 6**). In this test, minimum value of the oxygen index is obtained under controlled conditions response of heat and flame. Oxygen index is the mini-

Component	Content
B_2O_3	36.47% min
$Na_2B_4O_710H_2O$	99.90% min
Na_2O	16.24% min
SO_4	200 ppm.max.
Cl	70 ppm.max.
Fe	15 ppm.max.
Vapor pressure	<0.0000001 hPa in 25°C
pH	4 in 20°C
Density	1.489 g/cm^3
Distrubition coefficient	−1.09 in 22°C
Decomposition temperature	70°C
Shape	Powder
Particle size	−0.0063 mm

Table 3. Technical properties of borax decahydrate.

Component	Content
B_2O_3	56.50% min
Equivalent H_3BO_3	99.93% min
SO_4	300 ppm.max.
Cl	5 ppm.max.
Fe	5 ppm.max.
Vapor pressure	<0.0000001 hPa in 25°C
pH	4 in 23°C
Density	1.376 g/cm^3
Distribution coefficient	−1.18 in 20°C
Decomposition temperature	70°C
Shape	Granules
Particle size	−0.0128 mm

Table 4. Technical properties of boric acid.

mum concentration of oxygen that will just support flaming combustion of a material initially at 23 ± 2°C under the test conditions. However, these conditions do not hold true under actual fire conditions for fire hazard or fire risk materials. It is responsibility of the user of this standard to establish appreciate safety and health practices and determine the applicability of regularity limitations prior to use [20].

A small test specimen is supported vertically in a mixture of oxygen and nitrogen flowing upward through a transparent chimney. The upper end of the specimen is ignited and

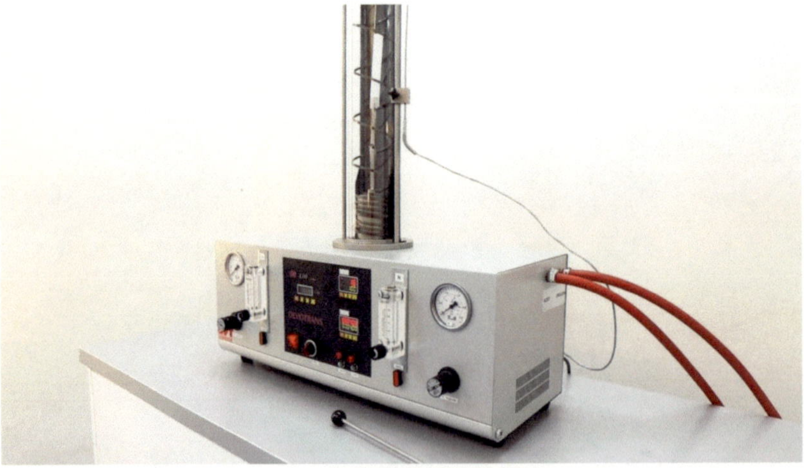

Figure 6. Devotrans LOI.

subsequent burning behavior of the specimen is observed to compare the period for which burning continues or the length of specimen burn with specified limits. Limited oxygen index is obtained by conducting using top surface ignition or propagation ignition. The test chimney consists of a heat resistant glass tube of 75–100 mm inside diameter and 450–500 mm height. The opening at the top of the chimney shall be restricted to provide an outlet of 40 ± 2 mm diameter. The bottom of the chimney, or the base to which the tube is attached, shall contain noncombustible material to evenly mix and distribute the gas mixture entering at the base. Glass beads 3–5 mm in diameter in a bed 80–100 mm deep have found suitable. The chimney shall be mounted securely on the base to prevent air leaks. Specimen holder is small holding device that will support the specimen at its base and hold it vertically in the center of the chimney. Gas measurement and control devices shall be suitable for measuring the concentration of oxygen in the gas mixture entering the chimney with an accuracy of ±0.5%, by volume, of the gas mixture and for adjusting the concentration of oxygen in the precision of ±0.1%, by volume, of the gas mixture, when the gas velocity through the chimney is 40 ± 2 mm/s at 23 ± 2°C. The system for gas measurement and control involves needle valves on individual and mixed gas supply lines, a paramagnetic oxygen analyzer that continuously samples the mixed gas, and a flow meter to indicate when the gas flow through the chimney is within the required limits [20].

The specimen was prepared by hand in 100 mm length, 10 ± 0.5 mm width, and 2 ± 0.25 mm thickness. Surface of the specimens shall be cleaned and free from flaws that could affect burning behavior, for example, peripheral molding flash or burrs from machining. The edges of the specimens shall be smooth and free from fuzz or burns of material left from scissors. Specimen was mounted vertically in the center of the chimney so that the top of the specimen is at least mm below the open top of the chimney and lowest exposed part of the specimen is at least 100 mm above the top of the gas distribution device at the base of the chimney. Gas mixing was set to flow controls so that an oxygen/nitrogen mixture. The gas flow purge is let for at least 30 s prior to ignition of each specimen, and the flow was maintained without change during ignition and combustion of each specimen. At the end of the test, burning behavior was noted as dripping, charring, erratic, burning, and glowing combustion or after-glow [20].

The limited oxygen index (LOI) was applied to investigate the ignition and ease of extinction of a yarn. LOI was performed according to the international standard ASTM D2863.

In addition to flammability test, mechanical properties of 100% PP carpet yarns have been tested according to ISO and ASTM Standards to finished and unfinished filaments. Yarn linear density is measured ISO 2060: 1994 Textiles; yarn from packages—determination of linear density (mass per unit length) by the skein method. Yarns are randomly selected from the BCF specimens. Test results are reported and tex are calculated. Yarn strength and elongation are measured according to ISO 2062:1993 standards titled textiles; yarns from packages—determination of single-end breaking force and elongation at break. Crimp contraction and shrinkage are measured textured yarn tester (TYT) by Lawson Hemphill by ASTM D-6774-02; crimp and shrinkage properties for textured yarns use a dynamic textured. In addition, longitudinal views of the samples were taken by scanning electron microscope (SEM) (**Figure 7**) and examined. For this examination, JEOL JSM 6390LV SEM is used in the University of Gaziantep.

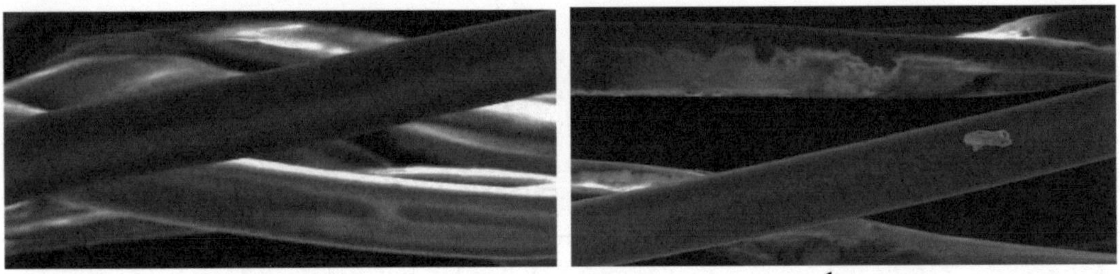

Figure 7. SEM view (a) without flammability chemical and (b) with flammability chemical.

Before examination, samples are coated with Au-Pd. It must be reflective material for surface display. The thickness of this reflective material must be at least 100 Å. First samples are located into two-sided adhesive on metal stabs tape. The coating is placed on the spray coating device. The coating takes place in 90 s with 5 mA, 8 Pa.

3. Experimental results

3.1. Flammability (LOI)

In **Figure 8** shows that limited oxygen index (LOI) versus four different yarn number (1300, 1600, 1900, and 2500 dtex) of BCF PP carpet yarn. When the figures are examined, it can be observed that LOI value of BCF PP yarns increase from 5.20 to 9.04 for non-boric acid and borax decahydrate, from 10.5 to 21.6 for boric acid and from 17.2 to 28.3 for borax decahydrate.

From the LOI data, LOI of BCF PP without boron is approximately 6.95 and increases with the increasing boric acid and borax decahydrate. It also shows that borax decahydrate has maximum beneficial effect in terms of flammability properties and highest fire performance.

3.2. Mechanical properties

Strength and Elongation: mechanical properties of fibers are probably their most important properties for end products. The purpose of tensile test is to estimate how strength pile yarns for carpet. Yarn properties and to estimate their suitability in terms of strength for pile to estimate how strength.

Crimp: crimp is a physical change in the structure of the individual filament. Crimp is a relatively unstable mechanical structure of individual filaments, which together with other filaments have a volume. Changes in crimp result in variations of the yarn's geometry (bulk). The crimp structure has a decisive influence on the appearance of the yarn in the carpet. The introduction of yarns with different crimp levels into the same fabric will change the reflectance of light off the surface of the fabric, producing dye shade differences. The shrinkage level influences the dye uptake or the dispersion of dyestuff particles into the molecular spacing, causing an increase or decrease of adhesion of the dyestuff particles into noncrystalline regions.

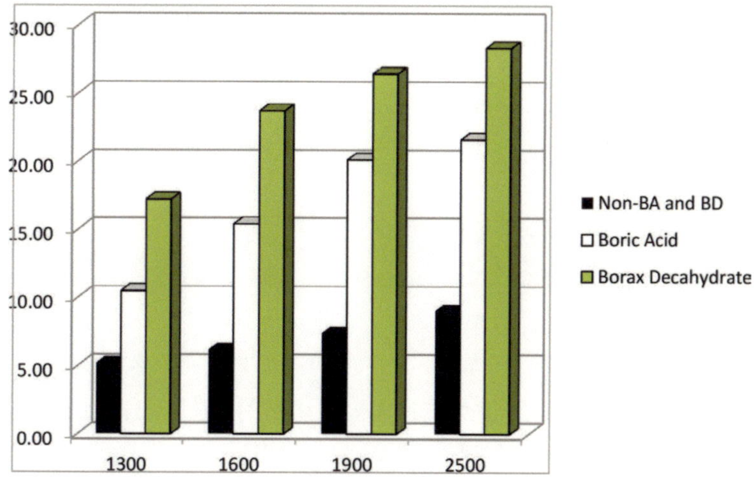

Figure 8. Limited oxygen index.

Shrinkage: nonheatsetting and/or nonuniform heatsetting can induce disorientation in the fiber morphology and is responsible for the fiber length change, leading to problems in applications, such as streaks in finished carpets. Therefore, shrinkage is a useful property for determining dimensional stability. The amount of fiber shrinkage will affect the processing behavior of yarns and dimensional changes in the fabrics made from these yarns. Differences in shrinkage behavior of threads may lead to cockled appearance in the fabric after finishing of carpets.

Yarn linear density: according to yarn count system yarn linear density is shown like that dtex = x g/10,000 m. dtex is depend on metric system.

In **Table 5**, the mechanical test results which performed according to standard is shown. As can be seen in the table, the mechanical properties of yarns cannot be affected with chemical additions.

		Without boron	Boric acid	Borax decahydrate
1200 dtex	Strength (cN/tex)	2.3	2.2	2.3
	Elongation (%)	41	47	46
	Crimp (%)	15.4	14.5	14.7
	Shrinkage (%)	2.6	2.8	2.6
	Yarn number (dtex)	1142	1136	1137
1600 dtex	Strength (cN/tex)	2.3	2.3	2.2
	Elongation (%)	47	47	42
	Crimp (%)	17	16.1	17.2
	Shrinkage (%)	2.9	2.6	2.8
	Yarn number (dtex)	1591	1611	1595

		Without boron	Boric acid	Borax decahydrate
1900 dtex	Strength (cN/tex)	2.6	2.5	2.9
	Elongation (%)	51	50	53
	Crimp (%)	18.6	17.9	16.8
	Shrinkage (%)	3.2	3.5	3.0
	Yarn number (dtex)	1984	1916	1910
2500 dtex	Strength (cN/tex)	3.8	3.9	3.7
	Elongation (%)	55	54	52
	Crimp (%)	20.0	21.5	22.6
	Shrinkage (%)	4.5	4.2	4.1
	Yarn number (dtex)	2493	2498	2517

Table 5. Mechanical test results.

4. Statistical analysis (ANOVA)

The experimental results have been statistically evaluated by using the design expert Analysis of Variance (ANOVA) software with *F*-values of the significance level of $\alpha = 0.05$, with the intention of exploring whether there is any statistically significant difference between the variations obtained. We evaluated the results based on the *F*-ratio and probability of *F*-ratio (prob >*F*). The lower the probability of *F*-ratio, it is stronger the contribution of the variation and the more significant the variable. The best models for each fabric were obtained and the corresponding regression equations and regression curves were fitted. The test results of the related samples were entered into the software for the analysis of the general design [11].

Table 6 summarizes the statistical significance analysis for all data obtained in this study. In the table, parameters yarn number (dtex) and flame-retardant (boric acid and borax decahydrate). **Table 6** also indicates that significant interactions between yarn number and fire retardant. The term *A* (yarn number-dtex) and *B* (fire retardant-boric acid and borax decahydrate) in this table is independent variables, where as the LOI value is dependent variable. And also, the linear design is suggested by the software for flammability behavior of yarns.

While ANOVA table is checked, yarn number and fire retardant have significant impact on LOI values. In addition, according to the table, the R^2 values of model come to be approximately 0.96. At this stage, the model accounts for this circumstances in the model at 96% ratio. This relation displaced that the model devised for response value can express with rather high accuracy the relation between independent and dependent parameters and that experimental studies are admissible as precise.

A normality test was also applied on the inputs derived from LOI by changing yarn number (dtex) and fire retardant (boric acid and borax decahydrate). The results are demonstrated in **Figure 3**. In general, probability plotting is a graphical technique for determining whether

Source	Sum of squares	DF	F-Value	Prob >F	
Model	7020.39	5	637.87	<0.0001	Significant
A	1155.37	1	524.88	<0.0001	Significant
B	5704.39	2	1295.75	<0.0001	Significant
$A \times B$	160.62	2	36.49	<0.0001	Significant
Residual	250.94	114	2.20		
Lack of fit	246.93	6	41.15	1108.75	Significant
Pure error	4.01	108	0.037		
R^2	96.55				
R_P^2	96.40				
R_{adj}^2	96.20				

Notes: A: yarn number (dtex), B: fire retardant (boric acid and borax decahydrate), R_p^2: predicted R^2, R_{adj}: adjusted R^2. Here, the P values of models must be smaller than 0.005 if it is accepted as significant.

Table 6. ANOVA.

specimen data confirm to a hypothesized distribution based on a subjective visual examination of the data. The estimation is very easy. As can be seen in **Figure 9**, data is distributed around the normality axis. This figure confirms the selected model.

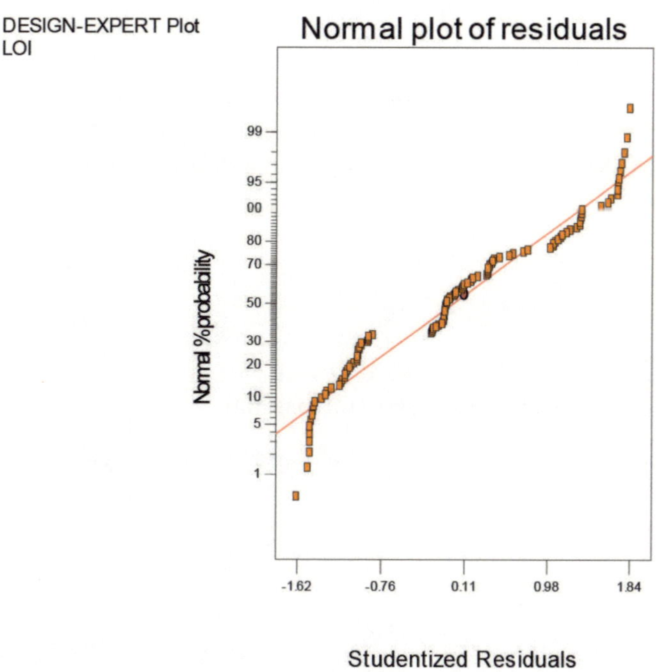

Figure 9. Normality test.

The regression equations of the model determined was found as:

$$LOI = +16.47 + 4.20A - 8.05B - 0.99AB \quad (1)$$

where A is the yarn number and B is the fire retardant in the equations.

5. Discussion and conclusion

One way to better protect combustible materials against initiating fires is the use of flame-retardants. Every day we come across new innovations, particularly, in the areas of flame-retardant fibers and chemicals. Boric acid and borax decahydrate are two new eco-friendly boron chemicals used for flame-retardant. Boron is highly resistant to fire due to the melting temperature of 2300°C. Because of this property, it is used as a flame-retardant material.

The research on the flammability behavior of samples can be summarized as follows:

- It can be inferred that borax decahydrate shows better flame-retarding effect rather than boric acid.
- Addition of flame-retardant changes the flammability behaviors of BCF PP yarns produced by conventional methods without flame-retardant.
- LOI value of the samples is also affected by the presence of additives.
- In the last few years, green or eco-friendly chemicals have been used because of increased human safety and the search for alternative environmentally materials. Efficient use of borax mines can contribute to the solution of waste accumulation problems of environment and flammability behavior by the aid of applying with adequate quantity.

Author details

Nazan Avcioğlu Kalebek

Address all correspondence to: nkalebek@gantep.edu.tr

Gaziantep University, Fine Art Faculty, Fashion and Textile Design Department, Gaziantep, Turkey

References

[1] Usenko V. Processing of man-made fibres. In: Chernyshova N, editor. Translated from the Russian. Moscow: Mir Publishing; 1979

[2] Corbman BP. Textiles fiber to fabric. 6th ed. McGraw Hill Inc.; 1983

[3] Elsasser VH. Textiles, 2nd ed. Fairchild Publication Inc.; 2007

[4] Wynne A. Textiles, Macmillan Education Ltd.; 1997

[5] Tortora PG. Collier BJ. Understanding textiles. 5th ed. Prentice Hall; 1997

[6] Klein W. Processing of man-made fibres. The Textile Institue; 1994

[7] Mclntyre JE. Synthetic fibres. Woodhead Publishing Limited; 2005

[8] Gupta VB. Kothari VK. Manufactured fibre technology. Chapman & Hall; 1997

[9] Fourne F. Synthetic fibers. Hanser Publishers; 1999

[10] Mostashari SM, Mostashari SZ. Verification concerning the relative superiority of a cellulosic fabric with regard to polyester in terms of flammability. Polymer-Plastics Technology and Engineering. 2008;**47**:313-317. DOI: 10.1080/03602550701870008

[11] Kalebek NA, Babaarslan O. Effect of Heat Setting Process of Polymers. In: Mendez-Vilas A, Salona A, editors. Polymer Science; Research Advances, Practical Applications and Educational Aspects. Spain: Formatex Research Center; 2016. pp. 199-207

[12] Shekar RI, Kotresh TM, Damodhara Rao PM, Kumar MN, Rahman S. Flammability behaviour of fiber-fiber hybrid fabrics and composites. Journal of Applied Polymer Science. 2011;**122**:2295-2301

[13] Shi Z, Fu R, Agathopoulos S, Gu X, Zhao W. Thermal conductivity and fire resistance of epoxy molding compounds filled with Si_3N_4 and $Al(OH)_3$. Materials and Design. 2012;**34**:820-824

[14] Yang CQ, He Q, Lyon RE, HuY. Investigation of the flammability of different textile fabrics using micro-scale combustion calorimeter. Polymer Degradation and Stability. 2010;**95**:108-115

[15] Horrocks AR. Flame retardant challenges for textiles and fibres; new chemistry versus innovatory solutions. Polymer Degradation and Stability. 2011;**96**:377-392

[16] Papaspyrides CD, Pavlidou S, Vouyiouka SN. Development of advanced textile materials: Natural fibre composites, Anti-microbial, and flame-retardant fabrics. Journal of Materials: Design and Applications. 2009;**223**:91-102

[17] Zhang S, Horrocks AR. A review of flame retardant polypropylene fibres. Progress in Polymer Science. 2003;**28**:1517-1538

[18] Charles AW, and Alexander BM. Fire retardancy of polymeric materials, CRC Press, Taylor Francis Group; 2010

[19] Web site. https://en.wikipedia.org/wiki/Boric_acid [Accessed: 12.02.2017]

[20] ASTM D2863-13, Standard Test Method for Measuring the Minimum Oxygen Concentration to Support Candle-Like Combustion of Plastic (Oxygen Index). West Conshohocken, PA: ASTM International;2013

Chapter 11

Textile Materials in Liquid Filtration Practices: Current Status and Perspectives in Water and Wastewater Treatment

Murat Eyvaz, Serkan Arslan, Ercan Gürbulak and Ebubekir Yüksel

Additional information is available at the end of the chapter

http://dx.doi.org/10.5772/intechopen.69462

Abstract

Filtration is considered the keystone of water and wastewater treatment and is used for various purposes, such as sludge dewatering and concentrating any solution. Moreover, as an advanced filtration technology, membranes can remove materials ranging from large visible particles to molecular and ionic chemical species. Proper selection of filter media/membrane material in filtration processes is often the most important consideration for assuring efficient separation. Filter media can be classified by their materials of construction, such as cotton, wool, linen, glass fiber, porous carbon, metals, and rayons. Recently, new polymeric materials have been used both individually and/or blended in filtration processes for the treatment of waters and wastewaters. The purpose of this chapter is to bring an overview on the textile-originated filter materials in filtration applications from conventional filtration to advanced membrane processes. Although many researches on filter media are available, very few researches have been carried out on the cutting-edge technologies about using filter materials on filtration processes from classical to advanced membrane processes. Therefore, in this part of the book, following major and minor titles are stated truly on the aforementioned new technologies and linked with conventional methods in water and wastewater treatment applications.

Keywords: filtration, membrane, textiles, woven, nonwoven, fibers, nanofibers, electrospinning, water treatment, wastewater treatment

1. Introduction

Physical, chemical, and biological methods are used in water treatment to convert raw water to potable water. The selection of the treatment method depends on the properties of the raw water. Water treatment methods can be based on simple physical processes, such as sedimentation processes or more complex physicochemical processes, such as coagulation. Among these purification methods, filtration is the process of removing solid substances from a fluid/liquid (water/wastewater) by passing them through a porous medium (filtration). Filtration is commonly used in water treatment to remove solids, including microorganisms (bacteria, viruses, etc.) and precipitated iron and manganese found in surface waters [1]. In addition to conventional filtration, direct filtration, which is a simple and economically attractive process in which the sedimentation phase is lifted, is often used. Direct filtration is suitable for raw waters with a turbidity value lower than 10 NTU. It does not require sedimentation tanks and, in some cases, floatation tanks, which leads to low installation and operating costs [2]. Filtration is a basic procedure widely used in environmental engineering applications for the removal of suspended solids, such as clay and silt particles, microorganisms, colloid and sediment humic substances, rotten plant particles, and calcium carbonate and magnesium hydroxide precipitates used in water softening [3]. Filtration is used in the treatment of drinking water, especially in high-quality surface water. For the treatment of wastewater, different kinds of filtration processes can be used at different stages of the process [4, 5]. In addition to solid-liquid separation, the filtration process is used for dewatering. A classification based on the operating mode of the filtration and the filtration unit used is shown in **Figure 1**.

Filters are selected in different qualities depending on the characteristics of the industries in which they are used. For example, the reverse osmosis process used to desalinate water and

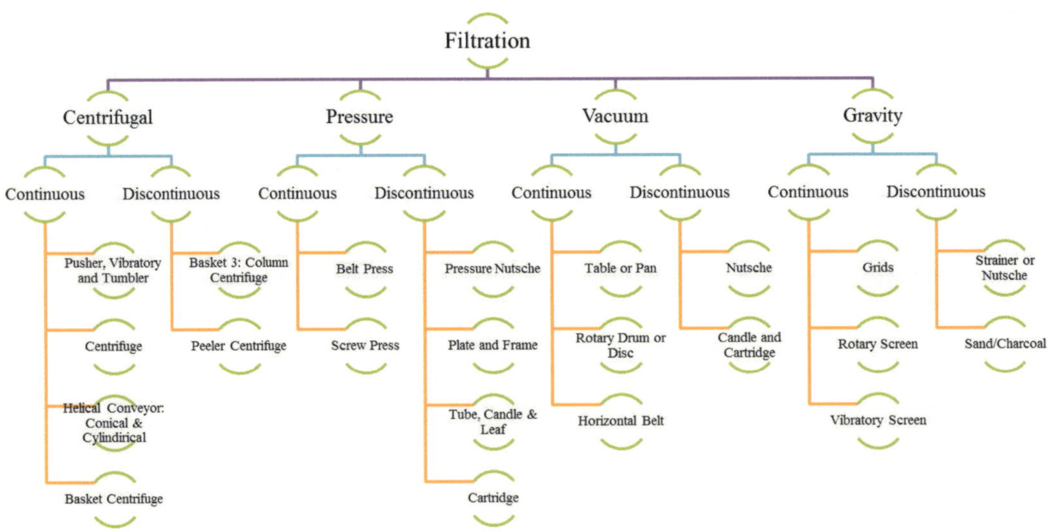

Figure 1. Operational modes of filtration (adopted from [6]).

the cellulose acetate and aramid hollow fiber membranes used are among the first applications of fiber-based filtration. The important properties of the fibers used in such a filtration are their hydrolytic nature, oxidative nature, and high biological resistance over a wide pH range. In addition, the fibers need to be resistant to temperature changes and chemicals used in different industries. Poly (phenylene sulfide), polysulfone, aramids, polyimide, PEEK (Victrex), fluorocarbon, and related fibers are examples of high-performance fibers effective for liquid filtration under extreme and rapid changing environmental conditions [7].

Proper selection of filter media/membrane material in filtration processes is often the most important consideration for assuring efficient separation. Filter media can be classified by their materials of construction, such as cotton, wool, linen, glass fiber, porous carbon, metals, and rayons. Recently, new polymeric materials have been used both individually and/or blended in filtration processes for the treatment of waters and wastewaters. The purpose of this chapter is to bring an overview on the textile-originated filter materials in filtration applications from conventional filtration to advanced membrane processes. Although many researches on filter media are available, very few researches have been carried out on the cutting-edge technologies about using filter materials on filtration processes from classical to advanced membrane processes. Therefore, in this part of the book, following major and minor titles are stated truly on the aforementioned new technologies and linked with conventional methods in water and wastewater treatment applications.

2. Principles of filtration

2.1. Filtration mechanisms

Filtration is based on the principle that water is passed through a porous medium at a certain rate. The filter does not allow particles to pass while allowing water to pass through. Particles (0.01–5 mm) retained in the filter medium are smaller than the size of the filter material (5–20 mm), and such retention takes the place in trapping in unfiltered extraneous materials, which strikes and adheres to the filter material due to the rapid flow of the fluid. Although the flow in the porous media was announced by Darcy in 1856, the first filters were designed by trial and error until the publication of the Carman-Cozeny equation (1937), which describes the flow in the filter bed. As research continues on the design of granule filters, the earlier empirical equations used in design are still widely used [8].

The trapping of particles in the filter bed consists of two phases, collision and attachment. In the first stage, the particles in the fluid approach the surface of the densities of the porous filter media by mechanisms, such as sedimentation, impaction, diffusion, and interception, and become accessible to hold. In the second step, the retention of particles in the filter depends on the balance between the superficial forces between the particles and the filter medium. These steps and the mechanisms that play a role in each step are explained elsewhere (**Figure 2**) [9, 10].

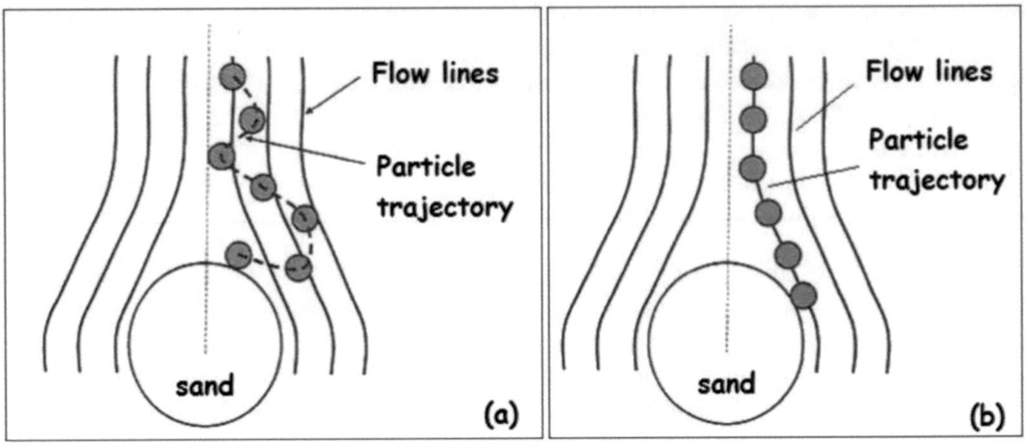

Figure 2. Filtration mechanisms in deep-bed filter, (a) diffusion, (b) interception [9, 10].

2.2. Types of filtration processes

2.2.1. Surface (membrane) filtration

Membrane filtration is the process of separating a material from a medium through which it can pass more easily than other materials in the same environment. In water and wastewater treatment, membranes are used to remove unwanted suspended or dissolved substances. However, in some cases, the membrane may move to remove contaminants from the wastewater or to transfer special components (such as oxygen) into the liquid medium. Extraction processes currently used include electrodialysis (ED), dialysis, pervaporation (PV), and gas transfer (GT). In such cases, the membrane is used to allow selective penetration of specific components dissolved in water. However, filtration processes such as reverse osmosis (RO), nanofiltration (NF), ultrafiltration (UF), and microfiltration (MF) are of much more industrial significance. In these processes, it is the bulk water that passes through the membrane under an applied pressure, leaving the pollutants in a concentrated form on the unpermeated side of the membrane [10, 11]. Membrane classification based on filtration is shown in **Figure 3**.

2.2.2. Depth filtration

Deep-bed filters are more than a century-old and are widely used in water and wastewater treatment applications. The theory of deep-bed filtration is the same as the formation of wells or springs in nature, which are drained from porous media, such as rocks and sand. In practice, deep-bed filters are obtained by placing the material to be filtered in a closed tank (pressurized) or in an open concrete pond. The filter tank is designed according to the properties of the liquid to be filtered or filtration method (pressure or gravity operation). Particles in the liquid are trapped in the filter bed by the abovementioned filtration mechanisms. During

http://dx.doi.org/10.5772/intechopen.69462

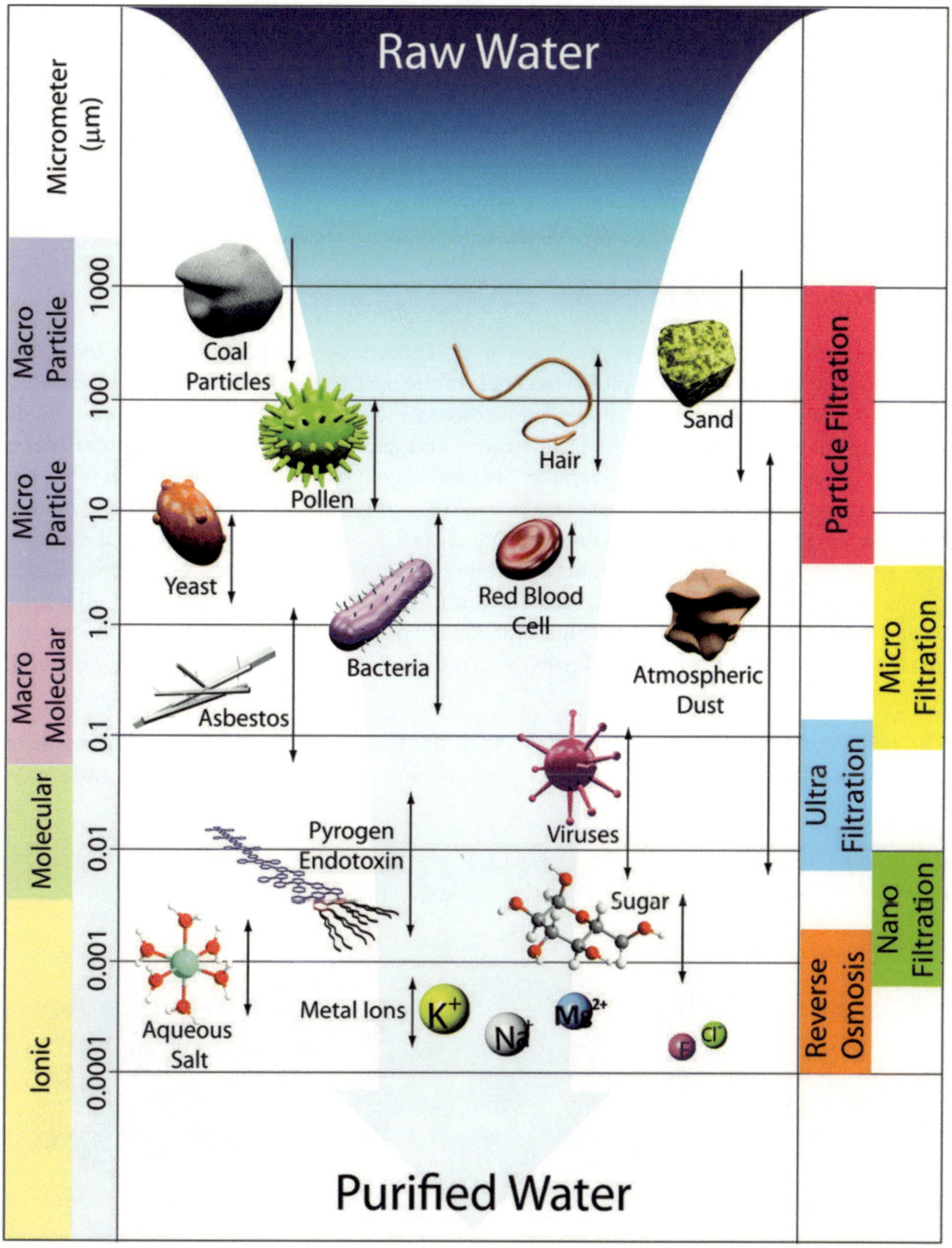

Figure 3. Schematic illustration of membrane filtration spectrum [11].

filtration, the particles that accumulate in the filter bed begin to clog the filter and reduce the fluid pressure. For this reason, solids are not required at high concentrations. Filtration is generally applied for concentrations lower than 0.5 g/L. Alternative or pretreatment systems should be considered as the solids concentration increases. Precoagulation and/or flocculation systems have to be applied before the filtration of submicron particles that are too small to be filtered or settled readily. The common types of deep-bed filtrations are slow sand filtration, rapid filtration, and direct filtration, which are explained in detail elsewhere [6]. Both surface and depth filtration are shown schematically in **Figure 4**.

2.3. Operating parameters of filtration

The filter medium is evaluated as the heart of any filtration process. Ideally, while the solids to be retained are concentrated on the feed side of the membrane, the liquid portion is forced to pass through the membrane and is transported to its other side. A filter medium, by its nature, is not homogeneous, and its dimensions and geometries come from irregular pores. These pores may also exhibit an irregular distribution over the membrane surface. Since the flow in the environment only takes place through the pores, the microfluidic velocity therein can cause large differences on the filter surface. This indicates that the top layers of the filter cake produced on the membrane surface are not homogeneous and are also formed based on the nature and properties of the filter medium. Since the number of passages in the filter cake is larger than the number in the filter media, the primary structure of the cake is strongly attached to the structure of the first layers. This means that the filter crayon and filter material are influenced by each other.

The pores containing passages extending along the filter medium can catch solid particles smaller than the narrowest cross section of the passageway. Such retention of the particles is

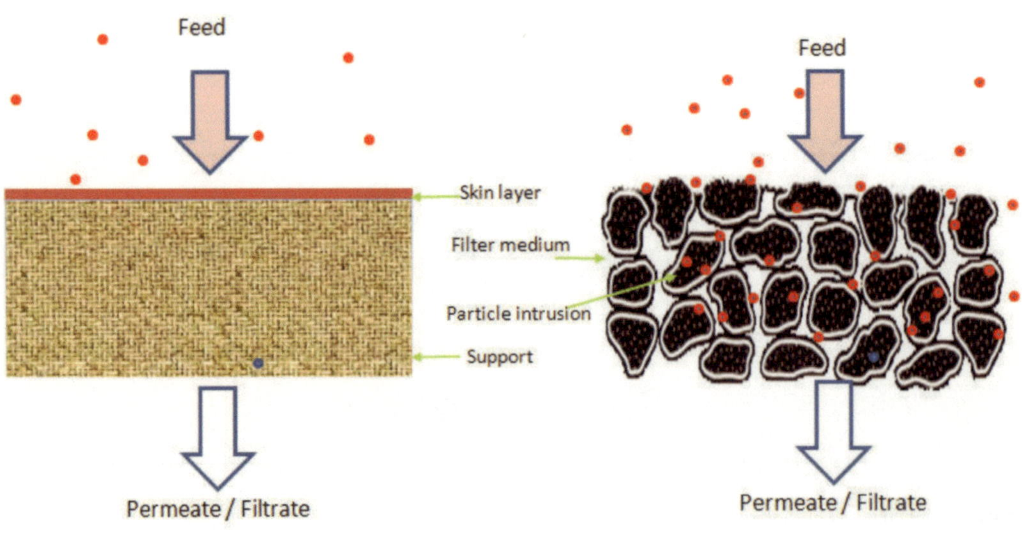

Figure 4. Schematic representation of surface filtration (on the left side) and depth filtration (on the right side) [12].

http://dx.doi.org/10.5772/intechopen.69462

generally explained by particle bridging or, in some cases, physical adsorption. Depending on the intended use, different filter media are used. The commonly used filter media are sand, diatomite, coal, cotton or wool fabrics, metal wire cloth, porous quartz plates, chamotte, sintered glass, metal dust, and powdered ebonite. The average pore size and configuration of the filter material (including tortuosity and connectivity) is due to the size and form of each element from which the material is produced. The manufacturing method of the filter material also influences the average pore size and shape. For example, pore characteristics vary when the fibrous media is first pressed. Pore properties also depend on the properties of the fibers in the woven fabric or on the methods of sintering glass and metal powders. In addition to all these, some filter media, especially fibrous layers, are subject to significant compression when subjected to typical pressures used in industrial filtration processes. Other filter materials, such as sintered plates of ceramics, glass, and metal powders, are stable under the same operating conditions.

The filtration/separation process also influences pore properties. Because as filtering continues, effective pore size decreases and flow resistance increases. This is because the particles penetrate into the pores of the filter media. The separation of solid particles from the liquid by filtration is a complex process. In practice, it is desirable that the filter pores are larger than the average size of the particles to be filtered. However, the selected filter medium must have the ability to retain solids by adsorption, and cohesive forces between the particles must be large enough to induce particle agglomeration around the pore openings [13].

2.4. Application of filtration

2.4.1. In drinking water treatment

The filtration process discussed in this section is used to remove particulate material from the water. Filtration is one of the unit processes used in drinking water treatment. The particles retained in the filter may be particles present in the spring water, or may come into play during the purification process. Particulate materials include clay and silt particles; microorganisms, such as bacteria, viruses, and protozoa cysts; humic substances and other natural organic particles; calcium carbonate and magnesium hydroxide precipitates used in softening processes; or alumina or iron precipitation used in coagulation processes [14]. High-quality drinking water production and filtration units are shown schematically in **Figure 5**.

About 30–40 years ago, lead was considered among the dangerous harmful pollutants in drinking water. Today, along with lead, pesticides, bacteria, viruses, coliphages, nitrates, chlorine, chlororganic substances, and aluminum have been added to the list of health threats, and the pollutant list is renewing day by day [15]. The contaminants typically found in water are shown in **Table 1**.

2.4.2. In wastewater treatment

The treatment of wastewater is a big deal when considering the volume of the water to be treated. Most of the water used for domestic, commercial, institutional, and industrial purposes is returned to the environment as waste. For this reason, a suitable treatment for a safe discharge is needed, which can manage wastewater treatment, from collection of waters to treatment, from

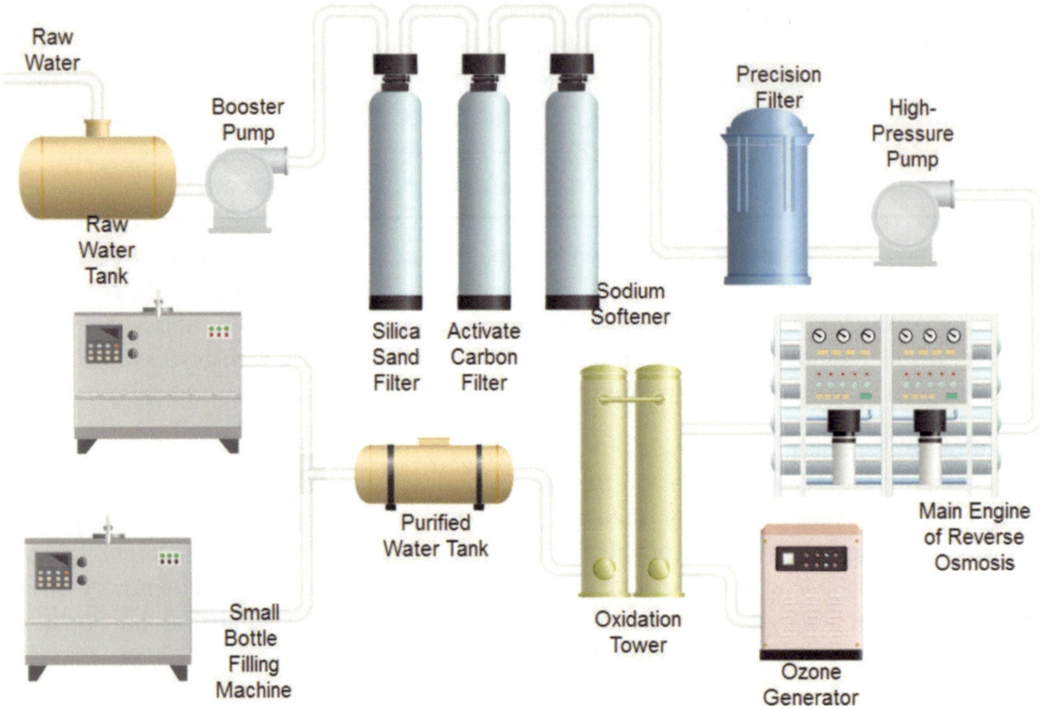

Figure 5. Drinking water production units.

Contaminants	Effects
Chlorine	• Reacts with organics and forms trihalomethanes
Bacterial diseases	• Cause cholera, the most widespread infection
The parasitic protozoan *Cryptosporidium*	• Get into the water supply from animal excrement
	• Can survive for long periods in water
	• Self-limiting gastroenteritis to life-threatening situations
Cryptosporidium	• Resist chlorine
	• Cause disease
	• Treatable with antibiotics
Giardia lamblia	• Cause disease
	• Treatable with antibiotics
Bacteria and viruses	• Can be removed by ultrafiltration and activated carbon

Table 1. Typical contaminants found in water (adopted from [15]).

equipment selection to process design. In most of the developed areas, wastewaters are collected by a municipality or a private operator and are directed to the treatment plant for collection by the sewerage system and surface runoff. The purpose of wastewater treatment is to convert

http://dx.doi.org/10.5772/intechopen.69462

these mixed wastes into a liquid stream, which will not harm the environment. Wastewater discharged without treatment threatens the life of plants and animals by consuming oxygen in the receiving environment. The contaminated receiving environment waters can be transported to the surface waters to be used for water supply and can adversely affect human health. Although most of the industrial wastewater treatment process is the same as domestic wastewater treatment, the characteristics of the industrial wastewater source should be taken into account.

Granular filtration is generally used for the treatment of water and wastewater containing suspended solids. The filter medium is composed of materials, such as sand and anthracite, which contain granular particles. The filter medium is contained in a basin, and underneath the material there is a layer that serves both as a support and as a drain. As the water or wastewater passes through the filter bed, the particles are trapped in and on the bed. When the filter is clogged, backwash is performed at high speed. The backwash water contains solids at high concentration and is recycled to another treatment step or plant inlet. Filtration is typically used for liquids with a solids content of 100–200 mg/L. As the concentration of suspended solids in the water-wastewater to be treated increases, the filter blockage accelerates and the frequency of backwash increases. Sudden and continuous flow changes in the filtration are another factor affecting the filtrate quality. Sedimentation is generally applied before filtration to reduce the suspended solids load. Filtration can also be applied to reduce suspended solids before biological treatment or before activated carbon process. The particles that can be removed by the granular filtration process are usually in colloidal size or may be in larger sizes such as floc. Although the floc is more likely to remain in the filter, it clogs the filter faster. Sometimes, however, flocculation of the particles may not be possible (as in many oil and water emulsions). In this case, some other equipment such as ultrafiltration may be needed. In a typical physical/chemical treatment plant, there are three parallel filters having three filter media layers (sand and anthracite) connected in parallel.

2.4.2.1. In municipal wastewater treatment

In most municipal wastewater treatment plants, the wastewater delivered to the treatment is not continuously characterized. For this reason, considering that other pollutants may also be transported to the plant wastewater inlet, the design of the treatment plant should be designed to cope with the pollutant in a wide range. These contaminants may be suspended or dissolved, organic or inorganic, or toxic or not. The treatment plant should be able to reduce the total amount of these pollutants below the limit values set by national and local regulators. These limits are regulated according to the structure of the receiving water environment. The units of a conventional municipal wastewater treatment plant are shown in **Figure 6**.

2.4.2.2. In industrial wastewater treatment

Industrial wastes differ from domestic wastes in three main ways:

(i) They contain more inorganic or biodegradable contaminants,

(ii) They generally have low or high pH, and

(iii) They often contain high amounts of toxic substances.

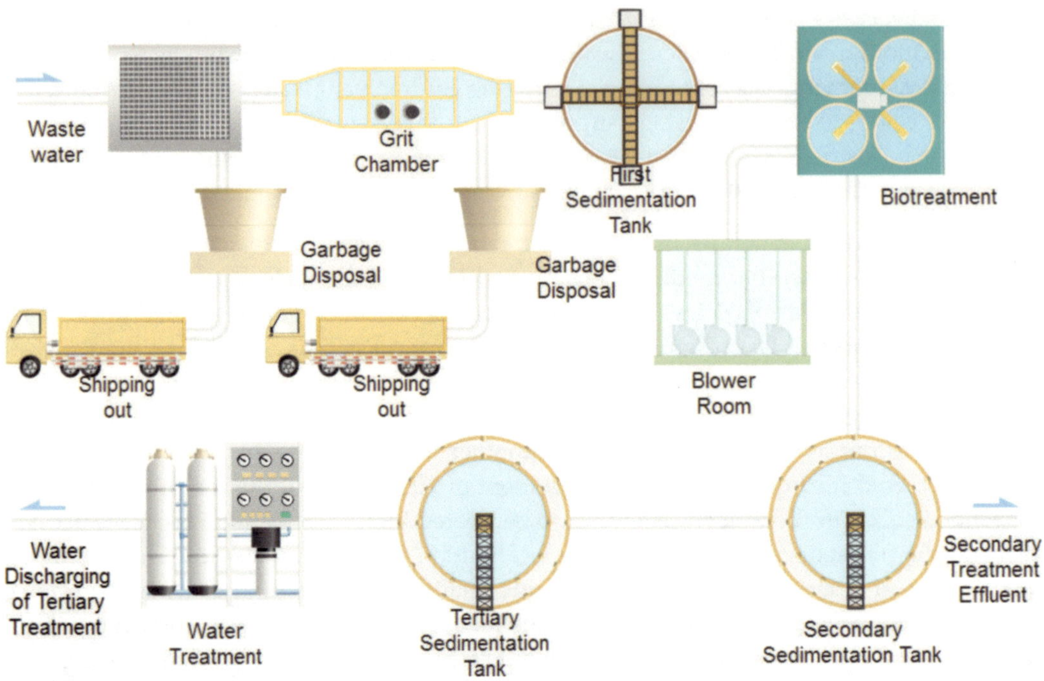

Figure 6. Conventional municipal wastewater treatment plant.

The oil compounds are also a problem for the treatment process. Although the waste amount of a plant that applies a certain production standard is generally constant in terms of compound and content, the purification cost is also very high since large volumes of wastewater and high amounts of sludge are formed. Although most industrialized countries have comprehensive legislation on industrial wastewater, this is not sufficient in wastewater management alone. It is also important that the industry owner knows the regulations and that the audits are carried out in sufficient frequency and extent. For this reason, wastewater treatment is a complex process in environmental management. In the majority of industrial wastewater treatment plants, there are two main parts: (i) to dispose of special wastes that are initially dependent on the raw material of the industry and (ii) to remove other general wastes. In the first part, the main function is to minimize the loss of any product material carried with the waste. If it is possible, the main thing is to recover the products in the waste [15].

2.4.2.3. Treatment of hazardous wastes

Membrane processes are used in many water and wastewater treatment applications. Most of these applications involve the separation and concentration of organic and inorganic substances. Wastewater can originate from industrial processes, contaminated ground or surface waters, or byproducts of other treatment processes. Membrane and filtration processes are used in water and wastewater treatment from visible particles to ionic species in many pollutant removal methods. Membrane processes are also preferred for the separation of hazardous wastes. In the organic pollutant removal, the separation takes place depending on the size (molecular weights) and polarities of the pollutants. For this, a suitable membrane having a different pore diameter is selected and used. The smaller the pores of the membrane, the higher the removal of small molecular weight compounds. However, as the pore size of the membrane decreases, the flux will also decrease. This will adversely affect the amount of product water obtained.

The polarity of an organic component is a measure of the ability of ionization in solution. Polar molecules include, for example, water, alcohols, and compounds having hydroxyl groups (e.g., phenols) and carboxyl groups (e.g., organic acids). Aliphatic hydrocarbons and polynuclear aromatic hydrocarbons are examples of nonpolar organic molecules. The chemical properties of the membrane can be used to separate nonpolar components from polar components in a waste stream. For example, a membrane that is surface hydrophilic will allow the passage of polar components while retaining nonpolar components. These membranes can be used to separate dissolved and emulsified oils from aqueous waste streams. Inorganic contaminants, such as salts and heavy metals, can be removed and concentrated by membrane processes from waste streams. Suspended inorganic materials can be easily removed using microfiltration membranes. These membranes have pore sizes ranging from 0.01 to several microns. Dissolved inorganics can be removed by reverse osmosis membranes or by precipitation followed by microfiltration. Only when reverse osmosis is used, a pretreatment is absolutely necessary, or it is used in the treatment of relatively clean waters. Chemical precipitation, on the other hand, provides higher flux with microfiltration membranes that are less sensitive to fouling [15].

3. Textile materials for filtration

3.1. Properties of filter media

There are a number of materials available for use in the filter medium that can be used to meet the needs of the user (**Table 2**). The material to be used must be easy to place on the filter tank/module. For this purpose, wool and nonwoven fabrics, natural or synthetic fibers may be preferred. In some cases, the weaving medium may be equipped with metal glands. The same material may be incorporated into the hard porous media (porous ceramic, sintered metal, woven wire, etc.), cartridge, and wax filter.

The proper selection of filter material is the most important factor in achieving efficient filtration. A good filter medium/filter should have the following characteristics: the filter medium should be able to retain the particles contained in the suspension in a wide size distribution; in order to obtain a high amount of filtrate in the desired quality, the filter must exhibit minimum resistance to flow; the cake deposited in the filter should be suitable for easy removal; it must be resistant to chemicals that can be transported to the filter medium and should not be a soluble material; during the filtration or backwashing process, the filter material should not swell; it must be sufficiently resistant to temperature changes in the fluid or environment; it must be resistant to the pressure applied in the filter and mechanical abrasion that may occur during the flow; and the filter material must be capable of preventing particles from being wedged in the pores [6]. The parameters to be considered in the filter selection are schematized in **Figure 7**.

Basic media format	Types of media
Loose granules	Deep bed
Loose fibers	Pads, felts
Structured granules	Bonded, sintered
Structured fiber	Needlefelts, spun
Sheet	Perforated, microporous
Woven/knitted	Spun yarn, monofilament
Tubular	Hollow fiber
Block	Rigid
Wound on core	Spun yarn, monofilament
Structured array	Ribbon, rods, bars
Extruded mesh	Netlon

Table 2. Filter media types [15].

Figure 7. Parameters in filter selection.

3.1.1. Classical filtration media

The filter medium is described as a porous (or at least semipermeable) barrier used to hold part or all of a suspension. When the pores of this barrier are much smaller than the diameter of the smallest particle to be filtered, the entire filtration process will take place on the surface, not the depth of the filter medium. If there is any particle smaller than the pore diameter, it will point toward the pores. Larger particles will clog the pores while smaller particles will clog the surface, thereby reducing the filtrate flow. At a certain point, the filtration process should be stopped and cleaned.

The mechanism by which the relationship between the particle size of the filtration process and the filter pore size depends strongly is called surface straining. As long as the particles are not deformed, the surface of the filter is separated by pore size with the help of tension. A second mechanism, referred to as deep stretching, occurs when a particle moves along the pore and is retained completely at a narrower point of the eye due to particle size. At this point, the clogged pore should be cleaned with backwash. The fine particles move on a tortuous path in filter bed and are trapped in the filter pores through a direct or inertial retention/diffusion mechanisms. This process is known as depth filtration. Congestion in the filter bed also occurs with these mechanisms. Particles retained in the filter material are compressed and do not completely cover the pores. Thus, the flow continues until the filter is completely clogged. In this case, the filter should be backwashed.

During the filtration process, particles in the pores are held together, and after a while they begin to function like filter material and assume the responsibility of trapping the particles that come after it in the fluid. A similar mechanism is also seen in the cake layer formed by the separated particles held on the filter media, and this separation is called cake filtration. More complicated mechanisms can occur in cake filtration. Because the cake compressed less or more by water pressure.

3.1.2. Membrane filtration media

Membranes are classified according to the size of the particles they separate. Macrofiltration is used to separate the intrinsic particles in the range of about 1 mm to 5 μm (screening is used for particles above 1 mm). Microfiltration is applied for particles from about 5 μm to about 0.1 μm, while ultrafiltration is used for lower dimensions. While ultrafiltration separates finer particles such as colloids, the lower limit of the particle size is usually determined by the term molecular weight, measured in Dalton. Further separation processes include nanofiltration and reverse osmosis. In both systems, a semipermeable membrane acts as a barrier in front of the fluid flow. However, the operating principle of NF and RO processes is different from that of UF. The liquid to be used in NF and RO is a solution which does not contain suspended solids or is prefiltered. The last two processes mentioned do not physically contain holes. The molecules are distributed across the membrane under high transmembrane pressure, and the liquid is removed from the other side of the membrane in pure form [15].

3.2. Textile media

Textile fibers are obtained from many natural and synthetic sources. Natural sources include materials of wood/cellulosic and vegetable origin, such as cotton, towel (flax), and jute, and materials of animal origin, such as silk, wool, fur, and hair. Synthetic materials are produced from natural sources such as glass, ceramics, carbon, metals, or reconstituted cellulose. They can also be obtained synthetically, extruded from thermoplastic polymers. The natural fibres are extremely long by comparison with their diameters, except in the case of wood cellulose, where the manufacturing process produces short fibres (in millimetres).

Fabrics are the largest component of filter materials. They are composed of fibers or filaments made of natural or synthetic material and are relatively soft. They are also not rigid like dry paper. For this reason, they need some kind of support when they need to be used as a filter medium. The fibers or filaments can be made into a fabric as is by a series of drylaying operations to produce a felt or the like. Such 'noninterlaced' fabrics are often referred to as 'nonwoven.' They are mentioned in the following sections of nonwoven fabrics [15]. A filter media classification (Ipurchas, 1967) based on rigidity is shown in **Table 3**.

3.2.1. Woven fabrics

Flexible and nonmetallic materials have been widely used as filter media for many years. These materials can not only be found in the form of fabric or as preformed nonwoven materials, but also as perforated plates. The fabric filter media is characterized by the number of weaves, mesh size, yarn size, and mesh type. The number of mesh or the number of thread of a fabric is the number of threads per inch. The number of yarns in the warp and weft direction are equal to each other and are represented by a single number.

The warp threads are placed longitudinally in a fabric and are parallel to the fabric edge. Weft or fill yarns also pass through the width of the fabric across the width of the warp. The space between the threads is the mesh opening and is measured in units of mm or inch. Different yarn sizes are normally defined as a diameter measurement in micrometers or mils (thousands of an inch). In warp and weft directions, the yarn sizes are normally the same and are represented by a single number. Fabrics are available in different mesh openings and different yarn diameters. The yarn diameter affects the amount of open space in the cloth to which it belongs, which determines the filtration flow rate or yield.

The diameter of the natural fibers varies according to their source, and is usually bigger than 1 mm. Synthetic fibers and filaments are formed by a kind of extrusion process that has a diameter that matches that of the extruded bending mouth. For this reason, their diameters may be in a much wider range than natural products, and in a wide range of sizes. The length and diameter of a natural fiber can be increased by turning the material into a yarn, but the

Type	**Example**	**Minimum trapped particles (μm)**
Edge filters	Wire-wound tubes Scalloped washers	5–25
Metallic sheets	Perforated plates Woven wire	100 5
Woven fabrics	Woven cloths Natural and synthetic fibers	10
Cartridges	Spools of yarns or fiber	2

Table 3. Filter media classification [6].

http://dx.doi.org/10.5772/intechopen.69462

yarns can also be made from fibers at the same time. Because the lengths are much longer, the fibers can usually be brought together to make a yarn, but the bundles are usually twisted to have a reasonably constant diameter. In order to impart sufficient strength to the resulting yarn, the shorter fiber filaments must be firmly twisted after being rotated for sequencing. Yarns made from fibers usually have a thin, smooth, and glossy appearance. Staple yarns are generally thicker, more hairy, and with less or no shine in appearance. Yarns can also be made from various types of tapes. For the filter medium, these bands are most likely fibrillated or made from other perforated material. The woven fabrics then consist of monofilaments or multifilament yarns or twisted staple yarns. The latter is normally used as a single yarn, but two or more spun yarns may be joined to the yarns which are twisted together; this is usually the opposite of the twist of each thread.

Fabric materials can be considered as a physically stronger alternative filter medium than paper materials and are used in a similar way for pleated elements. Fabric elements are in fact the most commonly used filter material for fine-size filtration and can be easily compared with modern paper filters in terms of filtration performance. Until the appearance of processed paper filters, fabric filters were more advantageous than conventional paper filters. While processed papers are now more commonly preferred as filter material due to lower cost, fabric filters can withstand higher working pressures with similar geometry. However, the fabric elements have a lower specific resistance than the paper elements. Though thicker than paper, fabric materials can carry a heavier pollutant load per unit area. However, when the same volume of packaging is taken into consideration, this advantage is offset by the decrease in surface area since the fabric material is thicker. Fabric filters may be preferred when large size filters are needed or when adsorption is required in addition to mechanical screening. The fabrics may contain a range of materials, woven and nonwoven, and may be modified by impregnation with synthetic resin or the like. In the same way, 'cloth' is often used to describe a natural or synthetic fabric media and even a woven wire cloth [15]. Some typical filtration performance curves for fabrics and papers are shown in **Figure 8**.

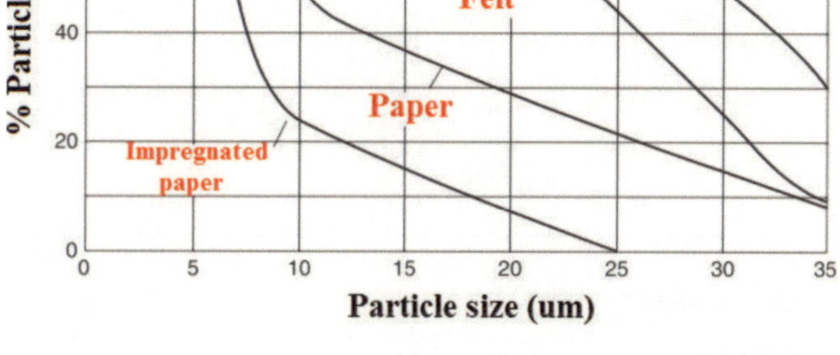

Figure 8. Filtration performance comparison [15].

Synthetic fiber-produced fabrics are superior to natural fabrics due to their resistance to swelling, acid-alkali, and various solvents and their resistance to the growth of fungi and bacteria, such as natural fibers. In addition, many synthetic fibers are resistant to high temperatures and can be easily cleaned due to their smooth surface. The physical properties of the most commonly used synthetic filter materials are shown in **Table 4** [14].

3.2.1.1. Woven yarn fabrics

Fabrics can be woven from a wide variety of yarns. Generally, the warp threads (extending longitudinally on the counter) are stronger, whereas the weft threads (running along the counter) may be more bulky and more tightly twisted. A weft is a thread of a very different material, while it is quite common that warp is a single, relatively stiff staple. Equally, it is normal to make both warp and weft from the same fiber or yarn. The properties of a fabric, especially as regards its behavior as a filter medium, are highly dependent on the way yarns are woven together. Many properties of the filter media are attributed to the natural properties of the fiber or filament or to the method of conversion to yarn. There are three basic yarn types commonly used for filter media: (i) monofilament, a single continuous filament composed of synthetic material (or silk); (ii) multifilament, a large number of filaments; and (iii) staples, spun or twisted filaments from natural materials such as cotton and wool, or synthetic ones cut from extruded filaments. The main feature of yarn type affecting filtration performance varies with monofilament and multifilament/staple fabrics. In monofiber fabrics, the filtration takes place in the spaces between the filaments, while, in multifilament and staple fabrics, yarn twisting is also important as filtration can also occur within the yarns as well as between them. The physical and chemical properties of a yarn are often due to the

Fibers	Acids	Alkalis	Solvents	Fiber tensile strength	Temperature limit (°F)
Acrilan	Good	Good	Good	High	275
Asbestos	Poor	Poor	Poor	Low	750
Cotton	Poor	Fair	Good	High	300
Dacron	Fair	Fair	Fair	High	350
Glass	High	Fair	Fair	High	600
Orlon	Good	Fair	Good	High	275
Saran	Good	Good	Good	High	240
Teflon	High	High	High	Fair	180
Wool	Fair	Poor	Fair	Low	300
Dynel	Good	Good	Good	Fair	200
Nylon	Fair	Good	Good	High	300

Table 4. Properties of woven filter cloth fibers [14].

physical and chemical properties of the fibers or filaments that make up the yarn. In addition to the number of natural fibers (mostly cotton, but some wool and silk) and a small but growing number of inorganic fibers, the bulk of filter fabrics is based on an increasingly wide variety of synthetic polymer fibers. The physical and chemical properties can then be adapted to the filtration needs by selecting the appropriate polymer.

In textile filtration, the basic material (fiber or filament type) of a woven fabric and its shaping are the most important parameters in the selection of the fabric. The variety of woven fabrics available to be used in filtration is virtually unlimited, even if only the way in which the materials of the fillers or yarns and the way the threads are touched is taken into account. Then the weaving process and the final process to be applied to the fabric after weaving should be added to the fabric structure. Woven fabrics consist of a specific and regular thread which is called knitting. There is no need for the warp and weft yarns to be parallel or perpendicular to each other, but this is true for most fabrics and this structure is also present in the filter medium. The basic properties of a woven fabric result from the geometric uniformity of its components and are retained by friction at the contact points, not by any solid connection.

The binding system or weave is the fundamental factor that determines the character of the woven fabric. Although there are many complex types of knitting in industrial textiles, three main types of knitting (plain, twill, and satin) are used (**Figure 9**). The differences between the wefts are dependent on the length of the weft threads formed when the threads are touched on or under the warp threads. In the plain weave, the weft thread passes over succeeding warp threads along the loom. The return weft then passes through the opposite direction of the subsequent warp threads, so that each weft is held firmly by engaging the warp threads together. Plain weaves can give the most dense fabric and the most robust woven fabric with the highest leverage efficiency. The texture braids are characterized by a strong diagonal pattern. The weft yarns are formed by being passed over two or more warps at one time, and then one or more underneath, regularly along the counter. The next weft thread follows the same and upper pattern, but is replaced by a warp thread. The essential feature of a twill is due to its regularity which leads to the diagonal pattern. In a twill weave, more weft threads can be crammed to a unit length fabric, which gives the fabric more bulk. Compared to a plain weave with the same yarns, the twill fabrics are more flexible and therefore it is easier to place them in a filter [14, 15]. A summary of the main types of weaves for wire cloth is given in **Table 5**.

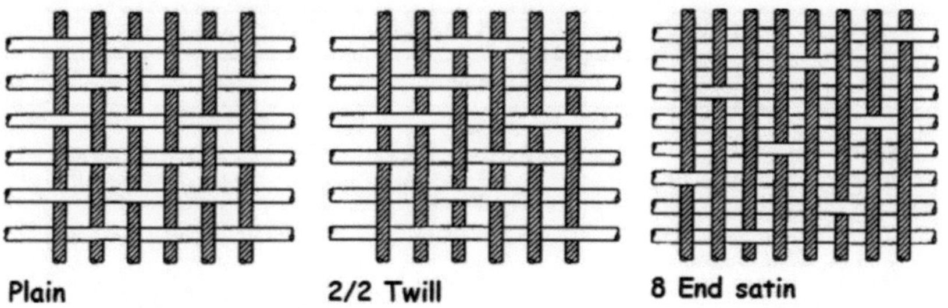

Figure 9. Weave patterns in woven cloths [6].

Name	Characteristics	Absolute rating range (μm)	Remarks
Square plain or twilled	Largest open area and lowest flow resistance. Aperture size is the same in both directions	20–300	Most common type of weave. Made in all grades from coarse to fine
Plain Dutch single weave	Good contaminant retention properties with low flow resistance	20–100	Openings are triangular
Reverse plain Dutch weave	Very strong with good contaminant retention	15–115	
Twilled Dutch double weave	Regular and consistent aperture size	6–100	Used for fine and ultrafine filtering

Table 5. Principle weaves for wire cloths [15].

Satin texture broadens the twill weave concept with wider spacing between touch points. Satin does not have the normal tissue pattern of the tissue, resulting in an irregular appearance with a smooth surface with relatively long layer of warp threads. Most satin fabrics are made from flat, slightly twisted yarn, so visual effects are enhanced. Satin weave fabrics are more flexible than the other two weave types because of the increased ease of thread twisting: this reduces the likelihood of trapped particles. The longer floats allow for the insertion of more warp threads in proportion, thereby further improving the smoothness of the surface resulting in easier cake drainage (**Table 6**). However, unless both warp- and weft-oriented yarns are compacted tightly, satin weaves generally do not provide high filtration effects, while long floats are more susceptible to abrasive wear. In addition to cleaning, all kinds of fabrics are subjected to a number of finishing processes, usually after weaving, to stabilize the fabric, modify surface properties, and regulate the permeability of the fabric. Calendering and singeing are two familiar surface processing methods that change the permeability.

3.2.1.2. Synthetic monofilament fabrics

Monofilaments are woven by extruded synthetic fibers produced with diameters from 30 μm to 2–3 mm. These fabrics are important as filter material in a wide range of industries and applications. Because they have corrosion resistance, vibration fatigue withstanding capacity,

Property	Weave			
	Plain	Chain	Twill	Sateen
Rigidity	1	2	3	4
Bulk	4	3	1	1
Initial flow rate	4	3	2	1
Retention efficiency	1	2	3	4
Cake release	2	3	4	1
Resistance to building	4	3	2	1

Table 6. Filtration requirements of weave [6].

uniformity, and economic resilience, they have taken the place of several other filter media types. Chemical and food processing industries, industrial hydraulics, and medical, automotive, and appliance markets are the main users of monofiber fabrics. These fabrics are available in a range of 5–5000 μm openings and are made from polymeric materials including nylon, polyester, polypropylene, and fluorocarbon. Synthetic monofiber fabrics, due to ductility and memory, can be flexed repeatedly without fatigue. Compared to a metal cloth, they can be folded or dented with less damage and they are lighter in weight. Some applications, at the same time, may require the filter medium properties of a synthetic monofilament and a metallized surface for static electricity dissipation. Accordingly, a metallized polyester monofilament fabric coated with nickel of 2 μm thickness is produced. Combined mono- and multifiber fabrics now have useful additional properties. Thus, such material is used in disc filter pieces that are elastic and will expand in the kickback phase to help release the cake. New belt press filters and large automatic filter presses are mainly used for fabric filters with heavy fibers [15].

In selecting the fiber to be used in the filtration process, the material with the highest chemical, thermal, and mechanical resistances should be preferred. For example, in **Table 7**, the resistance of different fiber materials to various chemical substances is roughly presented [14].

3.2.2. *Nonwoven fabrics*

The nonwoven medium is made from cotton, wool, synthetic and asbestos fibers, or mixtures thereof and from paper mass in the form of belts or plates. They can be used in different design filters such as filter presses, horizontal disc filters and rotary drum vacuum filters for liquid filtration. Most of these applications have low suspension concentrations; examples are milk, beverages, lacquers, and lubricating oils. The individual fibers in the nonwoven filter media are generally connected between them as a result of the mechanical treatment. A less common approach is the addition of binding agents. Sometimes, loose woven fabrics can be used on both sides of the filter to protect the filter media. Both absorbent and nonabsorbent raw materials can be used to produce nonwoven filter media with different materials and properties, different weights, and different filtration efficiencies. These filter media hold particles that are less scattered on their surface (more than 100 pm) or particles that are more dispersed in the depths of the filter media.

Wool felt is probably the oldest kind of textile fabrics and for many years is the only nonwoven fabric in practice. A strong adhesivebonded felt is developed, and drylaid synthetic fibers are collected and transformed to the shape of nonwoven media. Thus, nonwoven fabrics can sometimes be obtained by agglomeration of fibers, and sometimes continuous filaments are glued together to obtain desired flexibility.

The chemical properties of an untwisted fabric relate to the natural structure of the fiber that is used almost completely, as long as no binder is used that has significantly different properties. Nonwoven materials are classified into two main groups. These two classes are based largely on the methods and tools used to hold loose fibers together:

(i) The felts that utilize the basic fiber features to obtain mechanical strength or the mechanical processing (especially needling) and

(ii) Bonded cloths, which use some extra adhesive material to hold the fibers together.

Type of fiber	Insect proof	Resistance to aging	Acid	Alkali	Chlorocarbonic hydride	Ketone	Phenol	Benzene
Cotton	Medium	Low	Unstable	Low resistance, swelling	Resistant	Resistant	Resistant	Resistant
Silk	Medium	Low	Low R.	Unstable	Resistant	Resistant	Resistant	Resistant
Wool	Bad	Low	Low R.	Unstable	Resistant	Resistant	Resistant	Resistant
Glass	Good	Good	Low R.	Unstable	Resistant	Resistant	Resistant	Resistant
Steel fibers (Brunsmet®)	Good	Good	Low R.	Resistant	Resistant	Resistant	Resistant	Resistant
PA 6 (Perlon®)	Good	Good	Unstable	Resistant	Resistant	Resistant	Unstable	Resistant
PA 6.6 (Nylon)	Good	Good	Unstable	Resistant	Resistant	Resistant	Unstable	Resistant
PA 11 (Rislan®)	Good	Good	Low R.	Resistant	Resistant	Resistant	Unstable	Resistant
PA 12 (Vestamid®)	Good	Good	Low R.	Resistant	Swelling	Resistant	Unstable	Swelling
PA Nomex®	Good	Good	Low R.	Resistant	Resistant	Resistant	Unstable	Resistant
Polyester	Good	Good	Resistant	Low R.	Resistant	Resistant	Unstable	Resistant
Polyacrylonitrile	Good	Good	Resistant	Low R.	Resistant	Resistant	Resistant	Resistant
Polyvinylchloride	Good	Good	Resistant	Resistant	Resistant	Unstable	Unstable	Unstable
Polyvinylidenechloride(Saran®)	Good	Good	Resistant	Resistant except NH_4OH	Resistant	Resistant	Unstable	Resistant
Polyolefins								
Polyethylene								
High-pressure	Good	Good	Resistant	Resistant	Swelling	Resistant	Resistant	Resistant
Low-pressure	Good	Good	Resistant	Resistant	Swelling	Resistant	Resistant	Resistant
Polypropylene	Good	Good	Resistant	Resistant	Resistant	Resistant	Resistant	Resistant
Polytetrafluoroethylene	Good	Good	Resistant	Resistant	Resistant	Resistant	Resistant	

Table 7. Chemical resistances of fibers [14].

The more common method is to rely on the natural thermoplastic properties of the polymeric material to obtain adhesion when properly heated. Bonded cloths are divided into two groups based on whether the fiber formation is an integral part of the manufacture of the filter medium or not. The basic felt does not contain binders: some wooly fibers have the ability to assemble together to form a coherent mass due to protrusions on the fiber surface [14, 15].

The first step in making any felt is to scan the fibers so that the fibers are roughly aligned in one direction and drawn into a thin web. The pieces of this web are placed on top of one another to achieve the desired thickness of the felting. Consecutive layers can be placed in the same direction as the fibers or they can be aligned in different directions to increase strength. When sufficient thickness is achieved, the felt is compressed, heated, and often finalized after the dampening process has been carried out. The strength of this felt is basically weak. For this reason, the strength of most felts is reinforced by the inclusion of a woven layer called scrim.

Since the fibers in a felt are not tightly bonded to the mass of the fabric, there is a risk of fibers moving away from the filter media during filtration. For this reason, the use of various bonding elements, the integration of the thermoplastic fibers into the fabric, and the consolidation of the fibers in the filter medium by means of various mechanical joining processes based on needling or suturing braids are done. Modern felts are produced from synthetic or natural fibers or mixtures thereof. The fibers are mechanically or with the help of an adhesive bonded and passed through a controlled production to obtain a consistent density, pore size, and mesh geometry. Thus cutoff performance can be reasonably predicted. The structure of the felt is much looser than the paper, so depth filtration could be performed, the specific resistance is reduced, and higher flow rates can be achieved with smaller filter volumes and at lower pressure drop. The high-temperature-resistant meta-aramid filter has helped the industry to move one step closer to the zero emission target by providing a combination of high separation efficiency and low differential pressure of hot gas filtration technology.

3.2.2.1. Wool resin media

In solid-gas filtration and, more rarely, in solid-liquid filtration, the particles desired to be retained in the filter may carry an electrostatic charge. In this case, the use of filter material carrying an electrostatic charge opposite to that of the particles will provide a more effective filtration. For this purpose, many different filter media can be electrostatically charged. For example, long-term electrostatic effects can be obtained by adding a special resin to the wool felts used in the submicroscope aerosol filtration. Electrostatic charge is achieved by resin transfer by adding the resin powder into the wool matrix enabling charge transfer. The wool then has a positive charge and the resin has a negative charge. In this case, the filter is electrically neutral in general. The random distribution of resin powder on the wool fibers and the random arrangement of the wool fibers in the filter means that the electric field is not uniform and is therefore very effective for trapping both charged and unloaded particles. This electric charge gives wool resin a very low resistance to airflow, allowing submicrometer particles to achieve a higher than 99.5% efficiency in their filtration. Thanks to the resin's very high electrical resistance, the resin filter can maintain its filtration efficiency for many years despite the adverse effects of tropical conditions. Wool resins were first developed for use in respiratory devices for World War I and are still widely used in the respiratory industry, 90 years after its development. It is preferred against many new materials with low resistance to breathing and high filtration efficiency. Vacuum cleaners and other independent dust collectors also benefit from the high retention of wool resin against asbestos and other harmful dusts. In addition, wool resins are used as prefilters for high efficiency particulate air (HEPA) filters and for heating and ventilation in clean rooms for computer suites.

3.2.2.2. Needlefelts

For some simple filtration applications, it is possible to use the felt directly. In most other applications, however, mechanical or chemical treatment is required because of the low mechanical resistance of the felts and because of the disassembly of the fibers and mixing to the filtrate. Needle punching is one of the mechanical strengthening methods used for this purpose. This method has emerged as the most preferred mechanical strengthening technique for natural fibers in the 1880s, but since the early 1970s, it has been suitable for many synthetic fibers for processing the felts. In this method, carded fibers are pounded and compressed into a more dense structure by punching with a series of specially barbed needles moving back and forth at 2000 strokes/min and moving perpendicular to the felt layer. With 100 needle punches/cm^2, it is possible to circulate the fibers in the felt thickness both together and to reduce the felt thickness considerably. The punching operation can be carried out by one or both sides of the felt, so that the felt has a homogeneous structure. Needle felts are commonly used as bag filters for filtering dust and gases. Typical applications include cement industry, steel and aluminum plants, spray drying, coal grinding, sandblasting, food industry, detergent manufacturing, pneumatic conveying, and hot gas filtration using metal fiber felts and ceramic fibers. Some typical applications for filter fabrics of various kinds are shown in **Table 8** with their key characteristics. Felt is mechanically strengthened by needling, but alternatively the hydroentanglement method is used as a more professional technique. In this method, the fibers are tried to be fixed with the help of pressure water jet.

3.2.2.3. Meltspun materials

The use of new synthetic meltspun fibers has begun to spread quickly, while filtration applications are commonly used with needle felts and woven fabrics. These fibers are obtained by extruding a molten thermopolymer from a fine nozzle. As the fiber leaves the syringe, it is quickly quenched in an air stream and then collected on a moving collecting belt running underneath the nozzle. The filaments on the collector are then pressed at a certain temperature for consolidation. Thus, the fibers adhere to the points where they touch each other, and the fiber network is strengthened. This consolidation process is called spun bonding. If the airflow is placed immediately at the exit of the nozzle and along the line where the filament falls onto the collector, the fibers break off due to the air flow and fall on the collector in short pieces. When these fibers are pressed and sintered, this process is called melt blowing [15].

Material	Suitable for	Maximum service temp (°C)	Principal advantage(s)	Principal disadvantage(s)
Cotton	Aqueous solutions, oils, fats, waxes, cold acids, and volatile organic acids	90	Inexpensive	Subject to attack by mildew and fungi
Jute wool	Aqueous solutions	85	Easy to seal joints in filter presses	High shrinkage, subject to moth attack in store
Nylon	Acids, petrochemicals, organic solvents, alkaline suspensions	150	High strength or flexibility	Absorbs water; not suitable for alkalis
Polyester (Terylene)	Acids, common organic solvents, oxidizing agents	100	Easy cake discharge. Long life. Good strength and flexibility. Initial shrinkage	Not suitable for alkalis
PVC	Acids and alkalis	Up to 90		May become brittle. Heat resistance poor
PTFE	Virtually all chemicals	200	Extreme chemical resistance. Excellent cake discharge	High cost
Polyethylene	Acids and alkalis	70	Easy cake discharge	Soften at moderate temperatures
Polypropylene	Acids, alkalis, solvents (except aromatics and chlorinated hydrocarbons)	130	Low moisture absorption	
Dynel	Acids, alkalis, solvents, petrochemicals	110		
Orlon	Acids (including chromic acid), petrochemicals	Over 150		
Vinyon	Acids, alkalis, solvents, petroleum products	110		
Glass fiber	Concentrated hot acids, chemical solutions	250	Suitable for a wide range of chemical solutions, hot or cold (except alkalis)	Lacks fatigue strength for flexing. Abrasive resistance poor

Table 8. Typical applications for filter fabrics [15].

4. Recent processes in fabric filtration

4.1. Nanofiber spun membranes

Compared to other polymeric membranes, nanofiber membranes have attracted great interest in recent years due to their advantages such as high selectivity, hydrophilicity, and mechanical strength. Nanofibers are very thin polymeric fibers with a thickness of less than 100 nm which are preferred in various industrial fields, such as electronics [16], biomedical [17], textile [18], and environment [19]. Nanofibers stand out among similar polymeric membranes with high specific surface area, high porosity, and interconnected pore networks [20]. The nanofibers used in applications where microfiltration and ultrafiltration are used provide high water flux by reducing membrane resistance in water and wastewater treatment [21].

4.2. Electrospinning processes

Nanofibers can be obtained by one of the methods: drawing, template synthesis, phase separation, or electrospinning. Electrospinning is a frequently preferred method in recent times in obtaining high porosity nanofiber mat. In this method, nanofibers are obtained from a charged polymer solution under a high electric field. Parameters affecting the process (voltage intensity, feed rate of the polymer solution, nozzle-collector distance, polymer concentration and type, and duration of electrospinning) can easily be changed and controlled (**Figure 10**). Conditions such as room temperature and humidity are also factors that affect nanofiber morphology [22]. The molecular weight of the selected polymer directly affects the fiber properties. The uniformity of the pore size of the nanofiber mats is obtained when uniform and continuous collection of nanofibers from the nozzles to the collector is achieved [23]. The nanofiber layer, consisting of nanofibers ranging from 50 nm to 10 mm, offers many advantages such as high aspect ratio (length to diameter ratio), broad specific surface area, unique physicochemical properties, and design flexibility for chemical/physical surface functionalization [24].

4.3. Nanofibers in water and wastewater treatment

Nanofiber membrane processes are preferred in many industrial applications due to energy-saving use, environmental friendliness, operational simplicity, and flexibility during design. As the nanofiber production technology improves, the use of nanofibers as an alternative to membrane processes, such as conventional microfiltration, ultrafiltration, and nanofiltration, has opened the way [21]. In one study, nanofibers produced from the polysulfone polymer were used for prefiltration to remove microscale particles prior to the ultra/nanofiltration process, thus extending the ultra/nanofiltration membranes' life span [26]. The performance of

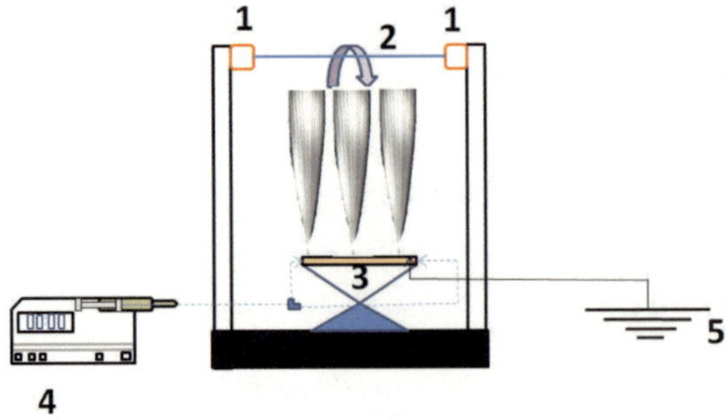

Figure 10. Schematic illustration of electrospinning setup: (1) rotating backing material, (2) conductive wire, (3) nozzle tray, (4) syringe pump, (5) high voltage power [25].

the membrane processes (retention of particles and the amount of permeate flux) strongly depends on the particle size. It has also been reported that the addition of nanofiber material and additives (such as nanoparticles and nanotubes) to the polymer solution affects the separation performance [27].

One of the latest nanofiber studies is the work of Aslan et al. [25]. In this study, nanofibers were obtained at the scale of ultra/microfiltration by means of electrospinning from the solution prepared by polyacrylonitrile polymer. For the first time in the literature, nanofibers were collected on a tubular support layer. The new membrane was tested with both standard particle solution and a surface water. The novel tubular nanofiber membrane removes 95% turbidity and 29% total organic carbon, which can be evaluated as high removal efficiencies when compared to the commercial microfiltration membrane. Membrane surface and cross section SEM images are given in **Figure 11**.

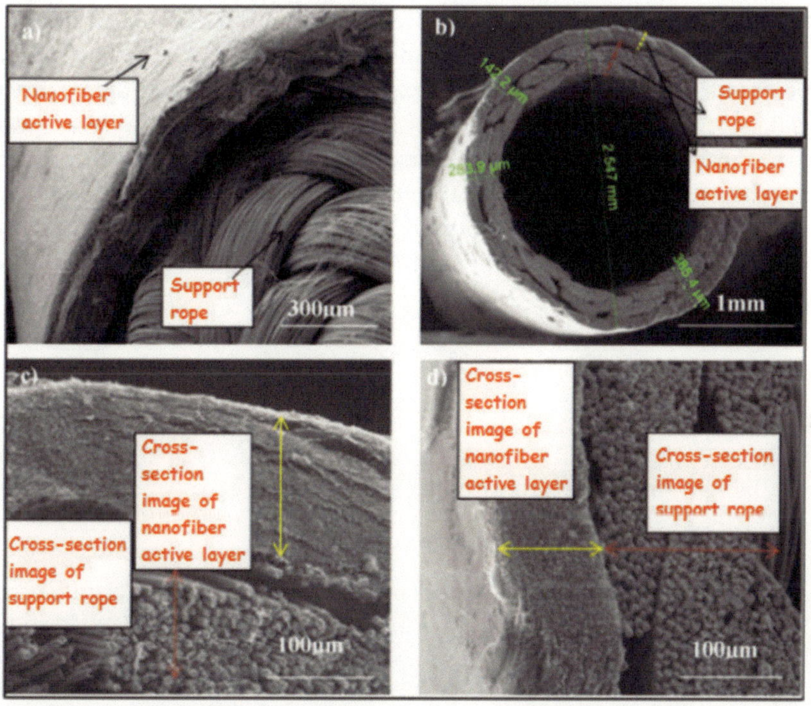

Figure 11. SEM images of tubular backing material and nanofiber layer at different scales [25].

5. Conclusions and recommendations

Filtration is considered the keystone of water and wastewater treatment and is used for various purposes such as sludge dewatering and concentrating any solution. Moreover, as an advanced filtration technology, membranes can remove materials ranging from large visible particles to molecular and ionic chemical species. Filtration performance depends on operating conditions, such as fluid characteristics, filtration rates, and filter media. Among them, proper selection of filter media/membrane material in filtration processes is often the most important consideration for assuring efficient separation.

Filter media can be classified by their materials of construction, such as cotton, wool, linen, glass fiber, porous carbon, metals, and rayons. Recently, new polymeric materials have been used both individually and/or blended in filtration processes for the treatment of waters and wastewaters. The purpose of this chapter is to bring an overview on the textile-originated filter materials in filtration applications from conventional filtration to advanced membrane processes. Although many researches on filter media are available, very few reports are represented on the cutting-edge technologies about using filter materials on filtration processes from classical to advanced membrane processes.

Textile materials and membranes are two important elements of surface filtration. The performance of surface filtration is closely related to the physicochemical properties of the filter surface. New materials are produced or the surface of the existing material is modified in order to improve the performance of the filter surface. These modifications may involve the use of different chemicals (e.g., polymer blends) in the production of the filter material, as well as the addition of additives to the base material (e.g., nanoparticles, nanotubes). In recent years, textile nanofibers have emerged in liquid filtration with their unique properties such as high aspect ratio, broad specific surface area, unique physicochemical properties, and design flexibility for chemical/physical surface functionalization, and they will attract more attention in near future as filter material in both liquid and gas filtration.

Author details

Murat Eyvaz*, Serkan Arslan, Ercan Gürbulak and Ebubekir Yüksel

*Address all correspondence to: meyvaz@gtu.edu.tr

Environmental Engineering Department, Gebze Technical University, Gebze, Kocaeli, Turkey

References

[1] Faust SD, Aly OM. Chemistry of Water Treatment. 2nd ed. Chelsea, MI: Ann Arbor Press; 1998. p. 273

[2] American Water Works Association (AWWA). Water Quality and Treatment: A Handbook of Community Water Supplies. 4th ed. New York: McGraw-Hill Inc.; 1990

[3] Duran M, Demirer GN. Fundamental Processes in Water Treatment. Ankara, Turkey: TMMOB-CMO Publications; 1997 [in Turkish]

[4] Masschelein WJ. Unit Processes in Drinking Water Treatment. Madison Avenue, New York, USA: Marcel Dekker Inc.; 1992. p. 270

[5] Keskinler B, Çakıcı A, Yıldız E. Lecture Notes in Environmental Engineering and Fundamental Processes. Erzurum, Turkey: Ataturk University Publications; 1994 [in Turkish]

[6] Rushton A, Ward AS, Holdich RG. Solid Liquid Filtration and Separation Technology. Weinheim, New York, Basel, Cambridge, Tokyo: VCH; 1996

[7] Polk M, Vigo TL, Turbak AF. High performance fibers. In: Encyclopedia of Polymer Science and Technology. Hoboken: John Wiley & Sons, Inc.; 2002

[8] Parsons SA, Jefferson B. Introduction to Potable Water Treatment Process. Oxford, UK: Blackwell Publishing; 2006

[9] Eyvaz M. Declining rate filtration [doctoral dissertation]. Turkey: Environmental Engineering, Gebze Institute of Technology; 2013 [In Turkish]

[10] Judd S. The MBR Book: Principles and Applications of Membrane Bioreactors for Water and Wastewater Treatment. Oxford, UK: Elsevier; 2003

[11] Lee A, Elam JW, Darling SB. Membrane materials for water purification: Design, development, and application. Environmental Science: Water Research & Technology. 2016;**2**:17-42

[12] http://onlinembr.info/membrane-process/classification-of-filtration-modes/ January 13, 2017]

[13] Cleasby JL, Logsdon GS. Granular bed and precoat filtration. In: Letterman RD, editor. Water Quality and Treatment: A Handbook of Community Water Supplies. Chapter 8. New York: McGraw-Hill; 1999

[14] Cheremisinoff NP. Liquid Filtration. 2nd ed. Boston: Butterworth-Heinemann; 1998. p. 319

[15] Sutherland K. Filters and Filtration Handbook. 5th ed. Oxford: Elsevier/Butterworth-Heinemann; 2008

[16] Yao W, Yang J, Wang J, Tao L. Synthesis and electrochemical performance of carbon nanofiber–cobalt oxide composites. Electrochimica Acta. 2008;**53**:7326-7330

[17] Venugopal J, Zhang YZ, Ramakrishna S. Electrospun nanofibres: Biomedical applications. Proceedings of the Institution of Mechanical Engineers, Part N: Journal of Nanoengineering and Nanosystems. 2004;**218**:35-45

[18] Deitzel J, Kleinmeyer J, Harris D, Beck Tan NC. The effect of processing variables on the morphology of electrospun nanofibers and textiles. Polymer. 2001;**42**:261-272

[19] Jeong EH, Yang J, Youk JH. Preparation of polyurethane cationomer nano-fiber mats for use in antimicrobial nanofilter applications. Materials Letters. 2007;**61**:3991-3994

[20] Pillay V, Dott C, Choonara YE, Tyagi C, Tomar L, Kumar P. A review of the effect of processing variables on the fabrication of electrospun nanofibers for drug delivery applications. Journal of Nanomaterials. 2013:1-22

[21] Balamurugan R, Sundarrajan S, Ramakrishna S. Recent trends in nanofibrous membranes and their suitability for air and water filtrations. Membranes. 2011;**1**:232-248

[22] Frenot A, Chronakis IS. Polymer nanofibers assembled by electrospinning. Current Opinion in Colloid & Interface Science. 2003;**8**:64-75

[23] Ahmed FE, Lalia BS, Hashaikeh R. A review on electrospinning for membrane fabrication: Challenges and applications. Desalination. 2015;**356**:15-30

[24] Cloete TE, De Kwaadsteniet M, Botes M, Manuel J, Walker A. Nanotechnology in Water Treatment Applications, Norfolk, UK: Caister Academic Press; 2010

[25] Aslan T, Arslan S, Eyvaz M, Güçlü S, Yuksel E, Koyuncu İ. Novel nanofiber microfiltration membrane: Fabrication and characterization of tubular electrospun nanofiber (TuEN) membrane. Journal of Membrane Science. 2016;**520**:616-629

[26] Gopal R, Kaur S, Chao YF, Chan C, Ramakrishna S, Tabe S, Matsuura T. Electrospun nanofibrous polysulfone membranes as pre-filters: Particulate removal. Journal of Membrane Science. 2007;**289**:210-219

[27] Desai K, Kit K, Li JJ, Davidson PM, Zivanovic S, Meyer H. Nanofibrous chitosan nonwovens for filtration applications. Polymer. 2009;**50**:3661-3669

Section 5

Smart and Functional Textiles

Chapter 12

Shape Memory Polymers for Smart Textile Applications

Suman Thakur

Additional information is available at the end of the chapter

http://dx.doi.org/10.5772/intechopen.69742

Abstract

Shape memory polymer (SMP) is a genre of smart materials which demonstrate a capability to fix in the temporarily programed shape and to recover the original shape upon exposure to external stimuli. Such unique and amazing properties of SMPs are applied to develop smart textiles by incorporating SMP into the fabric. Various types of stimuli-responsive SMPs such as thermal, light, pH, and moisture have been used in textiles for enhancing or attaining some smart properties. The integration of SMPs in fabric provides many fascinating and improved properties such as good aesthetic appeal, comfort, textile soft display, smart controlled drug release, fantasy design, wound monitoring, smart wetting properties and protection against extreme variations in environmental conditions. In this chapter, we discussed different types of SMP used for this purpose, their functions, and applications in textiles with their potential applications in the near future.

Keywords: shape memory polymers, stimuli-responsive materials, smart textile

1. Introduction

Shape memory polymer (SMP) is a genre of stimuli-responsive materials with ability to fix a programed shape and to return from the deformed shape to its original permanent shape while induced by an external stimulus such as heat, light, humidity, electric field, magnetic energy, pH and so on (**Figure 1**) [1–4]. This amazing property of SMPs has inspired many researchers and industrialists to develop smart textiles by incorporating them into the fabric. Nowadays, SMPs carved an attention for use in textile where they strongly respond to changes in heat and moisture levels, ensuring greater comfort for the wearer. Incorporation of SMPs also provides some smart features along with comforts such as aesthetic appeal, textile soft display, smart-controlled drug release, wound monitoring, smart wetting properties, and protection against extreme variations in environmental conditions. Therefore, smart functional textiles have developed swiftly in the past few decades [5–7]. Digital components,

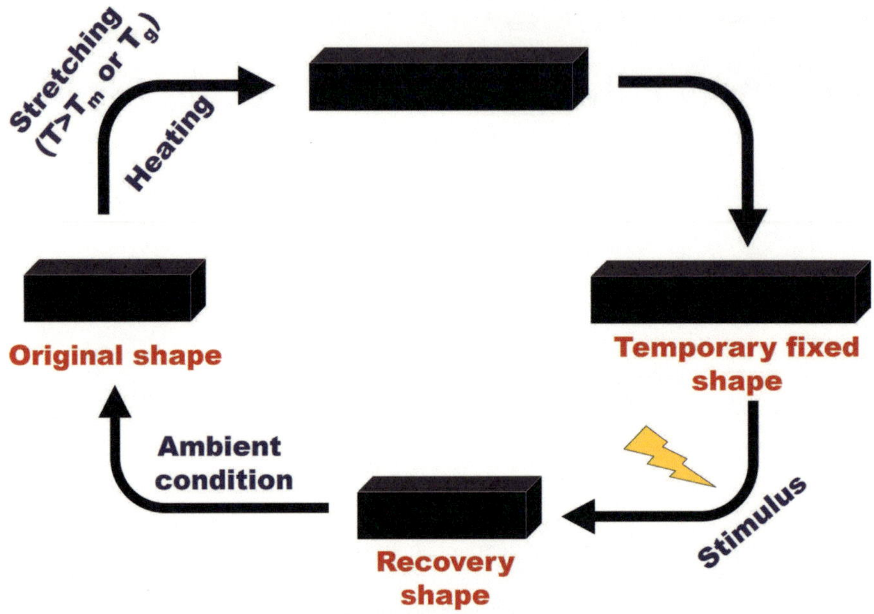

Figure 1. The schematic representation of shape memory behavior of SMPs.

computing, and electronics embedded e-textiles offered many novel functions to garments [8–10]. Various new and unique functions containing textiles are developed such as luminescent textiles, textile displays, emotion sensing dresses, self-cleaning textiles, temperature-regulated textiles, and self-moving textiles [11–14].

Different SMPs such as shape memory polyurethane (SMPU), polyester, poly-hydroxyproline, polysilamine, etc., and some responsive hydrogels including poly(N-isopropylacrylamide) (PNIPAAm) hydrogels, polythiophene gel, etc., are used to fabricate such smart textile. In addition to that, various SMP composites also prepared for different end applications. SMPs and their composites have several advantages for textile coatings:

(a) Different stimuli can be used which textiles commonly undergo, such as water, light, and heat

(b) They have highly flexible programming, which can be triggered by various stimuli by single- or multiple-step processes

(c) They exhibit tunable properties that can be engineered easily to be applied onto the fabric surface

(d) They have a light and adjustable modulus that is easily identical to a textile's softness

In this chapter, we discussed the different types of SMPs used for smart textile, their functions, and especially applications of such smart textile with their potential applications in the near future.

http://dx.doi.org/10.5772/intechopen.69742

2. Function of shape memory polymers in textiles

SMPs possess different versatile functions. The following contents introduce such functions of SMPs which use for the fabrication of smart textile.

2.1. Thermal and moisture controlling

Smart breathable garment is one of the most desired products in textile sector, which can be fabricated using SMPs. This garment can regulate the transfer of heat and moisture to wearer's body. The water vapor permeability (WVP) of the SMPs regulates with the human body temperature. The molecular free volume of the SMPs significantly increases when the body temperature is above the glass transition temperature (T_g) of SMPs. This aids the transfer of vapor and heat through perspiration for comfort in wearer's body. Again, the molecular free volume decreases when the body temperature is below T_g of the SMPs. This restricts the both heat loss and the moisture to pass through. Thus, the SMP is a good choice for maintaining stable body temperature. Textiles and garments containing SMPs are still being developed and investigated. Lamination, coating, knitting, and weaving are used to integrate SMPs into textiles [15].

In this regards, Wang *et al.* fabricated a garment which afforded thermal protection in too cold conditions [16]. In addition to that, Crespy and Rossi also incorporated the thermal-responsive hydrogels in textiles for smart heat and moisture management [17]. Kim *et al.* found that PNIPAAm-grafted polypropylene nonwovens possessed good thermo-responsive WVP [18]. SMPU-coated fabrics demonstrated a rapid increase in WVP at transition temperature. Additionally, Mondal and Hu fabricated a cotton fabric by using SMPUs containing a small percentage of multiwalled carbon nanotubes (MWNTs). This fabricated cotton demonstrated excellent WVP with good UV protectability [19].

Along with this, few commercial products also came into market. In this context, Mitsubishi Heavy Industries fabricated a SMPU membrane namely 'DiAPLEX-The intelligent material' [20]. It provides some special properties such as waterproofed, windproofed and breathability. Its working mechanism is based on the principle of thermal vibrations. These vibrations are generated by micro-Brownian movements of polymeric chains. SMPs can be attached in form of a laminate to a variety of fabrics, being capable of adapting to the surrounding environmental conditions (**Figure 2**). In addition, Toray Industries and Marmot Mountain Works® also developed a SMPU film MemBrain® laminated between adjacent layers of clothing. This fabricated cloth also demonstrates waterproof and breathable functions [20]. SMPU film responds when the temperature of the outer layer of clothing is sufficiently down; therefore, the air gap between the layers of clothing becomes broader. At low temperature, this air-gap broadening is attained if the film changes an out-of-plane deformation. This deformation need be adequately strong to resist the weight of the clothing, and the forces induced by the movements of the wearer. Again, the deformation should be reversal when the temperature of outer layer is subsequently hot. In addition to that, Apollo dress shirt from Ministry of Supply uses NASA space suit technology, which can also regulate body temperature and

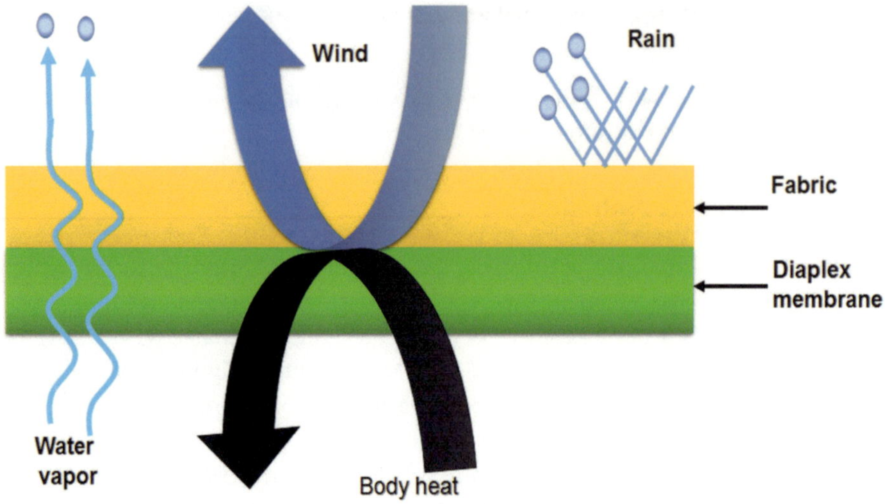

Figure 2. Structural illustration of twin layer of DiaPlex fabric.

also keeps wearers dry and odor free. Additionally, various types of thermo-regulated textile products, such as blankets, sleeping bags, underwear, jackets, sports garments, socks, ski boots, helmets, etc., have been on the market.

2.2. Self-adaptability of shape

SMP fibers are utilized to produce self-adaptable textiles which can easily regulate its structures with changes in environmental temperature. Though the shape memory effect (SME) of a fiber is basically considered as the change of its length, but this SME can take a range of forms after incorporated into fabrics, such as bending, shrinkage, and thickness increase, which is determined by the fabric structures. The SMP fibers fabricated garments can be properly expanded to fit in the wearer's body. Vertical pressure tests suggest that the garments made of SMP fibers possess a relatively low vertical tension stress in comparison with elastic fibers. This can be attributed to the deformability and fixability of SMP fibers into temporary shapes, which diminishes the adverse pressure sensation to wearers. The garment made of SMP fibers can enlarge and adapt to the wearer size, while no significant pressure is being exerted on the wearer due to the shape fixity of the SMP fibers. The garment made of Spandex fibers generates pressure on the wearer due to the high elasticity of the Spandex fibers in the fabric. The fabric made with SMP fibers with improved comfort sensation can be used especially for intimate apparel and low-pressure socks [21, 22].

2.3. Shape retention

The cotton fabrics treated with SMP possess a wrinkle-free effect [23]. In fabrics, especially cotton materials, wrinkles are easily generated during wearing or storage due to the debonding and slippage of hydrogen bonds. In order to avoid that wrinkles in cotton, dimethyloldihydroxyethylene urea (DMDHEU) is mostly used during finishing of cotton. The SMP-treated fabric does not release formaldehyde compared to DMDHEU, which is a benefit of usage of SMP. In addition to that SMPU finishing on cotton improves the mechanical strength of fabrics to some extent in contrast with that of conventional durable-press finishing process. The SMPU emulsion-treated fabric demonstrated adequate wrinkle-free effect after repeated washing, and it can easily last up to hundreds of laundering cycles [20]. SMPU-treated cotton fabrics also possess a greater crease and pattern retention ability due to presence of SME. The crease or pattern design on textiles can provide the esthetic appeal in garment. In addition to cotton, wool fabrics were also treated with the SMPU emulsion. SMP-treated wool garment has better dimensional stability than that of an untreated garment. This is due to the fact that SMPU covers wool fiber, and therefore, it reduces the wool directional frictional effect after the finishing process. The untreated wool garment shrinks to a small size, while the treated garment maintains its size after laundering. Wool fabrics and sweaters possess a daunting tendency to felting with the entanglement of scales by directional friction. The treated wool fabric showed significant reduction in felting.

2.4. Smart wettability

Smart wettability is one of the most desire requirements in garment sector. In most of the cases, it can be achieved by providing the hydrophobic surface on the fabric. Such surface with a water contact angle greater than 150° and a sliding angle (SA) lower than 10° can generally be obtained by periodic microsized or nanosized patterns. In addition to microparticle or nanoparticle coating or finishing on the textile surface, SMP coating provides the same effects. The micro- or nanopattern effect can be achieved by the phase transformation of SMPs. If a SMP coating is used on a fiber surface, it will demonstrate different shrinkage effects after heating and cooling. This reflects that SMP-coated fabric has a potential application to achieve a water-repellent effect or control water-spreading behavior. Besides hydrophobicity, controlling the movement of water on textile surfaces also attracts interest of many researchers. The behavior of water reverting from repellency (hydrophobic) to adhesion (hydrophilic) is basically governed by a combination of the contact angle and sliding angle. In this regard, Zhong et al. fabricated smart cleaning cotton fabrics using cross-linked thermo-responsive polymer, poly(2-(2-methoxyethoxy) ethoxyethyl methacrylate-co-ethylene glycol methacrylate) [24]. They found that both the wetting time and contact angle of the cotton fabrics significantly increased, when the temperature is above the lower critical solution temperature (LCST), indicating the cotton surface switches from hydrophilicity to hydrophobicity. As the performance of cleaning is generally enhanced when the surface is more hydrophilic, the cotton fabrics with cross-linked polymer can be used to prepare the fabrics with smart cleaning ability. Such fabrics can be cleaned at much lower temperatures as compared with common cotton fabrics. This can efficiently save a significant amount of energy which is in textile cleaning.

3. Different type SMPs used in textile application

3.1. Thermal-responsive SMPs

Thermal-responsive SMPs are one of the most extensively studied and used SMPs. Various forms of SMPs such as solution, emulsion, film, fiber, foam, and bulk forms are developed for suitable application under different circumstances. Generally, T_g or melting temperature (T_m) is selected as triggering temperature of this kind of SMPs. The triggering temperature of SMPs is adjustable and can be adjust around body temperature too. Several advantages of these SMPs make them suitable for textile applications such as superior processability, mechanical properties, high deformability, and high recoverability. Yang et al. developed an *N*-isopropylacrylamide (NIPAAm) hydrogel–treated cotton fabric that was able to absorb a large amount of water from humid air, which can be subsequently released at the higher temperature than the LCST of the polymer. The combined effect of the intrinsic phase transition of PNIPAAm around the LCST and the change between extreme superhydrophilic/superhydrophobic morphologies of the hydrogel is mainly responsible for this phenomenon [25].

3.2. Water/solvent-responsive SMPs

Water or other solvent is used for triggering this kind of SMPs. The original shape is recovered due to the plasticizing effect of solvent molecules, which enhanced the flexibility of polymeric chains. If an SMP has a hydrophilic or water-soluble ingredient, the shape recovery can be accelerated. A pyridine unit, which is responsive to moisture, is good choice to use for improvement of the moisture absorption in PU. Hu *et al* introduced a pyridine unit (*N*-bis(2-hydroxyethyl) isonicotinamine) into SMPU to achieve moisture-responsive SMPU with high strain recovery and recovery speed [26]. In addition to that, Hu and his team also fabricated cellulose nanowhiskers (CNWs)/elastomer and CNW/SMPU nanocomposites which exhibited good water sensitivity due to the reversible hydrogen bonding. They found that the fabricated CNW/SMPU showed triple SMEs upon exposure to sequential thermal and water stimuli [27].

3.3. Thermal/pH-responsive polymeric hydrogels

Thermal/pH-responsive polymeric hydrogel is another kind of SMPs which is used in textiles to produce smart textile. This polymeric hydrogel is a three-dimensional macromolecular gel network containing a large fraction of water within their structures. The degree of swelling of such gels increases or decreases at below or above a critical temperature. At a higher temperature, the hydrophobic interaction among hydrophobic segments is high, while the hydrogen bonding is less. The final result yields shrinkage of gel due to the hydrophobic interactions. PNIPAAm hydrogel is most frequently studied polymeric gel in textile. PNIPAAm hydrogel possesses a LCST in aqueous medium around 32–34°C which is close to the human body temperature. By adjusting the PNIPAAm copolymer composition and topology, the phase transition can be easily tuned. PNIPAAm hydrogel-coated fabrics showed reversible swelling/shrinkage (hydration/dehydration) and good water vapor permeability (WVP) of the fabrics. These properties enable the achievement of temperature-sensitive hygroscopic fabrics, temperature-sensitive deodorant fibers, and temperature-sensitive nutrient delivery fabrics. Yang et al. grafted PNIPAAm on cotton fabrics to cover the hydrophilic surface of the cotton with a polymer layer which would be able to absorb

http://dx.doi.org/10.5772/intechopen.69742

water from a humid environment, below the LCST, and release it upon a temperature change [25]. The coated cotton fabric shows temperature-triggered reversible and repeatable change in the wettability (**Figure 3**). Instead of such advantages of PNIPAAm, it possesses moderate cytotoxicity which hampers its application in few special field. Researchers have developed the copolymer composed of 2-(2-methoxyethoxy)ethyl methacrylate and oligo(ethylene glycol) methacrylate as an ideal substitute for PNIPAAm [28]. Lee and his team investigated the pH-responsive LCST behavior of polyethylene oxide-based functional polymers with different pendant amine groups and varying side chain lengths. Depending on the nature of the pendant amine groups (primary amine, dimethylamine, and diethylamine) and the hydrophobicity of the side chains (ethyl, propyl, and hexyl), LCST is easily tuned from 44 to 100°C under different pH conditions [28].

3.4. Light-active polymers

Light-active polymers are particularly attractive for various textile applications. Many light-active polymers have been synthesized and used in this field, such as azobenzene-based crystalline elastomer, anthracene-based polymers, and coumarin-based polymers. Even though some light-active threads and nanofibers have been reported, the application of light-active polymers in textiles is yet to explore comprehensively. Requirement of special wavelength light to activate such light-active polymers is the main restriction for their practical applications in garments. Esteves and coworkers fabricated a light-active polymer-coated cotton fabric by functionalizing a spiropyran–NIPAAm hydrogel [29]. This treated fabric is capable of dimensional changes upon irradiation with visible light or upon a temperature stimulus (**Figure 4**). This light active fiber may also pave an opportunities in other technological applications, such as breathable textiles and agricultural purposes.

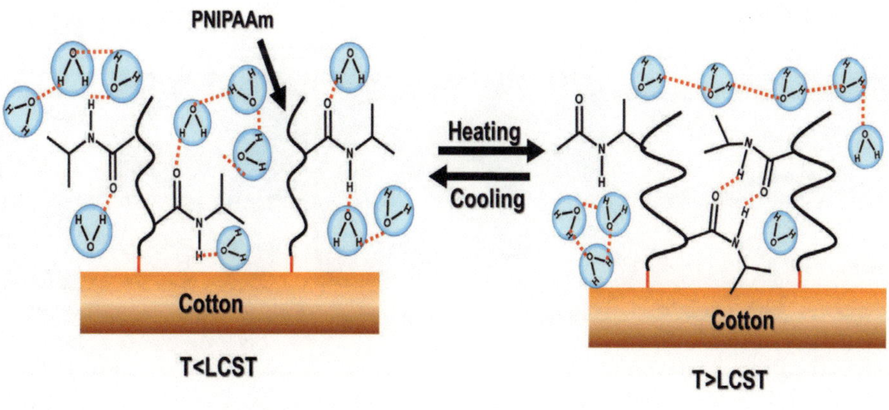

Figure 3. Temperature-triggered reversible wettability of PNIPAAm.

4. Application of SMPs in textile

4.1. Finishing fabrics

Thermo-responsive SMPs can be applied in the textile via garment finishing. Wrinkle-free, crease retention, anti-shrinkage and dynamic aesthetic textiles are fabricated by treating SMPs on fabrics. The cotton fabrics treated with SMP show good wrinkle-free effect. Cotton fabric treated with SMPU can recover to its original flat shape (wrinkle-free) within a minute upon blowing steam over it (**Figure 4a**). SMP-treated fabrics possess good crease and pattern retention ability (**Figure 4b**). As a result, good aesthetic appeal is achieved in such treated fabric.

4.2. Breathable fabrics

SMPUs are used for fabrication of breathable fabrics. WVP of the SMPU can be easily altered according to the wearer's body temperature. The overall WVP of SMPU films can be improved by incorporating hydrophilic segments such as dimethylpropionic acid and diol-terminated poly(ethylene oxide) in SMPUs. The overall WVP of SMPs can also be significantly increased by forming microfoams in the SMPUs.

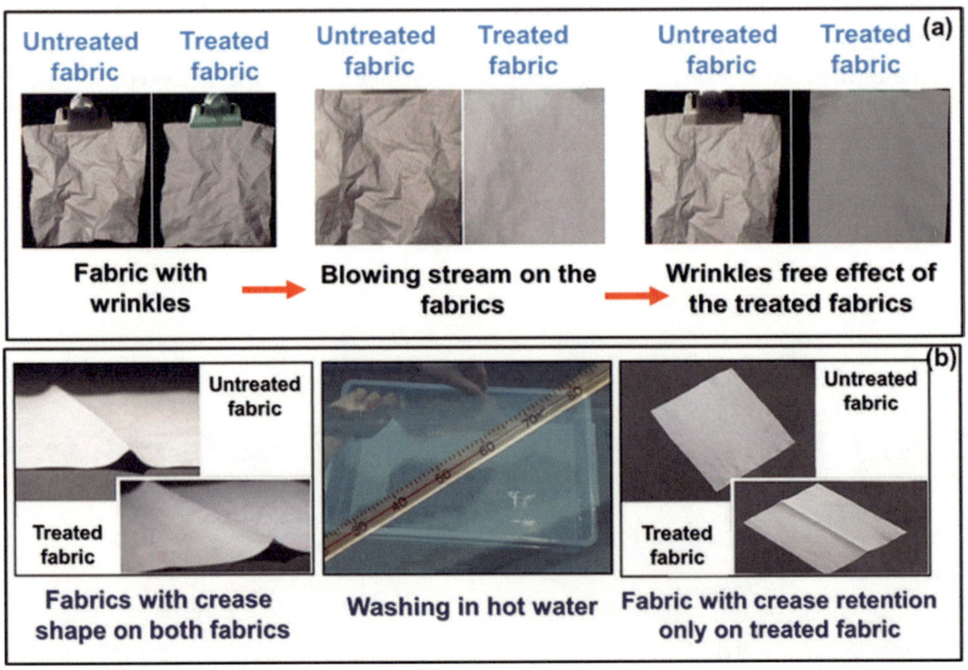

Figure 4. (a) Wrinkle-free effect and (b) crease and pattern retention ability of SMPU-treated cotton fabrics (reproduced with permission from Ref. [15]).

4.3. Damping fabrics

SMPs can absorb impact energy due to their good damping properties at around T_g. Allied Signal Inc. manufactured an automotive seatbelt fabric using SMP fibers (Securus fibers) which can absorb the kinetic energy and so effectively increase a passenger's safety by utilizing the damping effect of the SMPs [30]. The Securus fibers are made of shape memory poly(ethylene terephthalate)-poly(caprolactone) block copolymers. It is reported that the fiber can absorb energy from the body's forward motion and thus improves the safety of passengers during an accident. At first, the seatbelt holds the passenger securely in place; then, it elongates slightly and cushions the body as the belt absorbs the force from the body.

4.4. Skin-care products

The skin-care products prepared using stimuli-responsive hydrogel-treated textiles possess moisturizing, whitening, brightening and even anti-ageing effects on human skins. **Figure 5** demonstrates a facial mask which is made of nonwoven fabric treated with thermo-responsive polymeric gel [31]. This mask is responsive to human body temperature. Such smart mask can also utilize as a temperature-responsive carrier medium for controlled release of nutritious ingredients, perfumes, or other drugs to human skin. The study has revealed that the release of vitamin C easily could be controlled just by judiciously changing the temperature [32]. Even though such advantages, there are still some challenges such as the high stiffness and brittleness that are important to be resolved.

4.5. Wound-dressing products

Chitin/chitosan and chitosan derivatives are extensively used to fabricate wound dressing product due to their good antibacterial properties and wound-healing effects. Chitosan hydrogel as

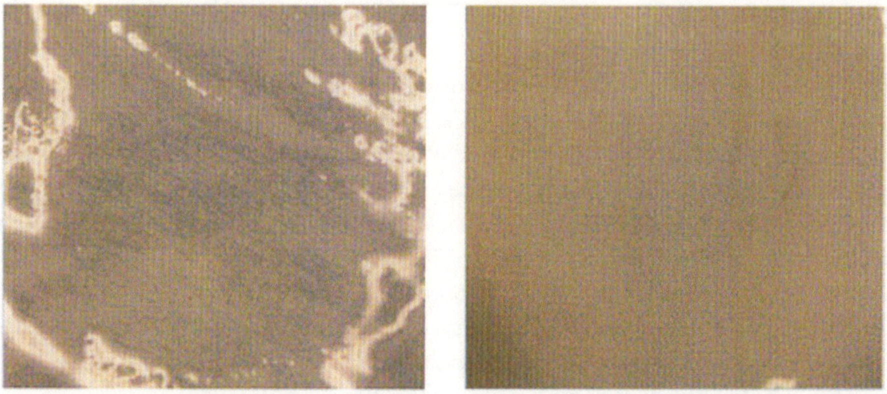

Figure 5. A facial mask made of the nonwoven fabric treated with TRPG at room temperature and body temperature (reproduced with permission from Ref. [15]).

a wound dressing material can help in the reestablishing of skin architecture. Chitosan-treated alginate filaments and cotton fabric have been fabricated for advanced wound dressings [33]. In addition to chitin/chitosan, various biopolymer-based hydrogel products have also been developed for wound dressing [34, 35]. Such smart wound dressing materials can deliver a novel drug release system in response to variations in pH/temperature; thus, the wounds can quickly heal.

4.6. Deodorant fabrics

Such special fabrics have an ability of releasing deodorant agents at certain temperatures. The smart fabric is fabricated by coating a stimuli-responsive polymeric hydrogel on the textile surface. The hydrogel is attached to the textiles surface through chemically cross-linking by using a functional monomer, such as acrylamide, and a cross-linking agent such as 2-(diethylamino)ethyl acrylate. Deodorant is generally incorporated into the hydrogel during or after the cross-linking reaction. Usually, β-cyclodextrin is loaded with polymeric gel for further enhancing the controlled release properties of hydrogel-modified textiles [36]. β-cyclodextrin has a hydrophobic interior cavity and a hydrophilic external surface in its structure [37]. Therefore, different types of hydrophobic guest deodorant moieties can be included in its cavity. Recently, a few body-responsive deodorant hydrogel products are also available on the market. Even though such advantages of deodorant fabrics, it is not fully explored. Few issues such as soft handle and high stability of the products need to be taken into consideration for development of such fabric. Another important factor is controlling of the hydration degree of the product in the open environment.

4.7. Smart energy storage textiles

Wearable electronics carve a distinct niche to our daily life by providing changes in outdated living habits to less consuming and healthier ones. Electronic circuits are generally integrated into flexible textile matrices to achieve such smart wearable electronics. Batteries and supercapacitors (SCs) are two archetypal energy storage devices, which is extensively used in our daily life. The capability of storing energy strongly relies on the properties of electrodes, including microarchitecture, conductivity, and specific surface area. SMPs have been applied in SCs to provide SME into SCs. SMPs in a SC usually serve as the substrate or core, with a conducting layer and an active layer attached successively to the surface of the shape-memory substrate or core to make a shape memory electrode. Huang *et al.* used shape memory material as the fiber core, which was consecutively coated with MnO_2 and PPy to fabricate temperature-sensitive electrodes [38]. Two of the such fabricated electrodes are twisted to make a capacitor with a specific capacitance of 198.2 F g^{-1} at a current density of 1 A g^{-1} for 20,000 cycles. They use shape memory material as the main skeleton to make wire-shaped supercapacitor possess a unique shape memory effect which has, indeed, a practical significance when it comes to risk management (**Figure 6**).

Deng et al. also developed a fiber-shaped supercapacitor (FSSC) by winding aligned carbon nanotube sheets on a SMPU substrate [39]. The fabricated FSSC can be transformed and "frozen" into user-required shapes and sizes; it can recover to the original shape and size automatically once the temperature exceeds the thermal transition temperature (**Figure 7**). This FSSC can be further woven in electronic textiles to fabricate smart clothes for flexible electronic devices.

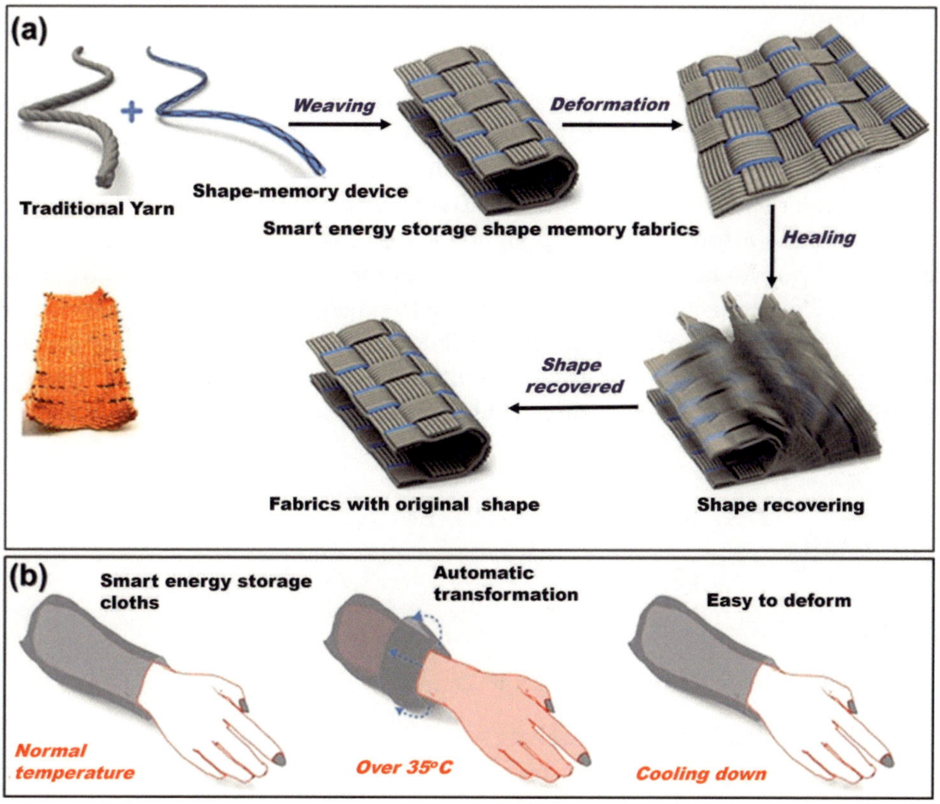

Figure 6. (a) Schematic demonstration of the shape memory SC woven with traditional fabric and the fabrication of a shape-recoverable smart textile and (b) schematic demonstration of this smart shape-memory textile used in a smart cloth (reproduced with permission from Ref. [38]).

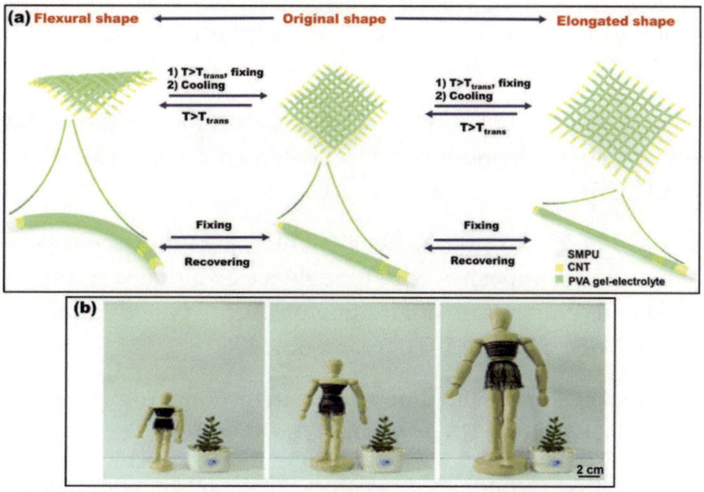

Figure 7. (a) Schematic of the FSSC and the fabricating textile that are reversibly altered into flexural or elongated states and recovered to the original shape and (b) photographs of the smart clothes fabricated by FSSC which were "frozen" into different shapes and sizes (reproduced with permission from Ref. [39]).

5. Conclusion

SMPs provide potential and enormous current opportunities in the textile industry. Textiles with SMPs can move or change their shapes, achieving different 3D forms in garments, enhancing their aesthetic appeal. Window curtains or screens with SMPs can open and close intelligently under environment stimulation. The microstructure or macrostructure changes in smart clothing in response to stimuli are a good means for achieving heat and moisture management of human bodies with a feeling of comfort. The change of fabric configuration can also be used for protection against extreme environments. With the rapid development of SMPs and novel strategies for integrating SMPs into textiles, it is anticipated that the research into smart textiles with SMPs will grow in multiple dimensions as a result of their promising potential applications. In the future, textiles may perform functions that are much more significant, far beyond what is being achieved at present.

Author details

Suman Thakur

Address all correspondence to: sumanchem89@gmail.com

Institute of Textiles and Clothing, The Hong Kong Polytechnic University, Hung Hom, Hong Kong

References

[1] Thakur S, Karak N. A tough, smart elastomeric bio-based hyperbranched polyurethane nanocomposites. New Journal of Chemistry. 2015;**39**(3):2146-2154. DOI: 10.1039/C4NJ01989J

[2] Yang B, Huang WM, Li C, Li L. Effects of moisture on the thermomechanical properties of a polyurethane shape memory polymer. Polymer. 2006;**47**:1348-1356. DOI: 10.1016/j.polymer.2005.12.051

[3] Yang G, Wan X, Liu Y, Li R, Su Y, Zeng X, Tang J. Luminescent poly(vinyl alcohol)/carbon quantum dots composites with tunable water-induced shape memory behavior in different pH and temperature environments. ACS Applied Materials and Interfaces. 2016;**8**(50):34744-34754. DOI: 10.1021/acsami.6b11476

[4] Jiang HY, Kelch S, Lendlein A. Polymers move in response to light. Advanced Materials. 2006;**18**:1471-1475. DOI: 10.1002/adma.200502266

[5] Cho G. Smart Clothing: Technology and Applications. 1st ed. Boca Raton: CRC Press; 2009

[6] Cho G, Lee S, Cho J. Review and reappraisal of smart clothing. The International Journal of Human-Computer Interaction. 2009;**25**:582-617. DOI: 10.1080/10447310902997744

[7] Gu JF, Gorgutsa S, Skorobogatiy M. Soft capacitor fibers using conductive polymers for electronic textiles. Smart Materials and Structures. 2010;**19**:115006. DOI: 10.1088/0964-1726/19/11/115006

[8] Tao X. Wearable Electronics and Photonics. 1st ed. Boca Raton: CRC Press; 2005

[9] Zheng NG, Wu ZH, Lin M. Infrastructure and reliability analysis of electric networks for E-textiles. IEEE Transactions on Systems, Man, and Cybernetics, Part C. 2010;**40**(1):36-51. DOI: 10.1109/TSMCC.2009.2031497

[10] Gu JF, Gorgutsa S, Skorobogatiy M. Soft capacitor fibers for electronic textiles. Applied Physics Letters. 2010;**97**:133305. DOI: 10.1063/1.3488351

[11] Kongolo D. Luminescent textiles. WIPO Patent Application WO/2008/148138; 2008

[12] Sayed I, Berzowska J, Skorobogatiy M. Jacquard-woven photonic bandgap fiber displays. Research Journal of Textile and Apparel. 2010;**14**:97-105

[13] Studstill K. Emotion Sensing Dress Releases Mood Driven Scents [Internet]. 2010. Available from: www.psfk.com/2010/01/emotion-sensing-dress-releases-mood-driven-scents.html

[14] Qi K, Chen X, Liu Y. Facile preparation of anatase/SiO2 spherical nanocomposites and their application in self-cleaning textiles. Journal of Materials Chemistry. 2007;**17**:3504-3508. DOI: 10.1039/B702887C

[15] Hu J, Meng H, Li G, Ibekwe SI. A review of stimuli-responsive polymers for smart textile applications. Smart Materials and Structures. 2012;**21**(5):053001. DOI: 10.1088/0964-1726/21/5/053001

[16] Wang, S. X., Li, Y., Tokura. Effect of Moisture management on functional performance of cold protective Clothing. Text. Res. J.. 2007;**77**:968-980. DOI: 10.1177/0040517507083552

[17] Crespy D, Rossi R. Temperature-responsive polymers with LCST in the physiological range and their applications in textiles. Polymer International. 2007;**56**:1461-1468. DOI: 10.1002/pi.2277

[18] Shanmuganathan K, Capadona JR, Rowan SJ, Weder C. Biomimetic mechanically adaptive nanocomposites. Progress in Polymer Science. 2010;**35**:212-222. DOI: 10.1016/j.progpolymsci.2009.10.005

[19] Mondal S, Hu JL. A novel approach to excellent UV protecting cotton fabric with functionalized MWNT containing water vapor permeable PU coating. The Journal of Applied Polymer Science. 2007;**103**(5):3370-3376. DOI: 10.1002/app.25437

[20] Hu J. Smart polymers for textile applications. In: Aguilar MR, Román JS, editors. Smart Polymers and Their Applications. 1st ed. Cambridge, UK: Woodhead Publishing; 2014. pp. 437-475

[21] Hu J, Lu J. Memory polymer coatings for smart textiles. In: Hu J, editor. Active Coatings for Smart Textiles. 1st ed. Cambridge, UK: Woodhead Publishing; 2016. pp. 11-34

[22] Liu Y, Chung A, Hu JL, Lu J. Shape memory behavior of SMPU knitted fabric. Journal of Zhejiang University SCIENCE A. 2007;**8**:830. DOI: 10.1631/jzus.2007.A0830

[23] Li Y, Chung SP, Chan L, Hu JL. Characterization of shape memory fabrics. Textile Asia. 2004;**35**:32-37

[24] Zhong Q, Chen YY, Guan SL, Fang QS, Chen T, Müller-Buschbaum P, Wang JP. Smart cleaning cotton fabrics cross-linked with thermo-responsive and flexible poly(2-(2-methoxyethoxy)ethoxyethyl methacrylate-co-ethylene glycol methacrylate). RSC Advances. 2015;**5**:38382-38390. DOI: 10.1039/C5RA03256C

[25] Yang H, Zhu H, Hendrix MMRM, Lousberg NJHGM, de With G, Esteves ACC, Xin JH. Temperature-triggered collection and release of water from fogs by a sponge-like cotton fabric. Advanced Materials. 2013;**25**:1150-1154. DOI: 10.1002/adma.201204278

[26] Chen S, Hu J, Yuen C, Chan L. Novel moisture-sensitive shape memory polyurethanes containing pyridine moieties. Polymer. 2009;**50**:4424-4428. DOI: 10.1016/j.polymer.2009.07.031

[27] Luo HS, Hu J, Zhu Y. Polymeric shape memory nano-composites with heterogeneous-twin-switches. Macromolecular Chemistry and Physics. 2011;**212**:1981-1986. DOI: 10.1002/macp.201100292

[28] Lutz JF, Hoth A. Preparation of Ideal PEG analogues with a tunable thermosensitivity by controlled radical copolymerization of 2-(2-methoxyethoxy)ethyl methacrylate and oligo(ethylene glycol) methacrylate. Macromolecules. 2006;**39**(2):893-896. DOI: 10.1021/ma0517042

[29] ter Schiphorst J, van den Broek JM, de Koning T, Murphy JN, Schenning APHJ, Esteves ACC. Dual light and temperature responsive cotton fabric functionalized with a surface-grafted spiropyran–NIPAAm-hydrogel. The Journal of Materials Chemistry A. 2016;**4**:8676-8681. DOI: 10.1039/C6TA00161K

[30] AlliedSignal. Introduces New Fiber to Boost Seatbelt Safety [Internet]. 1999. http://www.accidentreconstruction.com/news/jul99/070699a.html

[31] Hu J, Liu W, Liu B. Fabric-supported chitosan modified temperature responsive PNIPAAm/PU hydrogel and the use thereof in preparation of facial mask. US patent. 2016; **US20060286152**

[32] Hu J. Research and Scholarly Activities—Smart Fabrics for Magic Cosmetics [Internet]. 2010. Available from: www.polyu.edu.hk/cpa/profile/06feb/pdf/18-23.reseach.pdf [Accessed: March 12, 2011]

[33] Ribeiro MP, Espiga A, Silva D, Baptista P, Henriques J, Ferreira C, Silva JC, Borges JP, Pires E, Chaves P, Correia IJ. Development of a new chitosan hydrogel for wound dressing. Wound Repair and Regeneration. 2009;**17**:817-824. DOI: 10.1111/j.1524-475X.2009.00538.x

[34] Petrulyte S. Advanced textile materials and biopolymers in wound management. Danish Medical Bulletin. 2008;**55**:72-77

[35] Peng H T, Martineau L, Hung A. Hydrogel–elastomer composite biomaterials: 4. Experimental optimization of hydrogel–elastomer composite fibers for use as a wound dressing. The Journal of Materials Science: Materials in Medicine. 2008;**19**:1803-1813. DOI: 10.1007/s10856-007-3324-y

[36] Pathak H. Smart' textile materials: A paradigm shift. Journal of Textile Science and Eng ineering. 2012;**2**:107. DOI: 10.4172/2165-8064.1000e107

[37] Szejtli J. Past, present, and future of cyclodextrin research. Pure and Applied Chemistry. 2004;**76**:1825-46

[38] Huang Y, Zhu M, Pei Z, Xue Q, Huang Y, Zhi C. A shape memory supercapacitor and its application in smart energy storage textiles. The Journal of Materials Chemistry A. 2016;**4**:1290-1297. DOI: 10.1039/C5TA09473A

[39] Deng J, Zhang Y, Zhao Y, Chen P, Cheng X, Peng H. A shape-memory supercapacitor fiber. Angewandte Chemie International Edition. 2015;**54**:15419-15423. DOI: 10.1002/anie.201508293

Chapter 13

FELT: Communicating Emotion through a Shape Changing Textile Wall Panel

Felecia Davis

Additional information is available at the end of the chapter

http://dx.doi.org/10.5772/intechopen.70032

Abstract

FELT is a 5" × 6" (150 cm × 180 cm) computational textile panel, made up of four modules of frame and textile that was designed to understand what emotion is communicated to people using vision and touch from a still and shape changing textile. The purpose of FELT is to determine what still and shape changing, textural expressions of computational textiles can communicate emotion to people at the scale of an architectural wall. The central idea is that for both still and moving or shape changing textiles, there will be differences in what is communicated to people depending on whether the people experience the textiles via vision alone or via both vision and touch. In this chapter, two methods are described that were used to carry out the research. The first method is a design method used to produce shape changing textiles. The second method described is the user study. There are two user studies that will be discussed in this chapter to understand what emotion[s] are communicated by shape changing textiles and the FELT wall panel. If architects, artists, designers, engineers and scientists, and others could begin to understand the nature of what various textile expressions communicate, and what computational textiles communicate in transformation, then it would be possible to more clearly understand the role that texture of a computational textile plays in communicating emotion through a computational object. A textile that can move or change its shape could be used on a robot as robot skin, for example, for people who may benefit from some communication through vision and touch. A computational textile may be used on a wall, a pillow, curtain, furnishings, toys, and many other designed objects to communicate to people and children who do not have access to their emotions. This may be a child with autism or someone not in touch with what they are feeling. A textile can be used to have nonverbal communication with people through vision and touch.

Keywords: shape changing textiles, textile expression, computational textiles, emotion, communication

1. Introduction

Textiles used to create shelters have for centuries communicated narratives and stories to people and their communities. One only has to look at an example of Berber carpets used to make ancient tent shelters to learn about the communication of stories through images and symbols (**Figure 1**).

This figure shows tents erected by the nomadic Berber people of Morocco made from woven carpets knotted with sheep, goat, and/or other animals' hair. The carpets are richly patterned, and according to Barbatti, a Swiss scholar and writer, some of the symbols and patterns used date back to the hunter gatherer cultures of the European Mesolithic and Upper Paleolithic ([1], p. 16). For example, many of the hundreds of symbols used, including chevrons, x's, lozenges, Solomon's stars, and the floral motif called a "camel track," have to do with fertility.[1]

Yet, what of the information in textiles that can be carried through seeing and feeling texture and using the haptic system to receive communication from a textile rather than relying on symbolic form and symbolic color or symbols in general? Is it possible to communicate *intentionally* with people through textiles without using symbols? Is it possible to communicate emotion through textural expression of a textile? Emotion is selected as a quality to communicate in these studies as emotions are directly related to bodily experiences rather than sorting out things symbolically. In this chapter, the author will explore the potentials for texture to communicate emotion to people engaging in seeing and touching a still and shape changing, computational textile wall panel (**Figure 2**). The term computational textile is not a common term but one that has been in use for a decade or more. The term computational textile will be used

Figure 1. Shelter of Berber carpets. http://ethnographic-morocco.blogspot.com/2013_04_01_archive.html (Accessed December 2, 2017). Photo by Coralle Merier, 2009, Creative Commons License https://creativecommons.org/licenses/by-nc-sa/2.0/legalcode

[1](For a good description of the symbols and meanings on these Berber Carpets see Barbatti [1]:21).

in this chapter because typical computer commands to microcontrollers and other sensors are employed to make the textile behave or respond in a specific way. Computational textiles are textiles others often call e-textiles, electronic textiles, or smart textiles. If it is possible to communicate through texture and shape changing computational textures using vision and touch, then designers, engineers, and others may use computational textural expression in their designs for furnishings, robots, and other objects to communicate emotion to people. This opportunity may be especially useful in objects, places, and furnishings designed for people who are unable to communicate their emotions, or express their emotions to others and themselves. These computational objects can help those people get in touch with what they are feeling.

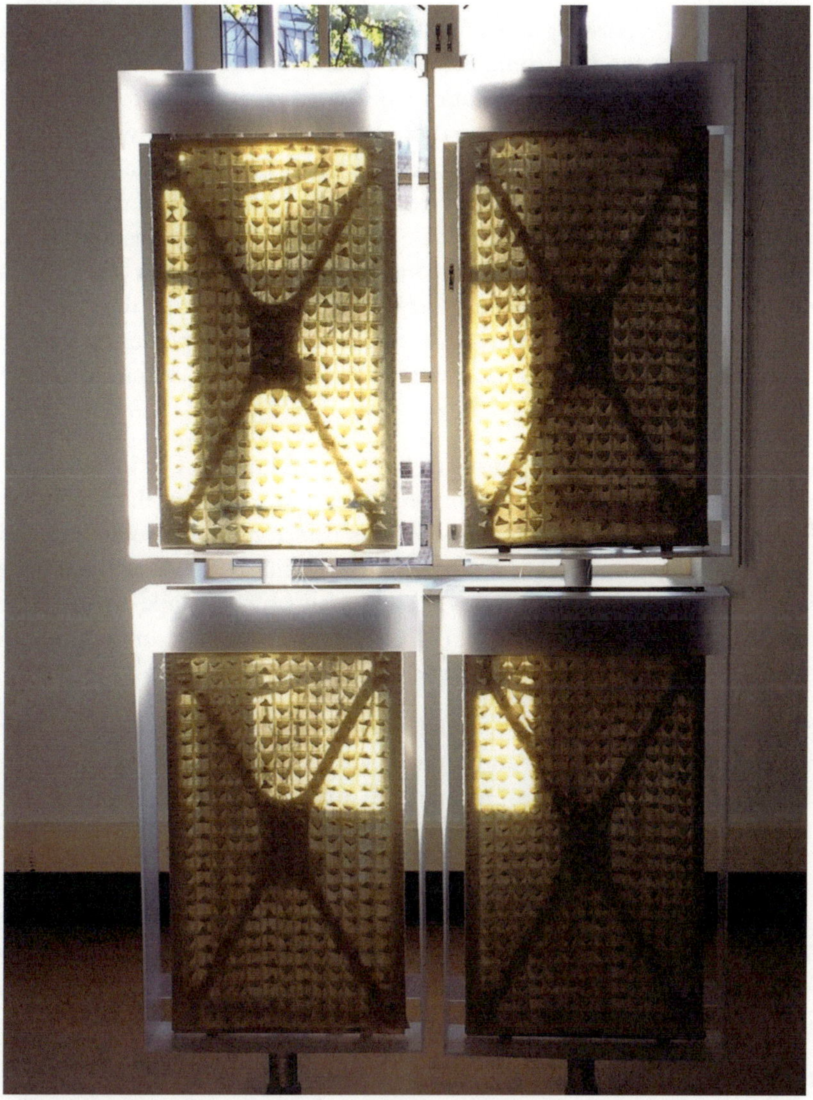

Figure 2. FELT wall panel.

1.1. Background: relating expression, communication, and emotion

Designers and artists have always been engaged with the problem of creating expressions that expand beyond their own experiences into something that can be shared, of finding ways to place viewers in a world that is relational. Expressions can communicate and are methods to connect and live with others, across boundaries. In this section, the relationship between the two words, i.e., expression and communication will be discussed. Expression and communication through the body, and the ways in which expression in a body, human or animal, is or is not communicated are discussed. If designers, scientists, and others understand that communication through the body is perceived through the expressions and habits formed by the body in its habitat, then it is possible to understand how the body communicates and is communicated via emotion.

If you look up the word *expression* in the Oxford English Dictionary, you will find two meanings ascribed to it. The second meaning has most relevance for this chapter. This meaning of expression designates it as a "representation" or "manifestation," for example, "the action of expressing or representing (a meaning, thought, state of things) in words or symbols; the utterance (of feelings, intentions, etc.)" (Oxford English Dictionary, accessed December 2, 2017) [2]. A critical word in this meaning or definition is that of intention. The word intention will be returned to later in this chapter. When the term "expression" is used in this chapter, no ontological claims are made. The word considered here is closer to the term "expressive power" as defined in computer science. "Expressive power" means the "measure of ideas expressible in any particular language" (Wikipedia "Expressive power (computer science)" accessed December 2, 2017) [3]. The language examined in this chapter is shape changing textile textures rather than symbols.

Communication has three primary senses in the Oxford English Dictionary [4]. The first of these senses is related to affinity or having something in common; the second is related to imparting or transmitting something: "The action of communicating something (as heat, feeling, motion, etc.), or of giving something to be shared." The last sense is that of having access, access between two people or places, having a shared physical link. Before discussing the methods and results of the experiments with FELT, it is important to understand the relationship between expression, communication, and emotion.

In *The Expression of the Emotions in Man and Animals,* Darwin starts to look at the issue of expression in the bodies of humans and animals (1890/2009). In this follow-up to *The Origin of the Species,* he describes hundreds of human and animal body expressions in minute, detail and maps them to different emotions and functions [5]. Yet Darwin made no comment on the causal connection between expression and emotion.

Darwin never investigated whether a person could experience an emotion without a corresponding bodily expression or whether bodily responses were a necessary condition for an emotional experience ([6], p. 100). Neither did Darwin make a connection between expression and communication.

In Darwin's account, many expressions of emotion such as the hair rising on the back of a person's neck when afraid are leftover responses related to an evolutionary function designed to maximize chances of survival in some prehistoric time. For example, writing about the behavior of a hen trying to chase a dog away from its brood, Darwin notes what he considers to be the principle cause of erect feathers, as follows:

> *(B)irds when frightened, as a general rule, closely adpress all their feathers, and their consequently diminished size is often astonishing. Though with birds, anger may be the chief and commonest cause of the erection of feathers ([7], p. 97).*

Summarizing from an extensive report on the expressive behavior of two classes of vertebrates, Darwin was trying to understand the function of emotional expression in the context of evolution. *Expression in this sense is about changing the body in relation to the environment.* In the case of the birds, by closing their feathers, they withdraw from the environment and by opening their feathers in anger they make an approach to it.

According to the psychologist Nico Frijda, the expressions that Darwin discusses are not intentional and have very low communicative value other than in the broadest sense of communication because those expressions are not intentional. In such cases, Darwin describes above the communication is a by-product of witnessing a message that was not intentionally sent although others received it ([8], p. 60). The message was inferred by an observer based on the subject's body in context.

Frijda argues that there are actually two senses of communication. A more restrictive sense refers to behavior produced in order to be perceived by another animal, and *"in order to influence the latter's behavior"* ([8], p. 103). This more restrictive sense is understood as true communication. Frijda proposes that to understand emotional expressions is to sense the impact of those expressions on the observer.

For Arnheim, a German perceptual psychologist, the meanings of esthetic expression can be understood through the expressions of the body. In dance, a dancer expresses happiness and joy by moving his/her body upward and an audience can recognize this as happiness. Similarly in dance, if a dancer wishes to communicate sadness to an audience, his/her body droops downward ([9], p. 261). People's ability to make inferences through these intentional clues is described as a process whereby people draw on their understanding of the relationship between patterns of bodily sensations and the emotions they produce ([9], p. 259).

This message is the kind of intentional, influential communication that artists and designers often want to achieve with their designs and is of interest in the FELT experiment.

1.2. Feeling things and being felt

James Gibson, a psychologist, argues that there is no separation between the body and the environment. In Gibson's account, therefore, energy flows into bodies from the environment and energy flows out from our bodies into the environment, such that it is impossible to understand the body without considering the environment ([10], p. 5). He writes about the environment as a place of flow and forces. The haptic system unlike any other perceptual system encompasses most of the body and literally puts people in touch with their world; it "lets us grab a hold of things" ([10], p. 97).

A specific quality of the human haptic system is that unlike the human visual system, it enables people to explore and alter their environment. Further, this ability means that people can also change what they perceive using their haptic system, which again is not possible with their visual system. Gibson gives a good example of this: "when we reach out to feel the edge of a table, we feel the edge, but simultaneously the table makes a dent in us." He argues that there are poles of experience that place the concepts of subject and object in a continuum. If you wish, you "can focus on the dent made in your hand or you can focus on the edge of the table" ([10], p. 99). The equipment people use to explore, feel, and alter their environment is the same equipment people use to feel and produce emotional experiences ([10], p. 99). As the philosopher Mark Johnson writes, "emotions are both *in us* and *in the world* at the same time (…) they are one of the most pervasive ways we are continually in touch with our environment" ([11], p. 67).

For Damasio, a neuroscientist, an emotion refers to a process in which initial expressive states of the body (that people can see) or physiological changes are associated with feelings. He defines feelings as the "idea of the body being in a certain way" ([12], p. 85).

John Dewey develops his own idea of emotion by connecting it to a person's lived situation. That is, for Dewey, there are emotional situations, not emotional states of mind. He writes: "I am happy" means "I feel happiness in me, and simultaneously in the situation I encounter" ([11], p. 67). By defining emotion as a transactional process, Dewey does not interiorize the concept of emotion that he argues has led to its banishment from the discussion of making meaning and creating knowledge ([11], p. 67). In his argument, Johnson picks up on a long history of abolishing emotion from scientific discourse, which happened by separating the brain from the body and by trying to maintain subjectivity and objectivity which he sees as abstractions rather than organism-environment transactions ([11], p. 67). In writing on Dewey, Johnson explains Dewey's position that hinges on connecting exterior environment with interior feelings in a way that is essentially symbiotic:

> *If emotions are merely private, interior, subjective responses then they tell us nothing objective about our world. (…) Once we see that emotions exist precisely because of the ways they are connected to our shared world and permit us to function within it, then it becomes possible to recognize their crucial role in our communal well-being ([11], p. 67).*

2. Methods used to determine emotion communicated from still and shape changing textiles

In this section, the author describes the design methods used to produce five shape changing textiles used as preliminary designs in a study which generated the texture for the final FELT panel. The chapter will also include the design methods to produce the final FELT panel.

In addition to discussing design methods, the author will also explain the user study with five textures, which informed the FELT panel. In both studies, participants were asked what emotion was communicated from photographed textures, textures that can be seen and touched, a preliminary group of five shape changing textiles as well as the final FELT panel.

2.1. Design of five shape changing textures

The textures for the studies were inspired from animal reactions some of which Darwin mentions. These reactions are seen in **Figure 3**. It was expected that movement of each animal's skin, fur or feathers would communicate or permit a person to infer a different emotion(s). Textures were developed based on the following animals and associations: a cat with its raised fur, which could indicate fear, alarm, or shock; a bird, whose feathers could indicate an angry state as discussed by Darwin ([7], p. 97); and an elephant whose skin wrinkles could indicate a number of emotions depending on how it was manipulated in compression and in tension. Only three of the five animals were selected as within this group as these three were deemed the easiest of the five to show expression through their fur or feathers and because these three showed all of the animal expressions seen in the five original animals.

A series of textile textures were fabricated from cotton muslin and felt and test-activated by pulling threads that were integrated in the fabric (**Figure 4**). Several of the textures were selected for their similarity to the reactions and the mobility seen in the skins of animals (**Figure 5**). In order to keep the focus on the transformation of the texture of the material rather than color, all five samples were made from a cream-colored felt. All the textures were 9" × 9" (23 cm× 23 cm) in size. There were varying textures, which were designed with the goal of communicating specific emotional qualities from excited to calm and either negative or positive. The samples included smooth textures, rough and triangulated textures, rounded textures, and textures with superimposed formal systems. All the samples, which were laser cut, were mounted on a box that hid the motors that actuated the texture (**Figure 6**). Some of the samples, i.e., Textures 2–4, required weights to return to their basic position. Each sample had a different motion and speed, which changed the texture of the textile. The motor speed for each textile was set so that the textures would transform in a way that mimicked the breathing of an animal in a relaxed state. Here is a link to a video that shows the five textures and the textures' range of motion and speed: http://vimeo.com/85620116.

In the end, five textures were settled on for the last study. During this study, the five textures were mounted on top of foam core boxes, which hid motors and wires inside. The motors were programmed to move the textiles slowly, at the pace of a calmly breathing creature. Small slats of wood weighted the textures so that they were pulled back down into their original position (**Figure 7**). The foam core boxes were mounted on the wall of a room at eye height so that people who participated in the study could see and touch them easily.

Figure 3. Animal reactions.

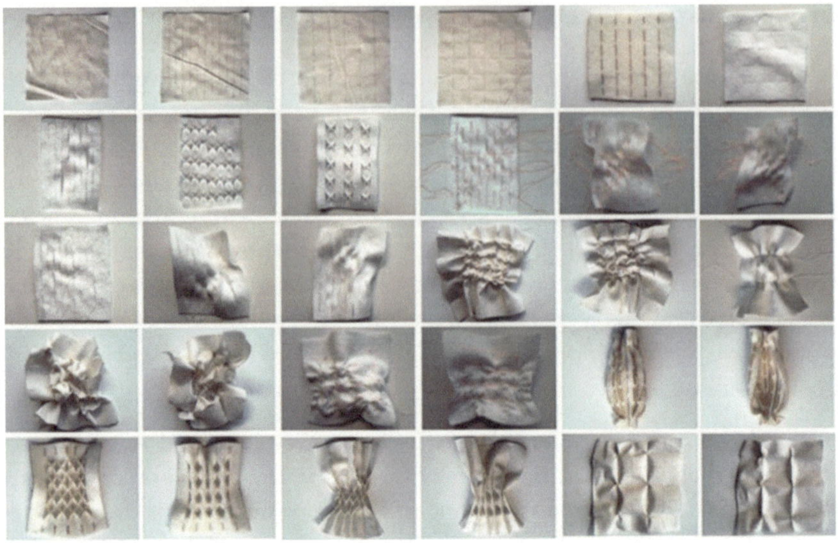

Figure 4. Transformational textures inspired by animal reactions.

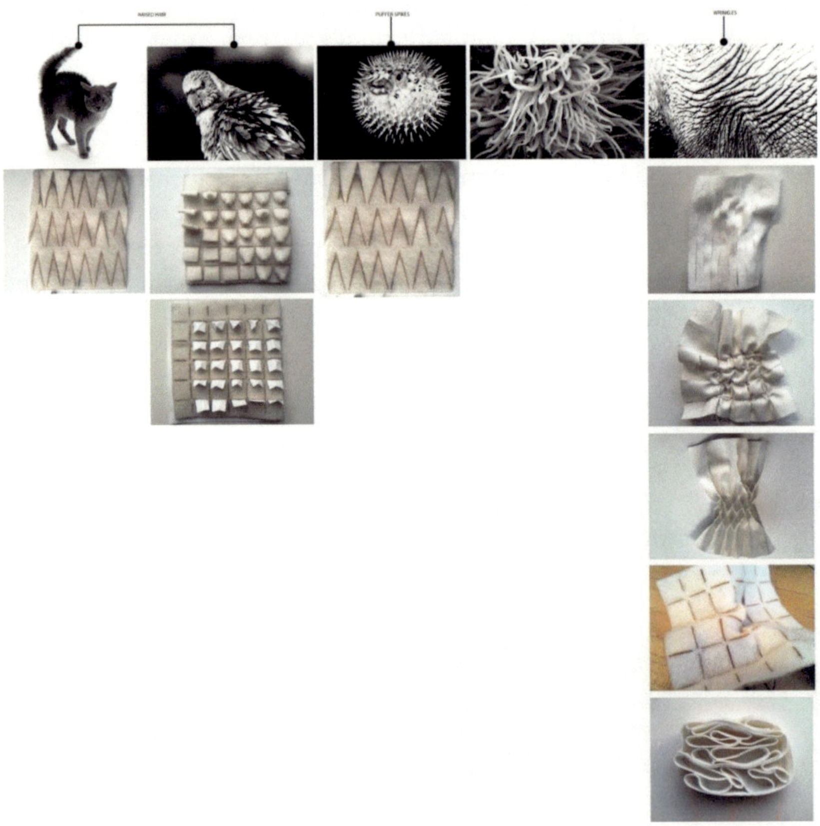

Figure 5. Selecting the textures that connected with animal reactions for the study with five textures.

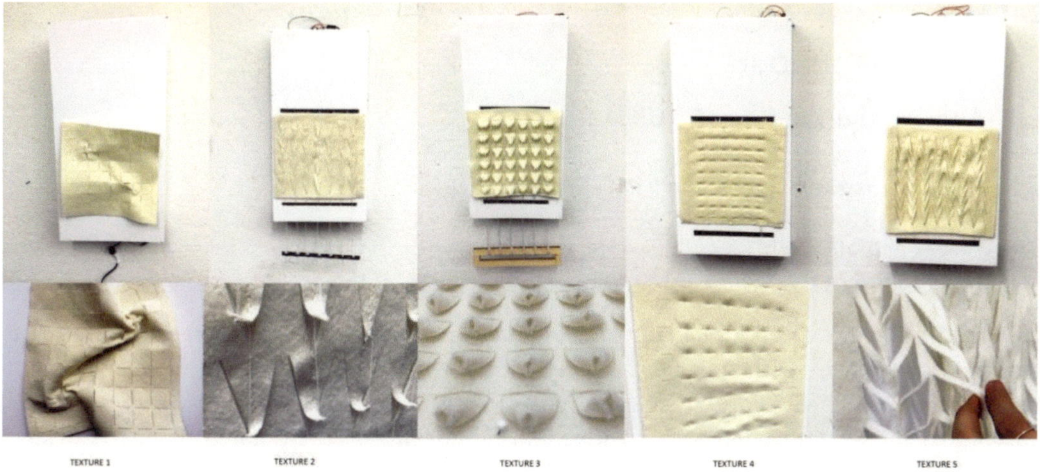

Figure 6. The five textile samples mounted on foam core boxes above, close ups photos below.

Figure 7. Face word graphics (The Makaton Charity: accessed December 2, 2017) [13].

2.2. User study with five textures: mapping emotion to still and moving textiles using vision and touch

The second method used in this experiment was the user study. The purpose of the study with five textures was to obtain participant feedback using vision and then vision and touch about emotions communicated from still and then moving textile expressions. An outline of the study is as follows:

1. Textiles are Still
   a. Looking **(ROUND 1)**
   b. Looking and Touching **(ROUND 2)**

2. Textiles are Moving
   a. Looking **(ROUND 3)**
   b. Looking and Touching **(ROUND 4)**

2.2.1. The hypotheses of the user study with five textures

There were actually four hypotheses for the study with five textures. Each hypothesis was tested in a round of the study. **Table 1** shows a summary of the hypotheses.

In Round 1, it was expected that using vision alone, people would consistently associate specific emotional states with specific characteristics of the textures of textiles in a state of stillness. Crisp, curvilinear shapes associated with positive, excited, and happy feelings; smooth curvilinear shapes associated with positive and calm feelings; triangulated shapes associated with negative and angry feelings; smooth triangulated shapes and superimposed systems, or a poorly defined combination associated with negative, depressed, and calm feelings.

In Round 2 of the study with five textures, it was expected that when people could see and touch a still texture, a negative emotional association using vision alone would change to a positive emotional association using both vision and haptic senses. When people used vision and the haptic senses together, the emotional associations changed.

In Round 3, participants were asked using their vision what emotion was communicated by textures in a state of motion. It was expected that this would raise or lower the prior rating of

Hypothesis number	Hypothesis description
Hypothesis 1, round 1	It was expected that using vision alone, people would consistently associate specific emotional states with specific characteristics of the textures of textiles in a state of stillness. Crisp, curvilinear shapes associated with positive, excited, and happy feelings; smooth curvilinear shapes associated with positive and calm feelings; triangulated shapes associated with negative and angry feelings; smooth triangulated shapes and superimposed systems or a poorly defined combination associated with negative, depressed, and calm feelings.
Hypothesis 2, round 2	It was expected that when people use vision and the haptic senses together the emotional associations would change. It was expected that when people could see and touch a still texture, a negative emotional association using vision alone would change when using both vision and the haptic senses to a positive emotional association.
Hypothesis 3, round 3	It was expected that when the textures were in a state of motion that this characteristic would raise or lower the participants' rating of what was communicated on a Circumplex grid based on what was associated with that texture motion.
Hypothesis 4, round 4	It was expected that the act of touching the moving textures would again change the ratings and what the textiles communicated.

Table 1. Study with five textures four hypotheses.

http://dx.doi.org/10.5772/intechopen.70032

that texture's position on a Circumplex Model of Affect for that texture in Round 1 based on what was associated with that texture motion.

Round 4 focused on what happened when the participants were permitted to see and touch the moving textures. It was expected that the act of touching would change the participants' responses concerning what emotion was communicated by the textile.

2.2.2. *Study procedure*

The study with five textures took place at M.I.T in January 2014, with 19 participants, i.e., 15 women and 4 men. The age range was 18–55. Participants were primarily faculty and students from the M.I.T. School of Architecture and Planning. There were also faculty and students from Aero Astro and Civil Engineering as well as from Harvard Graduate School of Design. Each session took 1 hour to complete four rounds of questions. The participants interacted with the textiles one on one with me in the room. All the participants were able to see the textiles and to touch them with their hands, and all were able to speak and write in English. The interviews were recorded by video with sound as well as by handwritten notes. All the textile samples were mounted onto foam board boxes, which hid servomotors and hung on a wall at about 46" (137 cm), which is approximately eye level for adults. The participants were standing as they viewed the textiles and were interviewed at each textile separately. All the textiles were uncovered for the entire interview.

In all four rounds, the participants were asked the same series of four questions. After the participants had the opportunity to see or see and touch the textiles, they were asked to freely associate for the first question.

1. For the first question, a set of three questions was asked in order to encourage the participants to give expansive answers. For example "What are some words that describe some emotions that you could attribute to this textile?" and "What are some adjectives that you could use to describe the mood of this textile?" Participants were told that the free association should focus on what the textile communicated to them in terms of emotional attributes. Participants were also told that it was Okay to talk about things that the textile reminded them of and to talk about any particular associations or memories that they attached to the textile. Their responses to this question were recorded by written notes and on videotape.

After this first question, the participants answered the questions presented on stapled 8.5" × 11" (22 cm × 28 cm) sheets of paper. The next two questions asked were those used in Studies 1, 2, and 3, which are as follows:

2. What does the texture communicate to you?

(Negative Mood) 1, 2, 3, 4, 5 (Positive Mood)

The participants were asked to circle a number between 1 and 5.

3. What does the texture communicate to you?

(Relaxed) 1, 2, 3, 4, 5 (Stimulated)

The participants were asked to circle a number between 1 and 5.

Last, the participants were given a sheet of faces which projected emotions and asked to pick a face in order to answer question 4.

4. Face: What mood would you associate with this texture?

Happy, Cross, Scared, Sad, O.K., Horrible, Worried, and Excited.

Then, they circled the words that they thought described one of the faces on the sheet. The faces are shown in **Figure 7**. The participants could circle as many of words as they wished.

After answering the four questions for one textile, the participants were asked to proceed to the next textile until all five textiles were reviewed for Round 1. Then, the participants were asked to begin at Textile 1 to start the next round or Round 2, then Round 3, and last Round 4.

2.2.3. The results of the user study with five textures

The results from the user study with five textures showed that emotions communicated to people using vision alone, then vision and touch together, and from still and moving textiles changed. Three analytical methods were employed to understand the results from each round of the study. These were Russell's Circumplex Model of Affect, face word cloud graphs, and the free association results.

2.2.4. Circumplex Model of Affect analysis

The most useful analytical tool was James Russell's Circumplex Model of Affect, which will be presented first. James Russell is a psychologist and professor of psychology at Boston University, who designed a two dimensional model of emotional affect in which people are asked to place words along an x axis where positive is happy on the far right of the axis. Negative is sad on the far left of the x axis. On the y axis, the highest point is excited or pumped. The lowest point on the y axis is calm [14]. **Figure 8** shows words Russell mapped to a Circumplex grid as an example. Note the position of words which express emotion on the grid. Happy excited words like "astonished" and "delighted" are in the upper right quadrant. Calm and sad words like "depressed" and "gloomy" are in the lower left quadrant. To create the Circumplex texture plot for this study, the numbers circled in questions 2 and 3 on the survey provided x coordinates and y coordinates respectively for each participant. Then the x coordinates and the y coordinates were averaged for each round to make the Circumplex plots in **Figure 8**.

During Round 1, the participants were invited to view the textures displayed in a still state. The results from Round 1 presented were not as expected. Textures 1, 2, 3, and 5 were all expected to start off in the upper-left quadrant, communicating agitation/anger, with Texture 4 starting in the lower-left quadrant, communicating boredom/depression. Instead, Textures 1, 2, 4, and 5 start in the right side of the quadrant, communicating positively and happiness. Texture 3 ends in the upper-left quadrant, communicating negativity/anger. **Figure 9** shows the Circumplex plots for all four rounds.

During Round 2, the participants were invited to view and touch the textures displayed in a still state. It was expected that all the textures would be on the positive half of the grid for

Figure 8. Circumplex Model of Emotional Affect for words, redrawn from Russell's Circumplex ([14], p. 1167).

this round, which was confirmed once Texture 3 had moved to the right. However, Texture 5 rather than going up in excitement once touched read as more subdued, indicating it was visually more interesting to see than to touch. Texture 1 moved up into the excited/happy quadrant and Texture 4 remained almost in the same place.

In Round 3, the participants were invited to see the textures in motion. It was expected that all the textures would end up in the upper-right half of the Circumplex, which the results showed was the case. However, surprisingly, Texture 1 moved into the upper left quadrant communicating agitation/anger.

In Round 4, the participants were invited to see and feel the textures in motion. It was predicted that the textures would stay in the upper-right quadrant, which proved to be the case. However, with one exception, all the textures rated lower on excitement and/or had lower y coordinates. The exception was Texture 3, which moved up considerably on the excitement scale. While this was not in the hypothesis 4, this result is understandable because participants had interacted with the textile samples3 times previously.

In the following analysis, Texture 3 alone will be looked at more closely because it had the highest variance compared to the other 4 textures. This meant people either loved or hated it. Texture 3 will become the texture used in the FELT wall panel.

2.2.5. The free association analysis for Texture 3

The next most useful analytical tool was the response to the free association questions. These responses helped to add nuance and understand why certain textures moved where they

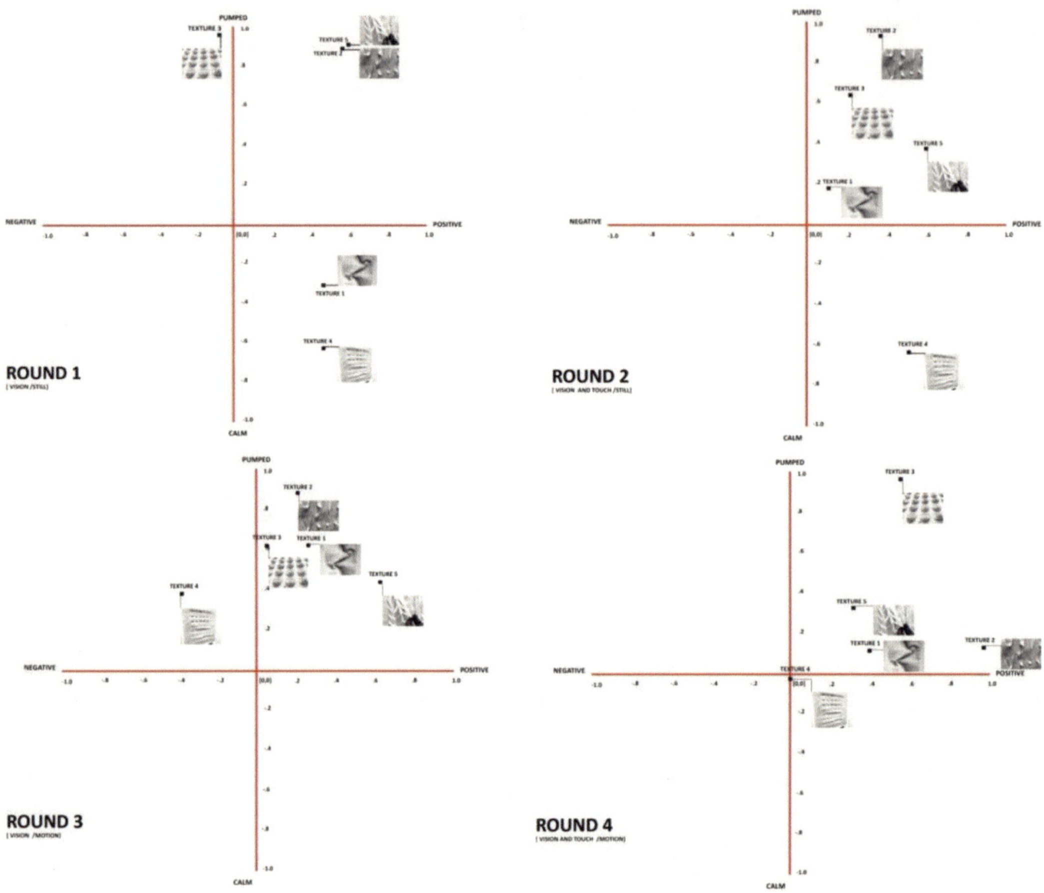

Figure 9. Circumplex Model of Affect for all four rounds.

did in the Circumplex analysis. The free association also showed change and transformation in communication of emotion as participants they stood in front of a texture. This showed that the communication of emotion was not static. In addition to this, several categories of responses emerged as the results were analyzed. These categories are often overlapping and it is certainly possible to describe other categories. The participants' comments are marked with a code (P), so that the comments can be connected to specific participants.

These categories include:

1. Memory and analogy relating to the texture via previous experiences.

This was by far the most popular response category.

2. Vitality or lack of vitality of the texture.

3. Expectation or comparison to imagination or previous texture.

4. Feels to skin.

5. Aggression or harm; understanding if texture will behave as if to harm.

Not every texture had all of these categories. A summary of the results from the free association is presented. Only three of the five categories mentioned emerged in the responses for Texture 3. These categories were memory/analogy; aggression or harm; and expectation or comparison.

The results for Texture 3 showed high variance. People reacted strongly to it either positively or negatively. For example, in Round 1, many people had good memories of this texture; these memories tended to be of in-animate objects like sweaters, bathmats, etc. For example, participant 18 said "pleasant anticipation; calls to mind a bath mat; soft to feet causes of cushioned feel; it might not be soft in this case difficult to tell; fabric looks soft; stiff protrusions like spikes or uncomfortable to touch." However, more were like participant 9 "ohh… uh scary, looks like aliens with boobs; cows breasts; got scared too many breasts; do not want to touch it; makes me tense; like a Dali piece or Indian Shiva sculpture." Participant 12 was "disgusted by biomorphic holes and aggregated forms; plant like; geometric natural form; like pores; my gut reaction is fear; negative; what it evokes is negative; it looks relaxed but if I get close to it would throw water at you."

In general, for all five textures, the results showed that participant's previous life experiences influenced what was communicated. In addition, people compared their experiences with previous rounds of the study, so expectation was a big factor. The results showed that textures which appeared livelier or more vital received more positive responses. Haptic manipulation made it possible for people to change what was communicated over the duration of the touching round, which either improved or decreased the positive aspect of communication. If the expectation was that something appeared soft and it was not soft to touch, this lowered the positive aspect of communication. If a texture seemed to respond to touch this increased positive communication. Negativity was increased by touch, specifically when the texture either became less vital under the pressure of touch or seemed to actively disregard touch. This made for a negative communication. In addition, if a texture was similar to a living breathing thing, people were worried about harming it or being harmed by it. Many people discussed animals in their responses, animals that were breathing, as expected.

2.2.6. Face Word Cloud analysis

The information from the Word Face question showed much less agreement between individual responses than seen in the Circumplex grid. Participants could select more than one face as a response. As a result, there was much less correspondence compared the Circumplex Model of Affect. However, it was expected that allowing this selection could show more nuanced communication. Many participants added their own word to the faces to express what they felt was communicated back from the texture. The results are included for Texture 3 (**Figure 10**).

Texture 3 shows variable responses in each round. This range should imply that the texture moved considerably on the Circumplex grid, which, indeed, it does. As expected, "worried" has the highest word count in Round 1, and Texture 3 starts on the negative side of the grid to the left of the y axis. In Round 2, the word "happy" received the most counts and the texture moved from the left to the right side of the y axis. In Round 4, "happy" and "excited" had the leading counts, which corresponded to its location on the Circumplex grid. In Round 3,

Figure 10. Face word clouds for Texture 3 Rounds 1–4 (left to right). Word graph generated using [15].

however, participants used the words "scared," "worried," and "excited" to describe Texture 3. One would, therefore, expect the texture to move to the left side of the *y* axis. Instead, its position in Round 3 is very similar to its position in Round 2. **Figure 10** shows the word graph results for Texture 3.

2.2.7. *Conclusion and discussion of the study with five textures*

The four study hypotheses given in **Table 2** were best answered by the Circumplex Model of Affect, the results of which are reported in Section 2.2.4. To sum up the results from the Circumplex Model of Affect, hypothesis 1 was not true. Hypothesis 2 is partially true because it is true when people were able to see and touch the textures, their responses changed, but not always for the better. Hypothesis 3 is true; all the textures received a higher rating of excitement than when they were still. Hypothesis 4 is true, all the ratings did change. However, using the Circumplex Model alone produces a static map, which shows only a slice of a

Hypothesis number	Hypothesis description	Result
Hypothesis 1, round 1	It was expected that using vision alone, people would consistently associate specific emotional states with specific characteristics of the textures of textiles in a state of stillness. Crisp, curvilinear shapes associated with positive, excited, and happy feelings; smooth curvilinear shapes associated with positive and calm feelings; triangulated shapes associated with negative and angry feelings; smooth triangulated shapes and superimposed systems or a poorly defined combination associated with negative, depressed, and calm feelings.	Not true
Hypothesis 2, round 2	It was expected that when people use vision and the haptic senses together the emotional associations would change. It was expected that when people could see and touch a still texture, a negative emotional association using vision alone would change when using both vision and the haptic senses to a positive emotional association.	Partially true
Hypothesis 3, round 3	It was expected that when the textures were in a state of motion that this characteristic would raise or lower the participants' rating of what was communicated on a Circumplex grid based on what was associated with that texture motion.	True
Hypothesis 4, round 4	It was expected that the act of touching the moving textures would again change the ratings and what the textiles communicated.	True

Table 2. Hypotheses for study with five textures results.

more complex series of events by which textures or things communicate emotions to people. The free association method, for example, showed that a participant's reaction to a texture changed over time, often starting from memory or analogy as a hook, which then developed into reflections pertaining to other ways in which the texture communicated to them.

The methods here were helpful in regard to determining how much more complex is the problem of understanding what is communicated to people by still and moving textural expressions. Further, these methods suggest that the problems of understanding emotions communicated by raw materials are different from communications from materials found in completed objects such as clothing or architectural screens. This issue calls for a different approach to answering the problem of what emotion is communicated by material expression. This is a problem of making meaning. On this point, the author argues that meaning is made in context in specific situations, as argued by [16, 17, 11] in Sections 1.1 and 1.2. The problem with studying textiles is that they are unfinished products that can be used to make some other product, which calls for other ways to consider how to frame what textile expressions mean. Textiles are an inherently unfinished product until they are selected and fashioned into something in a specific context. To use a textile before it has been fashioned into something is to imagine "what it could be."

In the FELT study, the textile shown to the participants was a full-scale 5" × 6" (150 cm × 180 cm) wall panel, which was on a much larger scale than the textures shown for the study with five textures. In addition to the change in scale, FELT is large enough to truly divide the space of a room and becomes a wall or room divider rather than just a textile sample. The author will describe the study for using the FELT wall panel in the next sections.

2.3. Design process for FELT wall panel

FELT is a large approximately 5" × 6" (150 cm × 180 cm) modular panel in **Figures 1** and **10**. The design for FELT was selected to replicate Texture 3 in the study with five textures because this texture produced the widest range of responses from love to intense dislike in terms of what the participants perceived it as communicating. FELT was an opportunity to explore how changing the scale of a textile could change what emotion was communicated to people from the textile expression through vision and touch. Four key steps were followed in order to make FELT. The first was making the fabric, the second was designing the framework to hold the textile, the third was connecting the electric motors to the frame and textile, and last was mounting the frames with textiles onto a rack, which allowed it to be used as a screen or a divider in space (**Figure 11**).

The textile texture for FELT used two sheets of white felt laser cut and sewn to create a 22" × 35" (56 cm × 89 cm) panel, thereby replicating Texture 3 at a larger scale. The felt used was a wool and polyester blend. The final size of the sheets of felt was determined by the maximum that could be cut on the laser cutting bed.

Figure 12 is an exploded axonometric drawing showing the various layers of the FELT panels in the Plexiglas box. **Figure 13** is an axonometric showing one Plexiglas box mounted on the supporting rack.

Figure 11. Close up of FELT Plexiglas frames and textile panels.

Figure 12. Exploded axonometric of FELT Plexiglas frame, aluminium frame and textile sheets 1 and 2 of the felt textile panel.

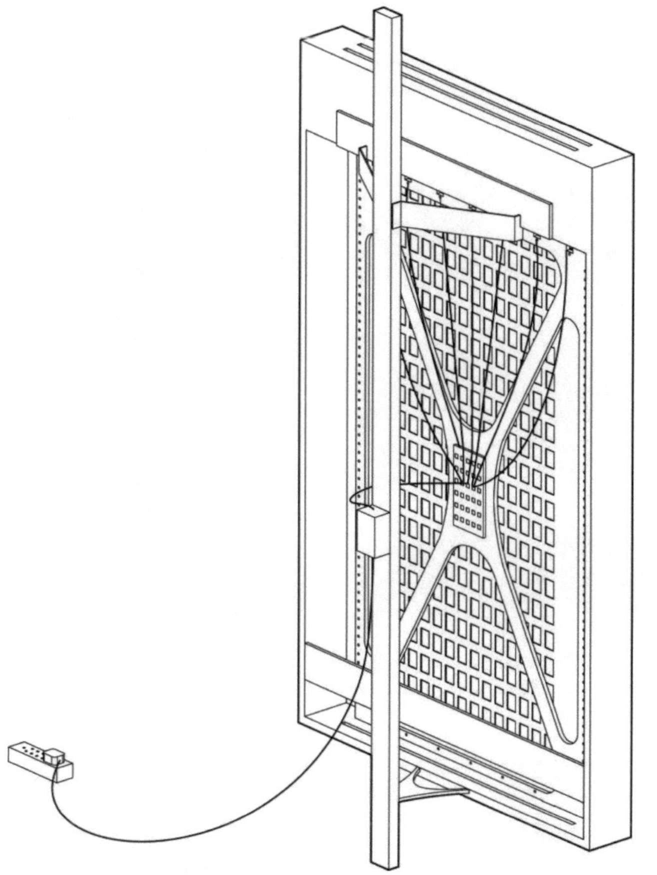

Figure 13. Axonometric showing one Plexiglas box mounted on the supporting rack.

The first step was making the textile panel. A first sheet of felt was cut into 1" wide × 1.2" flaps. A second sheet was cut with 1.2" long slits. The flaps on the first sheet were feed through corresponding slits on the second sheet to make the fabric. Second, the end of the flaps sticking out of the slit was sewn together making a cone shape. Then each cone was linked with a monofilament thread at the top. This monofilament thread slipped under the flap through the slit at the top and goes behind the fabric to be tied to a small digital 19 g/2.8 kg torque, 14/60° at 6 V servo motor at the top of the felt panel that pulled all the monofilament threads for that column of flaps. A second monofilament thread is linked to the cone at the bottom edge. This monofilament slips through the slit below and then runs behind the fabric to a weight that pulls the flaps back to their original position (**Figure 14**).

The second step was to sew the completed fabric to the aluminum frame (**Figure 15**). Once the fabric was on this frame, the motors were attached to the frame and to the monofilament lines from the fabric. Then, the entire aluminum fabric ensemble was snapped into a Plexiglas box, which held the panels (**Figures 1, 11–13**).

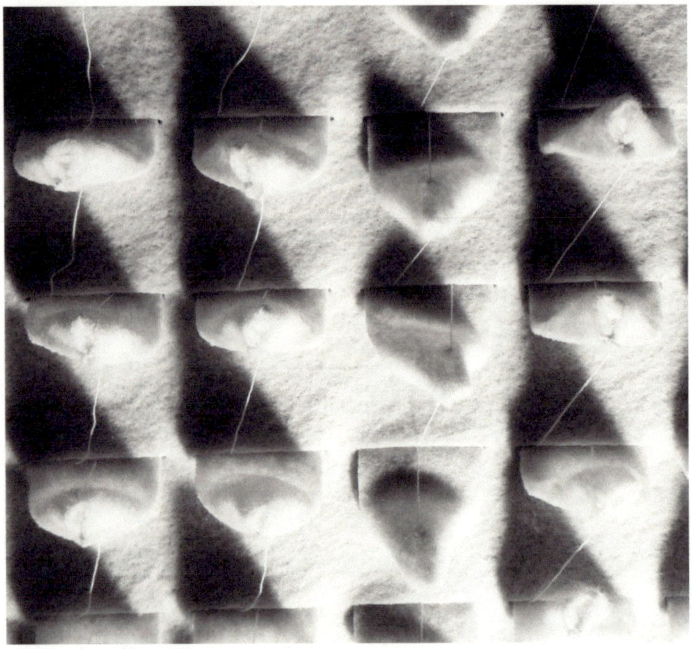

Figure 14. Close up of a FELT textile panel showing monofilament going up to be tied to a motor and another monofilament going down to be pulled by a weight.

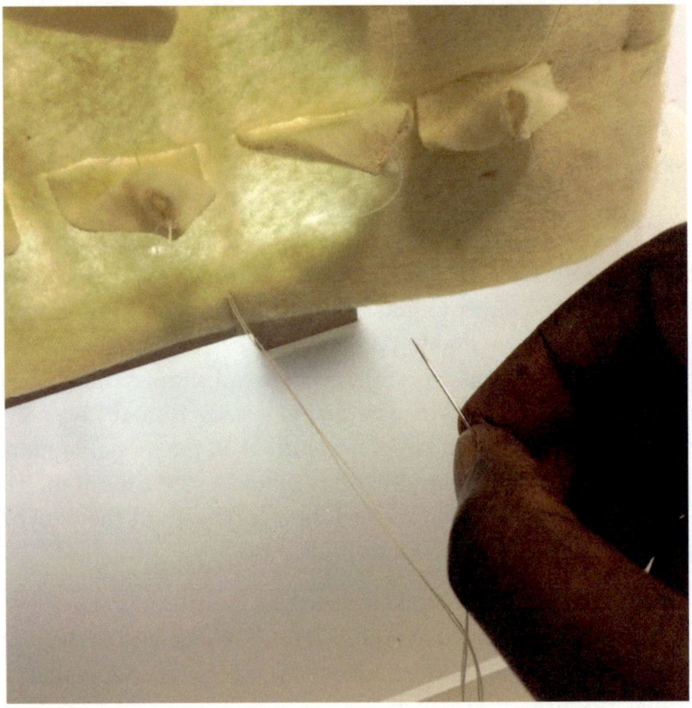

Figure 15. Sewing the felt panel to aluminum frame.

Figure 16 shows the wires from the motors going into breadboard and power source. **Figure 17** shows corner detail in Plexiglas frame.

2.4. User study with the FELT wall panel using vision and touch

The purpose of the FELT study was to obtain feedback from the participants in regard to the emotional attributes a textile or texture on a large-scale screen communicates to them through *vision* alone in a still state and in a moving state. The FELT study was also designed to obtain feedback on the emotional attributes from the same textile screen in a still and moving state when the participants used both vision and touch together. An outline of the FELT study is the same as the study with five textures in Section 2.2. An outline follows below:

1. FELT Panel is Still
    a. Looking (**ROUND 1**)
    b. Looking and Touching (**ROUND 2**)

Figure 16. Back of panel showing motors and wiring at the top as well as threads that lift the textile flaps.

2. FELT panel is Moving
   a. Looking (**ROUND 3**)
   b. Looking and Touching (**ROUND 4**)

2.4.1. Hypothesis for the FELT wall panel

Table 2 shows the hypothesis for the study with the FELT wall panel.

2.4.2. Study procedure for the FELT wall panel

This final FELT panel study took place at the Pennsylvania State University in August 2016. There were 17 participants, 13 of whom were women and 4 were men. The age range was 20–65 with the men's average age at 36 and women's average age at 35. Each session took half an hour to complete four rounds of questions. The participants interacted with the FELT panel one on one with me in the room. All the participants were able to see and touch the textile panels. All

Figure 17. Detail at corner of textile insert in Plexiglas frame.

the participants were able to speak and write in English. The interviews were recorded using handwritten notes. The panels used for the FELT project were four smaller panels making up one large 5" × 6" (150 cm × 180 cm) panel or the size of a room divider. The participants primarily stood as they viewed the FELT panel although one or two sat down to observe and interact with the panel. Each participant was interviewed separately.

The same four questions presented in Section 2.2.2 were asked of participants in the study for FELT. The same face/word chart was used as shown in **Figure 7**.

2.4.3. *The results of the user study with FELT*

The results from the FELT user study showed that emotions communicated to people using vision alone, then vision and touch together, and from still and moving textiles changed. Three analytical methods were employed to understand the results from each round of the study. These were Russell's Circumplex Model of Affect, face word cloud graphs, and the free association results.

2.4.4. *Circumplex Model of Affect analysis for the FELT wall panel*

The Circumplex Grid of Emotion plots show results that are consistent with the free association analysis. Question 1, which asked participants to rate 1–5 or negative to positive provided the x-axis data and question 2, which asked participants to rate from calm to stimulate provided the y-axis data. The Circumplex Grid was constructed with the averages of all the x data values and averages of all the y data values. Participants were much excited in Round 1, but when they touched the panels, they were disappointed in Round 2 when the visual appearance of the textile panels that appeared soft and pliable was not as soft and as pliable to their touch. Participants were further excited when the texture in the panels started to move or show motility in Round 3. In Round 4, when participants were permitted to touch the panels in full motion most, not all, participants reported a happy excited state as what was communicated back from the textile (**Figure 18**).

In terms of standard deviation, the highest negative/positive or x-axis data deviation was in Round 2, when participants could touch the textile texture. For the calm/excited or y-axis, the highest deviation was in Round 3 when participants could see the textile panels moving but not touch the panels.

2.4.5. *The free association analysis for the FELT wall panel*

There was some difference between the responses to the free association question in response to the study with the five sample textures and the large scale panel. Participants in the FELT study changed their positions to view the panel as expected to see the large- and small-scale features of FELT compared to no participants moving more than 3 feet away from the samples in the study with five textures, which were small textile samples mounted on the wall. Two people (P8 and P12) immediately and instinctively went to touch the large panel in spite of the verbal instructions issued not to touch it until requested. Before responding to the free association

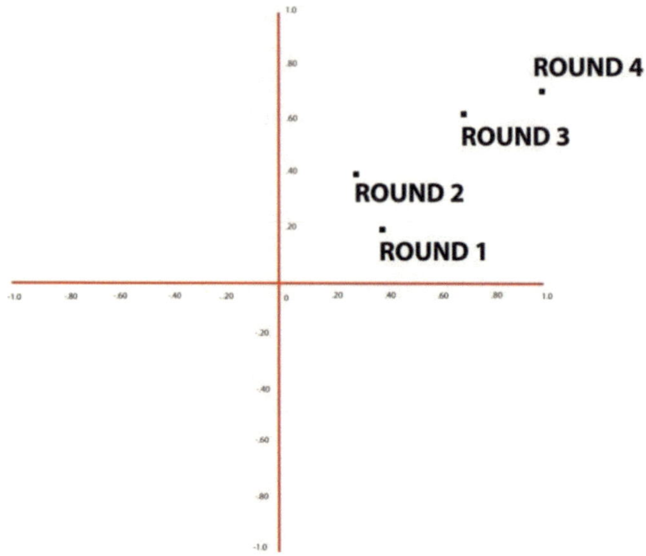

Figure 18. Circumplex Grid of Emotion for the FELT panel with the averages for all four rounds plotted.

question, four participants (P6, P8, P12, and P17) moved 8–10 feet away from the panel to have a look and then came back closer; in addition two participants (P13 and P17) sat down on a chair, which was near the panels or squatted to see what was going on in the lower panels.

In addition, several participants remarked on the light coming through the panels and that they were like windows when asked "what comes to mind?" One participant responded "A window, a curtain for a window, not a modern house, one like mom's or grandmother's house; this is like a cross and there is an extra band on one. It has a safe protected safe area. There was a war where I was growing up; windows were strengthened with tape to prevent glass from shattering" (P9). Also the large panel reminded people of walls as in this example, "what comes to mind is a pattern in a continuous wall; I am concerned with not knowing material which is just a physical aspect. It looks aggressive; like the thrown concrete on property walls in my country (Brazil). If you hit these walls and they are very sharp and you can be cut. But because the material is soft that will not happen here. But the pattern is not positive because it is like that concrete and because it is pointed and regular" (P6). Other participants connected it architecturally with descriptions like "high design corporate chic; an interpretation of 1960's modernism" (P3). In addition to connecting it visually with things associated with architecture people talked about the frames holding the textile which was unlike the study with five textures where no one mentioned the boxes hung on the wall. Participant 17 for example saw "duplicity, in the fact that there is more than one panel which means each panel not unique. I did not notice before but the backbone is different; the whole series of 4 is mirrored about vertical axis, probably unintentional. It is the idea of a hung textile in a frame with a sub-frame. I feel a lot of tension in backbone looks like it is pulling it apart and the fabric is trying to escape the frame more on the right side than left; it is like a person being stretched" (P17). This participant, an architect was reacting to the fact that the piece was made in four identical panels or frames, but the aluminum "backbones" that were seen through translucent felt showed different shadows, making the appearance duplicitous. Participants were asked about the lighting and often tried to figure out what was and was not a part of the study, for example, participant 8 asked if the shadow behind the panels was part of the project.

Outside of these comments in Round 1, many of the remaining comments fell into the categories that were seen in the study with five textures. As stated in the study with five textures, these categories are often overlapping and it is possible to describe other categories even as the ones selected do work. It maybe that these categories were similar to those in the study with five textures because when it came to really capturing their thoughts and words participants wanted to use for the free association, participants stood close or about 2 feet away from the panels to look at the texture, which is the same physical position participants took in the study with five textures. Thus, most of the responses were primarily about the texture up close. There are some of the categories raised in Round 1 of the FELT study, which are as follows:

1. Memory and analogy relating to the texture via previous experiences

This was by far the most popular response category

2. Vitality or lack of vitality of the texture
3. Expectation or comparison to imagination or previous texture
4. Feels to skin
5. Aggression or harm; understanding if texture will behave as if to harm
6. Noise
7. Color

In contrast to the free association for Texture 3, every round for FELT had all of these categories. A summary of the results from the free association is presented. A list of the hypotheses for FELT is provided your convenience in **Table 3**.

As in the study with five textures, analogy and memory were the most often used responses in Round 1 and in to a lesser extent in Round 3 when the textures were in motion (**Table 4**). There were no responses from the same participants that were the same across any rounds for the FELT wall panel. Every response was unique from round to round.

2.4.5.1. Free association results: general discussion

All but one of the 17 participants in the FELT study had general comments to offer at the end of the four rounds. Some of the most salient are included below.

Concerning the environment of the study, participant 3 raised the issue of the environment in which the FELT study was taking place. She stated "the environment is anti-sceptic; feels like a design crit; clean but welcoming environment; feeling like a hospital; ethos of cleanliness; curious how these response would have been in different context." Participant 12 also had his expectations framed by the location of the FELT study. Participant 12 stated, "We are in an iconic building on campus and so I anticipated something different from walking in the front door."

Hypothesis number	Hypothesis description
Scale and Context Hypothesis	*It was expected that participant responses would change if the textile was designed as a large architectural panel or space divider rather than small textile sample.*
Hypothesis 1, round 1	*It was expected that using vision alone, people would consistently associate specific emotional states with specific characteristics of the textures of textiles in a state of stillness. Crisp, curvilinear shapes associated with positive, excited, and happy feelings; smooth curvilinear shapes associated with positive and calm feelings; triangulated shapes associated with negative and angry feelings; smooth triangulated shapes and superimposed systems or a poorly defined combination associated with negative, depressed, and calm feelings.*
Hypothesis 2, round 2	*It was expected that when people use vision and the haptic senses together the emotional associations would change (as seen in Study 3). It was expected that when people could see and touch a still texture, a negative emotional association using vision alone would change when using both vision and the haptic senses to a positive emotional association.*
Hypothesis 3, round 3	*It was expected that when the textures were in a state of motion that this characteristic would raise or lower the participants' rating of what was communicated on a Circumplex grid based on what was associated with that texture motion.*
Hypothesis 4, round 4	*It was expected that the act of touching the moving textures would again change the ratings and what the textiles communicated.*

Table 3. Hypothesis used for the study with FELT.

Some participants were not sure what was and was not part of the FELT study. Some found the light shining through the felt panels, which made dark shadows on the panels distracting, for example, participant 13 states that "the X shadows at the back were a distraction, interrupts the way we are looking at it." In addition, participant 2 stated, "the light coming through pulls one's eyes to the darker areas until it starts moving; then one starts to notice the lighter areas."

Also in terms of the noises, participants heard one remarked "the panels were not moving as much as it sounded like they were moving." And, "it is difficult to classify in terms of sound and difficult to classify in general because of the noise. It is less obtrusive" (P2).

The narrowness of the face emoticons was another large topic for comment not just in the general comments section but also throughout the four study rounds. 10/17 participants (P4, P5, P7, P8, P9, P11, P12, P13, P15, and P17) drew in their own faces, wrote in additional emotions or mentioned that the faces did not offer enough variety in emotion to be adequate to describe what they thought was being communicated by the texture.

	Round 1	Round 2	Round 3	Round 4
FELT panel	16/17 (P1, P2, P3, P4, P5, P6, P7, P8, P9, P10, P11 ,P13, P14, P15, P16, P17)	9/17 (P2, P4, P5, P7, P9, P10, P11, P16,P17)	11/17 (P2, P4,P5, P7, P8, P10, P11, P13, P14, P15, P17)	9/17 (P4, P7, P9, P10, P11, P14, P15, P16, 17)

Table 4. Free association responses for FELT that included references to memories and/or analogies.

Some mentioned that if this panel was breathing wall that would be scary. This was somewhat a confusing statement made by participants because the panel was quite large.

Color emerged again as an important topic in the general comments section. Six out of seventeen participants had something to say about color at the end of the study. "It's not colored"(P2), "color changes how you think about this or a print would change my reaction"(P4) "I like that there is no color and it is pure white; the sound works for it, adds to the liveliness of it"(P10) "I thought change was going to be color, color is important and affects emotion and situation" (P13). "If the material could be smart then movement and color could change according to touch and kind of touch aggressive touch versus soft touch."(P15).

2.4.6. Face word cloud analysis for the FELT wall panel

The face word graphic analysis for all the four rounds for FELT is above in **Figure 19**. All graphs are at the same scale and are based on word counts from the responses given to question 4. As you can see in **Figure 19**, when looking at all the graphs together across all four rounds, "O.K.," "excited," and "happy" are words that stand out consistently. "O.K." starts off as the primary reading in Rounds 1 and 2, but "excited" and "happy" emerge almost equal in count to "O.K." in Round 2 and take over as equal counts in Round 3 with "excited" being the highest counted word in Round 4. This response is understandable as it is the first time the textile motors were turned on. "Excited" in Round 4 never becomes as strong as the "O.K." in Round 1. In addition to "O.K." fading from Rounds 1 to 4, the emergence of "worried," "curiosity," and "curious" exists at the beginning in Round 1 and 2 but gradually fade in Round 3 and almost entirely gone in Round 4. This was to be expected as participants became more familiar with the textile panel. In Round 2 "cross" appears as a second level of counts after "O.K.," "happy," and "excited." This is the only instance of the word "cross" rising to this secondary level in the four rounds. One speculation on the appearance of the word "cross" is that people were surprised by how rough the points and monofilament plastic strings felt when touched compared to the soft appearance. Fourteen out of seventeen participants indicated surprise, annoyance, or some negativity in their free association responses in Round 2.

In addition to the observations mentioned in the above paragraph, there are fewer words generated in Round 4, 14 words compared with Round 1, which have 17 different words. The density of words trails off as you look from left to right in **Figure 15**. This decreasing density seen in the word graph is the result of fewer counts for the smaller words in Round 3 and a

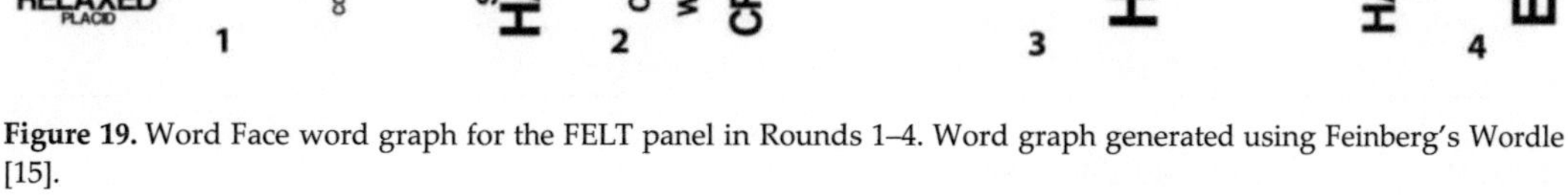

Figure 19. Word Face word graph for the FELT panel in Rounds 1–4. Word graph generated using Feinberg's Wordle [15].

coalescing of counts for fewer words. Thus, the scale of all the words in Rounds 1 and 2 is closer together than the scale of the words appearing in Round 4, for example. In Rounds 1 and 2, the word graph is made up of a blend of large-, medium-, and small-scale words. In Round 4, there are primarily large-scale words juxtaposed with tiny scale words with no intermediate scaled words.

2.4.7. Conclusion and discussion of the study with FELT

The scale and context hypothesis and the four study hypotheses given in **Table 5** were each best answered different methods. For example, the scale and context hypothesis was best answered by the free association responses. Participants did address the scale and context of the textile panel either in words or body movement as mentioned in Section 2.4.5. The scale and context hypothesis is true as expected.

The Circumplex Model of Affect most easily frames the remaining four hypotheses (**Figure 14**). However, the free association adds a level of explanation for why people select the choices they did for questions 1 and 2, which provide the x and y points for the model.

For the FELT study, hypothesis 1 is not true in the case of the large textile panel. In the FELT study Round 1, the textile texture is plotted low in the right quadrant of the model or positive trending toward excited. In the study with five textures, this same texture was seen as

Hypothesis number	Hypothesis description	Result
Scale and Context Hypothesis	*It was expected that participant responses would change if the textile was designed as a large architectural panel or space divider rather than small textile sample.*	True
Hypothesis 1, round 1	*It was expected that using vision alone, people would consistently associate specific emotional states with specific characteristics of the textures of textiles in a state of stillness. Crisp, curvilinear shapes associated with positive, excited, and happy feelings; smooth curvilinear shapes associated with positive and calm feelings; triangulated shapes associated with negative and angry feelings; smooth triangulated shapes and superimposed systems or a poorly defined combination associated with negative, depressed, and calm feelings.*	Not true
Hypothesis 2, round 2	*It was expected that when people use vision and the haptic senses together the emotional associations would change. It was expected that when people could see and touch a still texture, a negative emotional association using vision alone would change when using both vision and the haptic senses to a positive emotional association.*	True
Hypothesis 3, round 3	*It was expected that when the textures were in a state of motion that this characteristic would raise or lower the participants' rating of what was communicated on a Circumplex grid based on what was associated with that texture motion.*	True
Hypothesis 4, round 4	*It was expected that the act of touching the moving textures would again change the ratings and what the textiles communicated.*	True

Table 5. Results to the five hypotheses for FELT.

negative on the left side of the quadrant very high up on the *y* axis toward stimulated with a reading of agitated, angry, and aggressive as shown in **Figure 9**.

Part of hypothesis 2 is true. Responses did change when the textile texture was both seen and touched. However in the case of FELT, participant's touch did not produce a more positive response, in fact, it produced a more negative affect in Round 2.

Hypothesis 3 is true; the responses were more excited and trending toward positive and more stimulated for FELT when the textile texture moved.

Hypothesis 4 is true; the ratings did change in the final round. In this instance, the ratings were more positive and more stimulated than in Round 3.

The face word clouds supported the information from the Circumplex Model of Affect.

As in the study with five textures, the responses of people changed over the time of interaction, reflecting the changing information their eyes and sense of touch exchanged with the textile texture. Because the FELT study textile panel was much larger in size this seemed to produce more kinds of bodily exploration and thinking in the responses compared with the study with five textures.

2.5. Comparison of FELT wall panel and Texture 3

Figure 20 compares the plot averages in red for Texture 3, which was a 9" × 9" (23 cm × 23 cm) sample. Texture 3 was the exact texture pattern used to make the large textile panel FELT. The results of the plot averages for FELT are in black in **Figure 20**. The FELT Panel received more

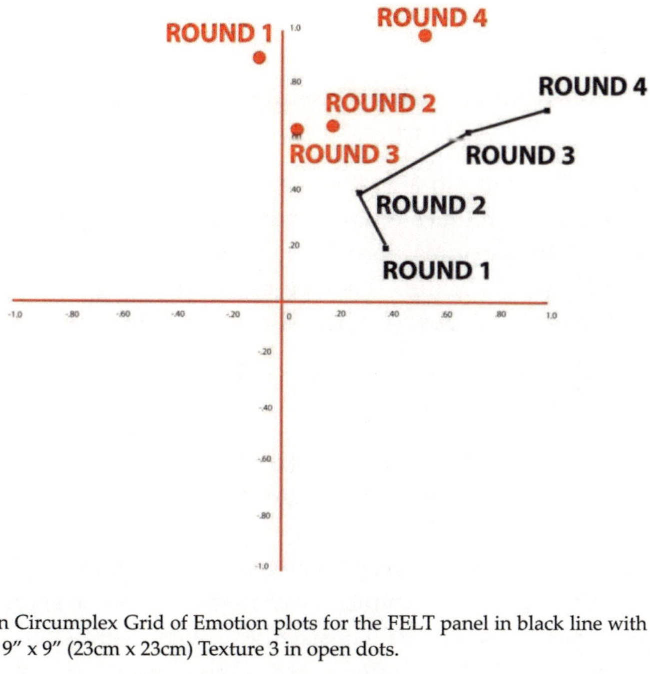

Figure 20. Comparison Circumplex Grid of Emotion plots for the FELT panel in black line with solid dots and the plot points for the original 9" x 9" (23cm x 23cm) Texture 3 in open dots.

positive ratings than those for Texture 3 and the FELT Panel received calmer ratings than those for Texture 3.

3. Conclusions and contributions

Hypothesis 1 was based on a belief that all humans are in some way hardwired or can understand certain shapes and textures in a very basic way through emotion. The experiments in this chapter did not prove this true, possibly because the context changes the messages of these shapes and textures.

Hypothesis 2 was based on the belief that once a person touches something that thing becomes known and thus is rated more positive. In addition, this hypothesis is based on the belief that if something was rated negatively using only vision that a negative visual rating could be overturned and rated positively using touch. Once a textile was touched, it did change its rating however, the change was not always positive.

Hypothesis 3 did affirm that when something in an environment changes, human reaction will typically, not in every case but most often, changes as well.

Hypothesis 4 is true. Yes, the ratings for positive/negative and stimulated/calm changed in every instance in study with five textures and the FELT study.

In closing, there are a few points of generalizable knowledge that demonstrated contributions of these studies. A summary of these salient points is as follows:

1. Using vision or vision and touch, the emotion(s) communicated to people changes during the process of exchange.
2. Emotions communicated through vision from a specific textile are different from those communicated using vision and touch.
3. Introducing motion or motility to a textile expression increases the stimulation and excitement of the emotion communicated by that textile. (The exception to this statement is seen in the study with five textures, specifically looking at Texture 3 in Round 3. In the Circumplex plot in Figure 20, Texture 3 shown in red ink decreases its positive rating in Round 3 compared to Round 2. It was expected to increase in positive rating.)
4. Analogy and memory are primary methods people use to understand what emotion is communicated by a textile.
5. The scale of the textile changed what emotion(s) were communicated from the textile because people must change their bodies to engage the larger scale.

Although these experiments did not prove that any specific live expression is related to any specific emotion, this work does demonstrate nevertheless that some expressions communicate in a range to most people somewhat consistently. Other expressions such as Texture 3 in the study with five textures had very strong, excited responses from different participants on opposite sides of positive and negative emotion communicated. These opposing types of responses went away with the FELT wall panel. There may be many reasons for this difference. For example, Texture 3 was compared to other textures during the study with five textures and was an unfinished sample of textile. FELT had a defined function as a wall panel and was much larger scale.

Much communication that people take from expression is context based and is perhaps inference rather than direct communication. There is much more work to be done, however, if designers, scientists, and others are to fully understand the relationship between esthetic expression and communication. In the end, these experiments have shown that emotion communicated from computational objects is woven between that object or space and an individual body. In these experiments, each individual was observed forming his/her own analogy or memory map searching for and connecting experience to something else he/she has learned or experienced before about that material and emotion. In these experiments, people were witnessed in the moment of seeing or seeing and touching making new analogies and memories to keep with them about the material and emotion. There is more work to be done, too, in the field of neuroscience to further our understanding of how these analogy maps are created, their location in the brain, and how, whether, and to what extent certain aspects of certain emotions and certain kinds of experiences have the same or similar maps in the brains of most people.

The author would like to thank the four graduate Research Assistants in SOFTLAB@PSU, at Penn State University. They are Niloufar Kioumarsi, Arman Esfahani, Niloofar Nikookar, and Nasim Motalebi, who with great patience and care helped bring the FELT wall panel to fruition and Xiao Han for his line diagrams of FELT.

Author details

Felecia Davis

Address all correspondence to: fadav@psu.edu

Stuckeman Center for Design Computing, Stuckeman School for Architecture and Landscape Architecture, Pennsylvania State University, Pennsylvania, USA

References

[1] Barbatti B. Berber Carpets of Morocco: The Symbols Origin and Meaning. Courbevoie, Paris: ACR Edition Internationale; 2009. Available from: from www.acr-edition.com

[2] Oxford English Dictionary. Expression [Internet]. n.d. Available from: http://www.oed.com/

[3] Expressive power. Computer Science. In Wikipedia [Internet]. n.d. Available from: https://en.wikipedia.org/w/index.php?title=Expressive_power_(computer_science)&oldid=741557765

[4] Oxford English Dictionary. Communication [Internet]. n.d. Available from: http://www.oed.com/

[5] Darwin C. The Origin of Species: 150th Anniversary Edition (Rep Anv edition). New York, NY: Bantam Dell; 1859

[6] Arnold MB. Emotion and Personality. Vol. 1. New York, NY: Columbia University Press; 1960

[7] Darwin C. The Expression of the Emotions in Man and Animals. 2nd ed. London, England: Penguin Group Ltd; 1890

[8] Frijda N. The meanings of emotional expression. In: Key MR, editor. Nonverbal Communication Today: Current Research. Berlin: Mouton Publishers; 1982. pp. 103-119

[9] Arnheim R. Expression. In: A Modern Book of Aesthetics: An Anthology. 3rd ed. New York: Holt, Rinehart and Winston; 1960. pp. 258-270

[10] Gibson J. The Senses Considered as Perceptual Systems. Oxford, England: Houghton Mifflin; 1966

[11] Johnson M. The Meaning of the Body: Aesthetics of Human Understanding. Chicago, I.L.: University of Chicago Press, 2008, 66 p

[12] Damasio A. Looking for Spinoza: Joy, Sorrow, and the Feeling Brain (Reprint edition). Orlando, FL: Harvest; 2003

[13] About Makaton [Internet]. n.d. Retrieved from: https://www.makaton.org/aboutMakaton/

[14] Russell JA. A circumplex model of affect. Journal of Personality and Social Psychology. 1980;**39**(6):1161-1178. https://doi.org/http://dx.doi.org.libproxy.mit.edu/10.1037/h0077714

[15] Feinberg J. Wordle—Beautiful Word Clouds [Internet]. 2008. Available from: http://www.wordle.net/

[16] Arheim R. Art and Visual Perception: A Psychology of the Creative Eye. Berkeley, and Los Angeles, CA: University of California Press; 1954. p. 426

[17] Dewey J. Experience and Nature. Vol. 1 of the Later Works, 1925-1953, In: Jo Ann Boyston, editor. Carbondale, IL: Southern Illinois University Press; 1925/1981

Chapter 14

ESD Knitted Fabrics from Conductive Yarns Used as Protective Garment for Electronic Industry

Gabriela Telipan, Beatrice Moasa, Elena Helerea,
Eftalea Carpus, Razvan Scarlat and
Gheorghe Enache

Additional information is available at the end of the chapter

http://dx.doi.org/10.5772/intechopen.69843

Abstract

Nowadays, interest in protection against electrostatic discharge (ESD), regarded as an important issue, is still growing. ESD may lead to serious economic losses and can also be hazardous to humans. In zones where substances with low values of ignition energy occur, ESD may involve a high risk of fire or explosion. Protective clothing is used to reduce the risk of electrostatic discharge (ESD) in electronics industry as well as in explosive atmospheres. In the manufacturing process of electronic products, significant percentage-related failures due to ESD phenomena that occur on the production line, especially where they are not installed systems and rigorous protection rules, are still recorded. Protective clothing with antistatic properties is designed to prevent the transfer of electrical charge from human operator to electronic device during the manufacturing process. In this chapter, a study was conducted about the electrical parameters representative for the control of electrostatic discharges as well as the surface and volume resistivity by applying the current absorption-resorption method and parameters of charge decay using the standardized induction method on several knitted conductive fabrics with carbon yarn. Current absorption-resorption method is applied for the first time in research under antistatic and dissipative fabrics.

Keywords: ESD garments, electrostatic discharge, absorption-resorption current method, surface, volume resistivity

1. Introduction

Textile materials that provide comfort, antimicrobial, antiallergenic, and anti-stress protection or shielding for working people in environments with high-risk factors are increasingly required to protect consumers' health, prevent diseases, and reduce environmental impact [1, 2].

A new concept has been created to transform the passive role of textile materials to an active role with all their influence for industry and consumers.

To keep up with the strategy for upgrading the textile companies, the researchers' main objective was to achieve high-performance technologies with low impact on the environment and human body and promotion of raw materials using superior hygienic, antibacterial, antiallergenic, and protective properties for people who work in hazardous environments.

In modern electrostatic discharge (ESD) protection systems, protective clothing is provided to prevent accidental ESD events caused by an operator and which can partially or totally destroy electronic devices from polychlorinated biphenyl (PCBs). Neither currently these problems are not completely eliminated. Thus, researchers have identified the need for protective clothing to have two properties at the same time but this is possible only if the garment is made of two layers.

The first property is linked to the need for a low conductivity of material for limiting electric charge processes and transferring related and the second property is linked to the need for an increased conductivity to facilitate dissipation of electrical charges processes on the material [3]. In order to prevent the occurrence of electrostatic fields which produce and induce electric charges in materials [4, 5], protective clothing and work surfaces should have electrostatic shielding properties [6]. ESD processes are dependent on temperature and humidity [5].

The conductive fabrics have been considered from their electromagnetic shielding and anti-electrostatic properties for various applications in the defense, and electrical and electronic industries [7–11].

The surface resistivity is a principal parameter which determines the ESD protection applications [12]. Generally, the synthetic fibers used for textiles fabrics have an electrical insulation character, with a surface resistivity about 10^{15} $\Omega/\square$. This value of surface resistivity is very higher for anti-electrostatic and electromagnetic shielding applications.

Many researches show that to obtain good and efficient anti-electrostatic material, a surface resistivity in the range of (10^{9}–10^{12}) $\Omega/\square$ is necessary, and for statically dissipative material, the surface resistivity values must be in the range of (10^{4}–10^{9}) $\Omega/\square$ [13].

In order to improve the electric conductivity of textiles, different techniques are applied, for example, the introduction in textile material of conductive yearns, carbon or metal [13–15], or treatment with conductive polymer, polyaniline, polypyrrole, poly(3,4-ethylenedioxythiophene)-polystyrene sulfonate PEDOT–PSS [13, 16, 17].

The research conducted by authors in the field of technical textiles was on knitted bilayer that provides good protection from accidental electrical discharges on the dielectric layer and a drainage of accumulated charges through conductive layer. An image of the bilayer concept is illustrated in **Figure 1** [18].

The external layer is predominantly dissipative and it provides protection to short circuit and the limitation of the electrostatic energy that can be dissipated to the work environment, and the internal layer is predominantly conductive, providing controlled drainage of cumulated

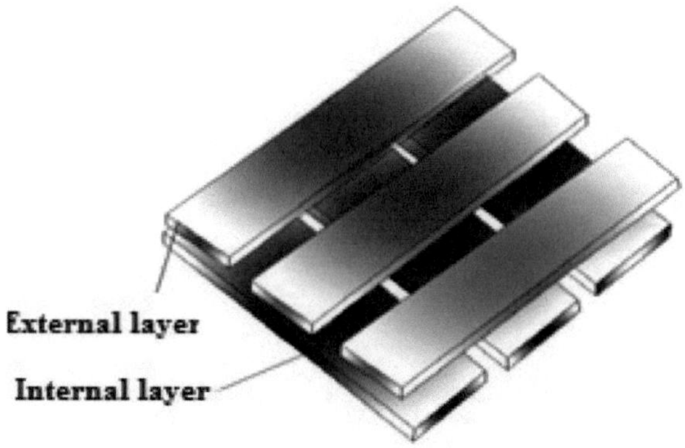

Figure 1. The bilayer concept of ESD fabric.

electrostatic charges. Also, an additional requirement for the internal layer is to provide comfort to the user.

The principal objective of our study consists in the development of a comparative analyzer for several ESD garments with carbon-conductive thread used for electronic industry and the characterization regarding their anti-electrostatic properties. The absorption-resorption measurement method is applied for obtaining the charge decays and surface and volume resistivities, as main parameters of ESD static and dissipative properties of garments.

2. ESD characteristics and measurement methods

2.1. Surface and volume electrical resistivities and the absorption-resorption current method

The procedure for determining the surface and volume resistivities is based on the current absorption-resorption method. This method was earlier applied to highlight the aging processes of insulating materials [19], but it can significantly also characterize the electrostatic charging-discharging processes in ESD protection materials.

The procedure consists of applying a DC voltage V on a sample, disposed in the measuring cell (with three electrodes, e.g., Keithley 8009 cell) and recording the values of current intensities to coupling source (absorption currents) and then to decoupling source (resorption currents).

Absorption current i_a is determined by applying a continuous voltage V across the capacitor (measuring cell), having as dielectric the material sample with conductivity σ and permittivity ε, an absorption current will flow through the circuit.

The absorption current intensity $i_a(t)$ is variable in time t and has specific components:

$$i_a(t) = i_0(t) + i_p(t) + i_{ch}(t) + I_\sigma \qquad (1)$$

where: $i_0(t)$ is the charging current of vacuum capacitor, $i_p(t)$ is the polarization current, $i_{ch}(t)$ is the space charge current, and I_σ is the conduction current [20, 21].

The charging current of vacuum condenser $i_0(t)$ decreases rapidly to zero, as the transitional regime is exceeded. Expression of this current is dependent on the time of variation, and the electric field intensity is

$$i_0(t) = \varepsilon_0 . \frac{\Delta E}{\Delta t} . S \tag{2}$$

where ε_0 is the permittivity of vacuum, S is the electrode surface, and E is the electric field intensity.

Polarization current $i_p(t)$ is determined by the polarization processes occurring in the material. Depending on the mechanism of polarization (electronic, orientation, interfacial/dishomogeneity polarizations), this current tends to zero after reaching permanent regime of DC power.

Space charge current $i_{ch}(t)$ stems from the movement of electrically charged structural units (ions, impurities, etc.) under the action of the electric field intensity E established in the material. The electric space charges can occur due to technological processes, in thermal degradation, electrical stresses, and so on. The space charge current can distort readings for conduction current intensity I_σ values. Therefore, it is necessary to determine the time constant τ feature for transitory regime of current flowing.

Steady-state regime is characterized by conduction current I_σ, given by the free electric charges flow under the action of the applied electric field and showing no change value in time. Usually, steady-state regime is considered to be achieved after a time of (3–5)τ, where τ is the time constant of the circuit. So, the measurement of conduction current intensity should be taken only after achieving the steady-state regime [22], and the average value of I_σ is obtained taking into account the measured values in steady-state regime of current flowing.

Depending on the arrangement of the measuring electrodes, the value of the volume current intensity I_V or the value of the surface current intensity I_S is determined and, as in the volt-ammeter method, the volume and surface resistivities are obtained.

When this no longer applies to voltages and the electrodes that are short-circuited, the discharging process begins, and in the material sample a transient current, named resorption current, flows. The intensity of resorption current $i_r(t)$ depends on time t and has specific components too:

$$i_r(t) = i_d(t) + i_{dp}(t) + i_{ch}(t) \tag{3}$$

where $i_d(t)$ is the discharge current, $i_{dp}(t)$ is the depolarization current, and $i_{ch}(t)$ is the space charge current.

A simplification of the time dependence of the absorption and the resorption current intensities in a polar dielectric is shown in **Figure 2**.

http://dx.doi.org/10.5772/intechopen.69843

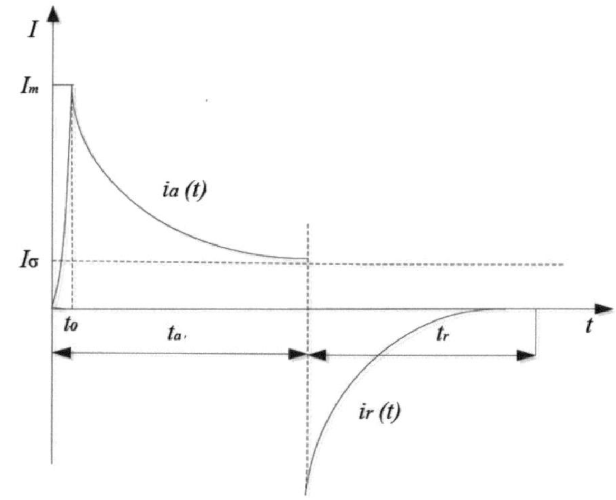

Figure 2. Time dependence of the absorption and resorption currents in a polar dielectric.

The absorption current decreases asymptotically toward the steady state, due to dielectric polarization and the sweep of mobile charges to the electrodes. For materials of high resistivity (up to 10^{13} Ωm, and 10^{12} Ω/□) the steady state is reached in several minutes, hours, and days. For these materials, the time dependency of the volume resistivity is taken into account.

For an antistatic material—surface resistivity is in the range of (10^9–10^{12}) Ω/□ [13]—the steady-state regime in general is reached within 1 min (**Figure 3**), and the resistance is then determined after this time of DC voltage supplying.

Comparison of **Figures 3** and **4** shows that the steady-state regime in antistatic polymers is established earlier than in polar dielectrics, the absorption current slope $i_a(t)$ is steeper and the resorption current $i_r(t)$ also has a steep slope.

For dissipative materials with surface resistivity of the order of 10^6–10^{10} Ω/□, the wave forms of absorption and resorption currents differ from those of antistatic or insulating materials [23–25].

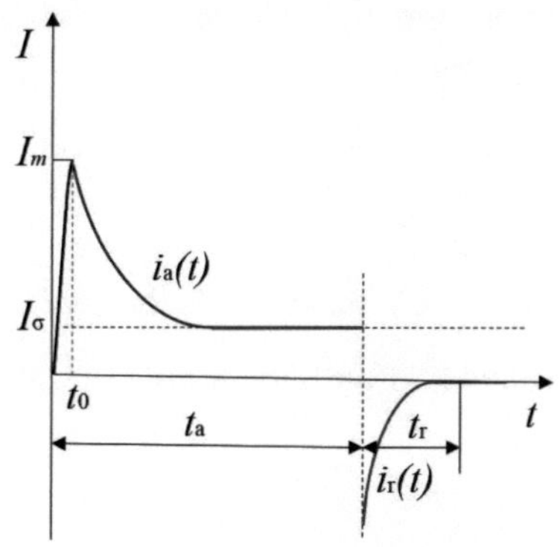

Figure 3. Time dependence of the absorption and resorption currents in antistatic polymers.

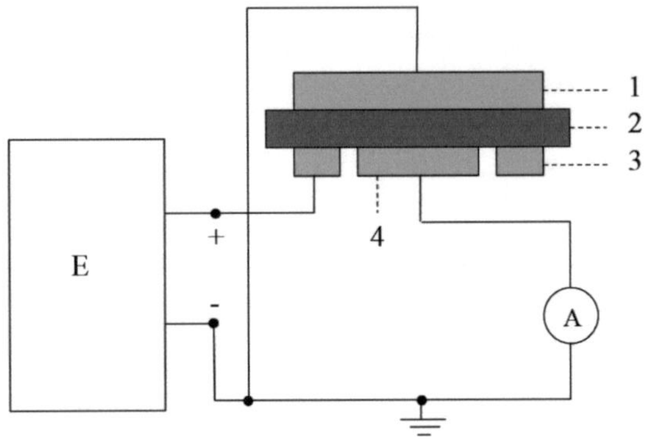

Figure 4. Scheme for measuring surface resistivity: (1) upper electrode; (2) sample material; (3) guard ring; and (4) the measuring electrode.

In materials, the free electric charges—both conduction processes—in volume of material and on the surface—are present, even at low electric field stresses.

Based on these processes, the volume and surface resistivities can be good indicators of ESD antistatic comportment of materials: the surface resistivity of material can characterize the ability of a material to dissipate electrostatic charges; the volume resistivity of a material is useful for evaluating the relative dispersion of a conductive additive throughout the polymer matrix, and it can related to electromagnetic shielding effectiveness in certain conductive fillers.

To determine the volume resistivity ρ_V and surface resistivity $\rho_{S,}$ with absorption-resorption current method, the same scheme is used, described in standard SR HD 429 S1 CEI 60093 [26]. The measurement principle of the method is similar to the volt-amperometric method, where single difference positions guard ring electrode.

Measurement scheme for determining the surface resistivity is shown in **Figure 4**.

In this case, the applied voltage reaches the sample through the guard ring, the read signal by the ammeter coming from the measuring electrode.

Surface resistivity is defined as the quotient of a DC electric field applied between two electrodes on a surface of a specimen and the filiform current density of the current flowing between the electrodes at a given time of electrification:

$$\rho_s = \frac{E}{J_s} \tag{4}$$

where E is the electric field intensity, in V/m, and J_S is the surface current density, in A/m.

The surface resistivity shall be calculated with the following relation:

$$\rho_s = \frac{R_s \cdot p}{g} = K_s \cdot \frac{U}{I_s} \tag{5}$$

where is the surface resistivity, in Ω/□, R_S is the surface resistance, in Ω, p is the effective perimeter of the guarded electrode, in m, g is the distance between the electrodes, in m, U is

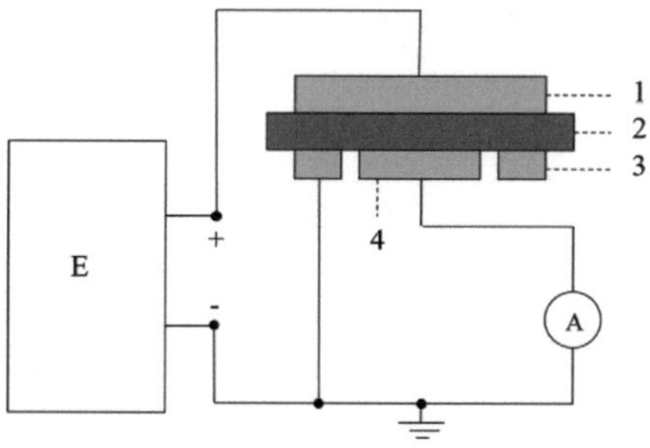

Figure 5. Scheme for measuring the volume resistivity: (1) upper electrode; (2) sample; (3) guard ring; and (4) measuring electrode.

the applied voltage, in V, and I_s is the surface current intensity, in A. The coefficient K_s is calculated in the function of the geometry of measurement cell and the particular electrode arrangement.

Measurement scheme for determining the volume resistivity is shown in **Figure 5**.

In this case, the applied voltage reaches the sample through the upper ring, the signal read by the ammeter coming from the measuring electrode.

Volume resistivity is defined as the quotient of a DC electric field strength applied to a material specimen disposed between two electrodes and the current density J_v of the current flowing between the electrodes at a given time of electrification:

$$\rho_v = \frac{E}{J_v} \tag{6}$$

where E is the electric field intensity, in V/m, and J_v is the volume current density, in A/m^2.

The volume resistivity shall be calculated with the following relation:

$$\rho_v = \frac{R_v \cdot A}{h} = \frac{K_v}{h} \cdot \frac{U}{I_v} \tag{7}$$

where ρ_v is the volume resistivity, in Ωm, R_v is the volume resistance, in Ω, A is the effective area of the guarded electrode, in m^2, h is the average thickness of the specimen, in m, U is the applied voltage, in V, I_v is the volume current intensity, in A. The coefficient K_v, given by the effective area A of guard electrode, is calculated in the function of the geometry of measurement cell and the particular electrode arrangement.

In the case of the measuring cell Keithley, the measuring electrode and guard ring are at the bottom of the circuit arrangement.

2.2. Parameters of charge decay and measurement methods

A way to characterize the rate of dissipation of electrostatic charge of garment materials is to study charge decay. Charge decay is the process of migration of charge across or through a material leading to a reduction of charge density or surface potential at the point where the charge was deposited.

To monitor the mobility of electrostatic accumulated charges, two methods of measurement are used [28]. In both methods, the charge is monitored by the observation of the electrostatic field it generates and this is done using non-contacting field-measuring instruments. The principal difference between the methods is the technique used to generate the electrostatic charge: (1) triboelectric charging, when the material specimen is in contact with a special rod, rubs together and subsequently separates. The electrical field strength from the charge generated on the test material is observed and recorded using an electrostatic field meter connected to a graphical recording device; (2) induction charging, which involves a field electrode, placed near the surface of a specimen, and which is energized with high voltage. If the specimen contains mobile electric charges, the charge of opposite polarity to the field electrode is induced on the specimen. Induced charge on the test material influences the net field, and this effect is measured and registered with a field-measuring probe, positioned above the test surface.

In **Figure 6**, the scheme for determining the discharge time of the material specimens, which are charged through electric induction [18], is shown. The main components of the measuring stand are signal generator (G), electronic voltmeter (VE), oscilloscope (OSC), sample (1), support sample (2), field electrode (3), measurement electrode (4), and guard ring (5) [18].

A step high voltage is applied on field electrode. By electrical induction, in the sample the electrical charges are separated and an electric field of intensity E_R is generated in the surrounding area. The electric field generated by electric induction is compared to the electric field generated by the electric field in the absence of a sample.

The electric field strength indicated on the recording device in the absence of the test specimen has maximum value and could be obtained with the electric flux law:

$$E_{\max} = \frac{Q}{\varepsilon_0 \cdot A} \tag{8}$$

where E_{max} is the intensity of the electric field near the measurement electrode in the absence of the specimen, in V/m; $\varepsilon_0 = 8.855$ pF/m is the vacuum permittivity; A is the effective area of the measurement electrode; Q is the accumulated electric charge in the specimen, respectively, in measurement electrode, obtained with the relation:

Figure 6. Installation for testing the capacity to dissipate electrical charge for textiles.

$$Q = C \cdot U \quad (9)$$

In relation (9), C is the capacity of the condenser without the specimen, and U is the voltage applied at the terminals. Capacity C is obtained usually through measurement with bridge measurement systems.

The electric field strength E_R indicated on the recording device with the test specimen in the measuring position is also registered in time.

The half-decay time t_{50} is defined as the interval of time in which the electric field intensity decreases to the value $E_R/2$. This parameter indicates the electric field strength decay to $E_R/2$ and is an indicator for classification of textiles as static or dissipative.

The shielding factor S is defined by the relationship between E_{max} and E_R and is calculated as: [18, 27, 28]

$$S = 1 - \frac{E_R}{E_{max}} \quad (10)$$

The advantage of using the induction method of charging the specimen is that it allows to determine both parameters—half-decay time and shielding factor, which characterize the ability to dissipate electrical charge of any antistatic or dissipative textile.

3. Experimental determinations

3.1. Conductive fabrics specimens manufacturing and examination

In order to assure the bilayer structure, the chosen knitting structures were plated plain jersey and plated rib. The knitting technology presumed electronic flatbed knitting machines (Stoll, 7E and 12E gauge). The bilayer knitted variants were made in plaited structures, with parallel evolution of two or more yarns with strictly determined relative position as a result of their settling at different angles (plaiting yarn V at an angle smaller than ground yarn F). In case of jersey structure, the plaiting yarn V appears on the foreground on the front and the ground yarn F, on the foreground on the back of the fabric. In case of rib structure due to alternating of front-back wales, both the plaiting yarn and the ground yarn will be present on the foreground, on each side of the fabric, **Figure 7**.

The used yarns were:

Base yarn: Nm 50/3, 100% cotton; Nm 30/2, 100% wool.

Conductive yarns, as follows:

- for inner layer: 75% cotton + 25% epitropic yarn (Nm 34/1 carbon coated polyester)—named yarn type 2.
- for outer layer: multifilament yarn made from surface-saturated nylon with carbon particles—named yarn type 5.

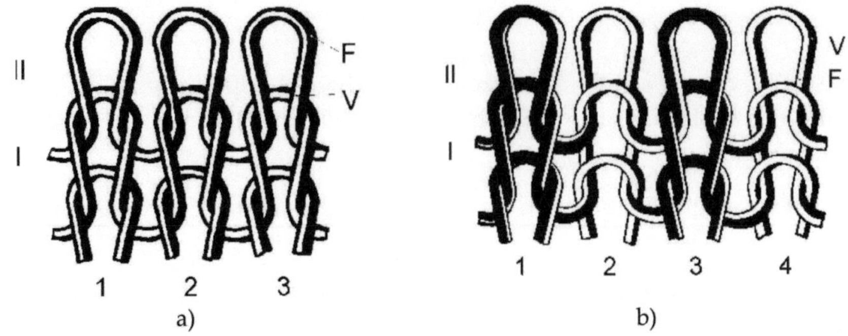

Figure 7. Structural representation of plaited: (a) plain jersey structure; (b) plaited rib structure.

Figure 8 shows the image for ESC knitted fabrics supposed to experiments.

In order to characterize the nine knitted fabric samples, complete sets of analyses have been realized and conducted for the following parameters: weight (g/m^2), density (wales/10 cm, rows/10 cm), thickness (mm), air permeability (l/m^2/s), water vapor permeability (%), thermal conductivity (mW/mK), and thermal resistance (m^{2}KW). Experimental data are presented in **Table 1**, and in **Figure 9**, few images with analyzed samples are shown.

In **Table 2**, the composition and structure of knitted fabrics for surface and volume electrical resistivity test and for electrostatic shielding factor are presented.

From the data presented in **Table 2**, it can be observed that the percentage of conductive wire varies depending on the number of yarns and their composition.

3.2. Measurement setups

A. Surface and volume resistivity measurement have a setup composed by Electrometer Keithley 6517A (1–1000V), with Keithley 8009 measure cell, and computer with control soft and data processing shown in **Figure 10** [30].

Figure 8. ESD knitted fabrics.

http://dx.doi.org/10.5772/intechopen.69843

Sample code	Weight [g/m^2]	Density [rows/10 cm]	Density [wales/10 cm]	Thickness [mm]	Air permeability [l/m^2/s]	Water vapor permeability [%]	Thermal resistance [m^{2}K/W]	Thermal conductivity [mW/mK]
P10	597	47	81	1.62	374.4	36.2	0.03269	49.55
P11	589	47	82	1.66	395.0	35.9	0.04444	37.35
P12	634	47	75	1.74	401.4	34.0	0.03395	51.25
P20	562	47	69	1.65	601.0	34.8	0.04099	40.25
P19	603	46	69	1.66	647.6	30.06	0.03567	46.55
P14	828	36	59	3.58	576.6	28.5	0.05732	62.45
P13	878	36	59	3.61	609.8	32.4	0.05726	63.05
P17	802	34	60	3.53	800.8	29.5	0.07801	45.25
P18	846	33	59	3.57	872.2	28.6	0.07786	45.85

Table 1. The characteristics of knitted samples [29].

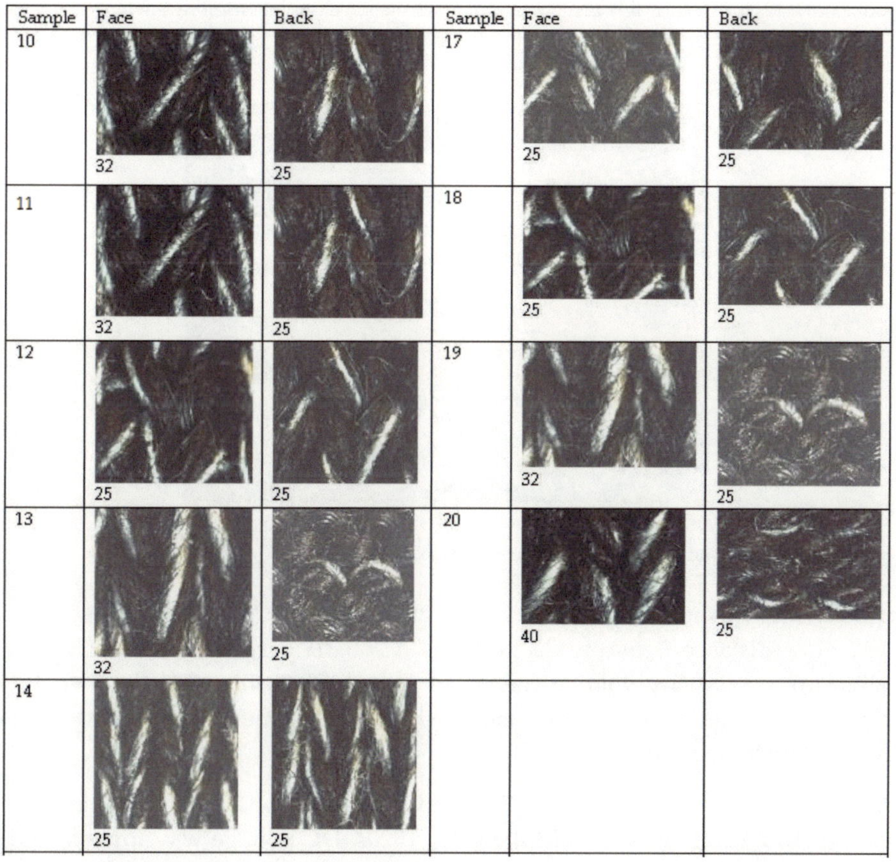

Figure 9. Optical images for knitted fabrics. The parameters 25, 32, and 40 represent the image magnitude.

Knitted structure	Knitted fabric		Percent of conductive yarn [%]	Indicator
	Outer layer/front side of fabric	Inner layer/back side of fabric		
Plaited plain jersey	One cotton yarn + one yarn type 2*	One cotton yarn + one yarn type 5**	4.5	P10
		One cotton yarn + 2 yarns type 5	6.0	P11
		One cotton yarn + 3 yarns type 5	7.5	P12
	One wool yarn + one yarn type 2	One wool yarn + 2 yarns type 5	6.0	P20
		One wool yarn + 3 yarns type 5	7.6	P19
Plaited rib	One cotton yarn + one yarn type 2	One cotton yarn + 2 yarns type 5	6.0	P14
		One cotton yarn + 3 yarns type 5	7.5	P13
	One wool yarn + one yarn type 2	One wool yarn + 2 yarns type 5	6.0	P17
		One wool yarn + 3 yarns type 5	7.5	P18

* Yarn type 2 content 75% cotton + 25% epitropic yarn (Nm 34/1 carbon coated polyester).
** Yarn type 5 content multifilament yarn made from surface-saturated nylon with carbon particles.

Table 2. Composition and structure of knitted fabrics for surface and volume electrical resistivity test and electrostatic shielding factor [28].

Figure 10. Stand for surface and volume resistivity measurements.

The characteristics of the measurement cell is as follows: (1) the size surface coefficient is K_s = 53.37 Ω/□ (distance between electrodes is d = 0.3175 cm, and the average diameter D = 5.3975 cm); (2) the size volume coefficient is $K_v = 20.25802 \times 10^{-4}$ cm^2, where ρ_v is in [Ω·cm], g is the average thickness of the sample in [cm]; U is the applied voltage, in [V], I_V is the volume conduction current, in [A].

The application time of the signal is t = 20 s and the read time after cutoff signal is t = 10 s. The total measuring time was 30 s. In the early 1920s, DC voltage was applied, and for the remaining 10 s the signal was cut off and the remaining currents were measured on surface material.

The minimum value of applied voltage to textile samples is 1 V and the maximum value applied varies from sample to sample; limitation criteria for applied voltage are given by electrometer. Keithley electrometer used can highlight the current of fA-mA (10^{-15}–10^{-3}) order.

From the values obtained for surface conduction current, values of steady state, within the time interval (12–18) s, have been extracted. This was done similarly for volume conduction current.

The values of surface and volume resistivity are calculated using relations (5) and (7).

The average values and mean-square dispersion values have been calculated and corrections have been done: for dispersions greater than 5%, the erroneous values have been eliminated.

In all the samples, it was observed that after about 12 s, the values of absorption current intensities are stabilized.

B. Time decay of electric charge and shielding factor evaluation has a setup built in accordance with standard SR EN 1149-3:2004 used. The image of measurement stand is shown in **Figure 11** [18, 28].

The test facility includes the following elements:

1. Electrostatic generator for inducing electric charge in the sample.
2. Base support.
3. Flange support.
4. Electrodes for the measurement of electric field intensity.
5. Stretching system of textile samples.

The apparatus used for electric induction charging is an electrostatic machine type, and for measurement, the following are used: Bridge ESCORT ELC-132A for measuring capacity; Oscilloscope-type Fluke 1968; kilovoltmeter Fluke HP 0-30 kV. The procedure for experimental determination of the half-decay time and the shielding factor using the induction charging method has been applied. For obtaining the electric field intensity value without fabric sample

Figure 11. Image of the ESD setup for assessing the ability to dissipate electrical charge.

E_{max} and with sample E_R, the following data have been obtained: (1) from generator, a voltage pulse is applied, the measured value is U_0 = 6.9 kV; (2) capacitance between the electrodes without the sample is C_0 = 59 pF; (3) the value of the accumulated electric charge, calculated by relation (9), is Q_0 = 407.1 nC; (4) electric field intensity value determined by relation (8) is E_{max} = 119.54 kV/cm. The electrodes have a diameter of 70 mm and section A = 38.465 cm.

The procedure for establishing the electric field intensity with sample between the electrodes is the following: (1) electrostatic generator induces electric field in a sample fixed on support; (2) the voltage value is recorded using an electronic voltmeter VE; and (3) the capacity of the sample system is measured using the bridge ESCORT ELC-132A.

4. Results and discussion

4.1. Surface and volume resistivity

The authors for each weaving technique have chosen a representative sample of the material. For vanish glad-weaving technique, the sample P11 experimental determinations are analyzed, and for vanish patent-weaving technique the sample P14 results are analyzed. The two samples have the same layer of the surface and the same type of conductor wire.

4.1.1. Textile sample P11

The image of P11 is shown in **Figure 12**.

For different values of DC voltage *U* applied to the sample, listed in **Table 3**, the average values of surface absorption current intensity of sample P11 in the steady-state regime are listed.

In **Figure 13**, the time dependences of the surface absorption current intensities for P11 sample, for different DC applied voltages, are shown.

For different values of DC voltage *U* applied to the sample, listed in **Table 4**, the average values of volume absorption current intensity of sample P11 in the steady-state regime are listed.

(a) (b)

Figure 12. Macroscopic image of P11 bilayer sample: (a) face and (b) back.

A. Voltage between 1 and 5 V					
U [V]	1	2	3	5	
$I_{S\ med}$ [A]	3.2005E − 06	7.0300E − 06	1.0572E − 05	1.7705E − 05	
ρ_s [Ω/□]	1.6678E + 07	1.5186E + 07	1.5147E + 07	1.5075E + 07	
B. Voltage between 8 and 25 V					
U [V]	8	10	15	20	25
$I_{\sigma\ med}$ [A]	3.0106E − 05	5.1182E − 05	8.9558E − 05	1.3911E − 04	2.9869E − 04
ρ_s [Ω/□]	1.4159E + 07	1.0429E + 07	8.9406E + 06	7.6524E + 06	4.4678E + 06

Table 3. Average values of surface absorption current intensity and surface resistivity for P11 sample.

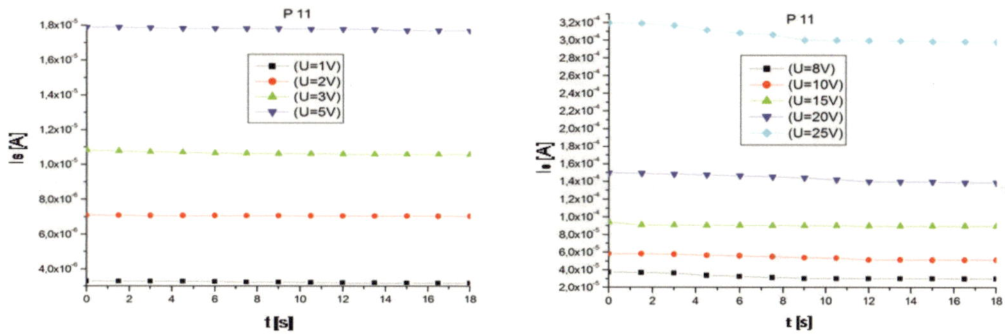

a) b)

Figure 13. The time variation of surface absorption currents intensity for P11 sample, for DC applied voltages: (a) voltage between 1 and 5 V; (b) voltages between 8 and 25 V.

U [V]	1	2	3
I_{med} [A]	6.7346E − 05	1.5310E − 04	2.6073E − 04
ρ_v [Ω·cm]	2.6323E + 06	2.3158E + 06	2.0398E + 06

Table 4. Average values of volume absorption current intensity and volume resistivity for P11 sample.

In **Figure 14**, the time dependences of the surface absorption current intensities for P11 sample, for different DC applied voltage, are shown.

4.1.1.1. Discussions

Data in **Table 3** show that the values of the measured surface current intensities increase with increasing the applied voltage. The order of magnitude of currents measured is between 10^{-6}A for an applied voltage of 1 V and 10^{-4} A for an applied voltage of 25 V.

Surface resistivity varies with voltage values having a tendency to decrease with the increase in the applied voltage. The order of magnitude for surface resistivity is between (10^6–10^7) Ω/□ depending on the voltage applied. This sample has static dissipative behavior.

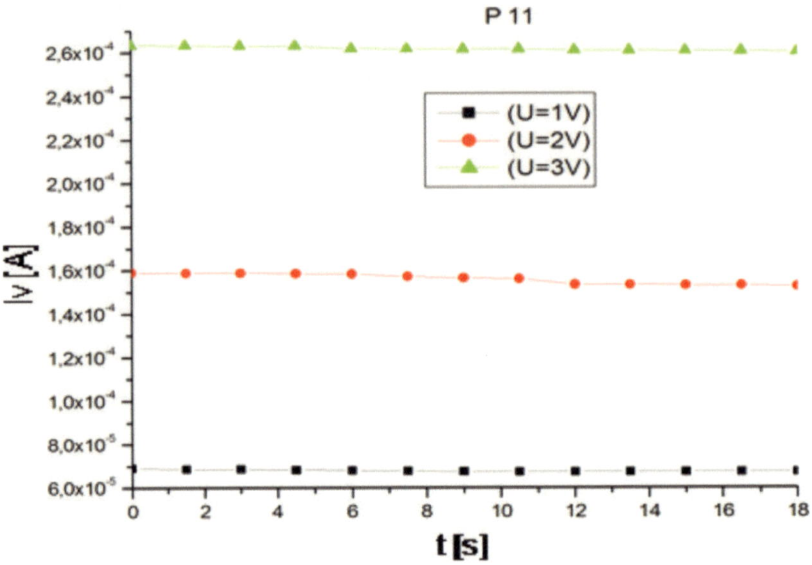

Figure 14. The time variation of volume absorption current intensity for P11 sample for DC applied voltages between 1 and 5 V.

Data in **Table 4** show that the order of magnitude of the volume absorption current is of 10^{-5} A for an applied voltage of 1 V and reach the values of 10^{-4} A for an applied voltage of 3 V.

Volume resistivity value decreases with the increasing applied voltage as in the case of surface resistivity. The order of magnitude for volume resistivity is 10^6 Ω·cm.

4.1.2. Textile sample P14

The image of P14 is shown in **Figure 15**.

For different values of DC voltage *U* applied to the sample, listed in **Table 5**, the average values of surface absorption current intensity of sample P14 in the steady-state regime are

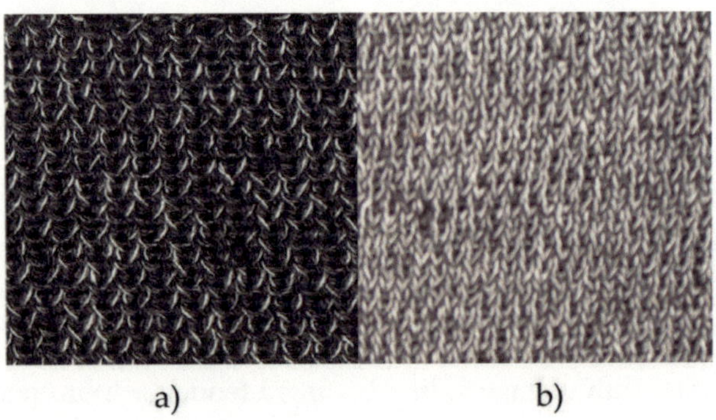

Figure 15. Macroscopic appearance of sample bilayer P14: (a) face and (b) back.

U [V]	1	2	3
I_{med} [A]	2.8207E − 04	6.6364E − 04	1.0680E − 03
ρ_s [Ω/□]	1.8924E + 05	1.6087E + 05	1.4994E + 05

Table 5. Average values of surface absorption current intensity and surface resistivity for P14 sample.

U [V]	1	2
I_{med} [A]	4.0038E − 04	1.4890E − 03
ρ_v [Ω·cm]	4.0224E + 05	2.1632E + 05

Table 6. Average values of volume absorption current intensity and surface resistivity for P14 sample.

listed. In **Table 6**, the average values of volume absorption current intensity of sample P14 in the steady-state regime are presented.

In **Figures 16** and **17**, the time dependences of the surface and volume absorption currents intensities for P14 sample, for different DC applied voltage, are shown.

4.1.2.1. Discussions

Data in **Table 5** show that the values of the surface measured current intensities increase with the increasing applied voltage value. The order of magnitude for surface measured current is between 10^{-4} A for an applied voltage of (1–2) V and 10^{-3} A for an applied voltage of 3 V.

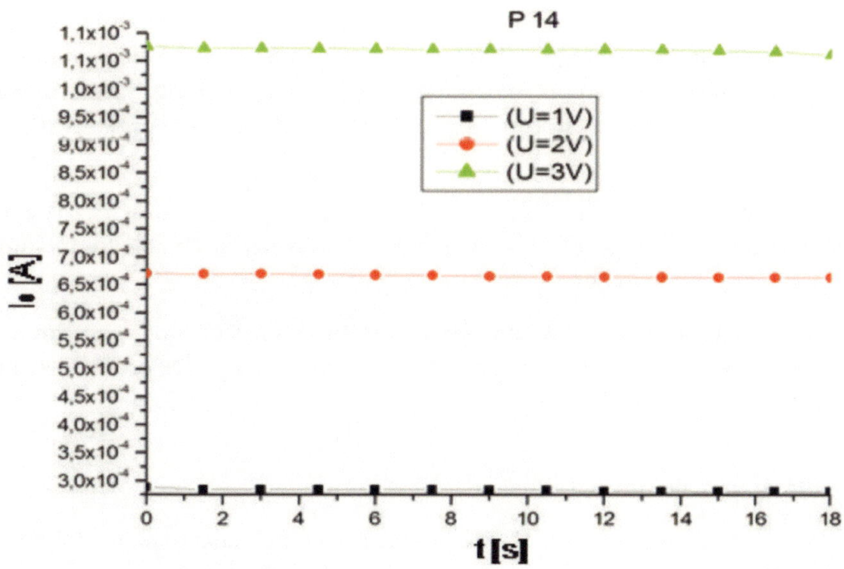

Figure 16. The time variation of surface absorption current intensity for P14 sample, for DC applied voltages between 1 and 3 V.

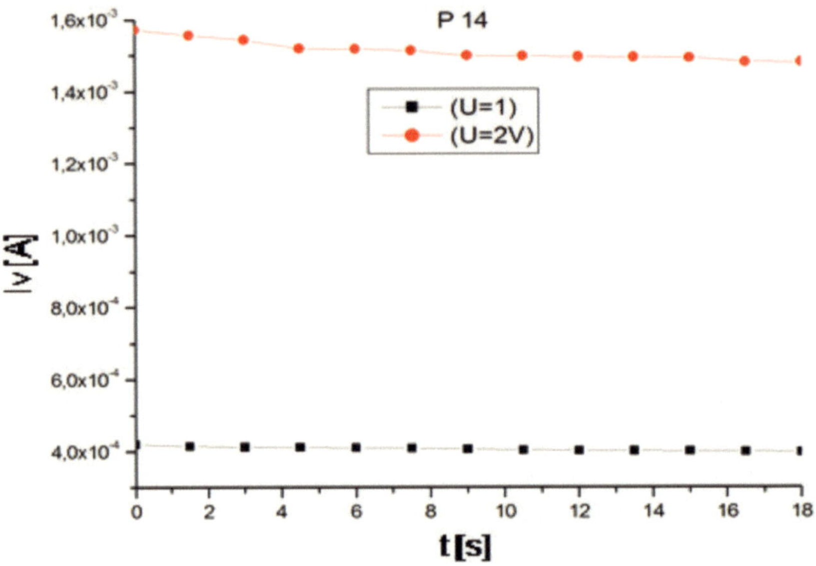

Figure 17. The time variation of volume absorption current intensity for P14 sample, for DC applied voltages between 1 and 2 V.

Surface resistivity decreases with increasing applied voltage, and the order of magnitude for surface resistivity is 10^5 Ω/□. This sample does not belong to the category of electrostatic dissipative materials ($\rho_s < 10^6$ Ω/□).

Data in **Table 6** show that the order of magnitude for volume measured current is 10^{-4} A corresponding to an applied voltage of 2 V. For voltage greater than 2 V, pico-ammeter cannot measure volume absorption currents for sample P14. Volume resistivity value decreases with the increasing applied voltage as in case of surface resistivity, and the order of magnitude of volume resistivity is 10^5 Ω·cm.

Analyzing the composition in terms of the two textile weaving techniques, it can be concluded with certainty that the technique of weaving plaited plain jersey is suitable for making electrostatic discharge protection clothes.

Plaited rib-weaving technique is not appropriate for these types of protective clothing. Resistivity values of surface and volume are not suitable for static dissipative properties of materials rather static conductive [31].

4.2. Charge decay and shielding factor

Experimental results obtained for capacity *C*, voltage *U*, electric charge on the fabric *Q*, electric field intensity E_R, half time decay of electric charge, and shielding factor are presented in **Table 7**. A comparative analysis regarding the half-decay time and shielding factor is shown in **Figure 18**, for the nine textile samples.

Sample number	Sample code	C [pF]	U [kV]	Q [nC]	E_r [kV/cm]	E_0 [kV/cm]	t_{50} [ms]	S
1	P10	32	37	1184	346.9	260.1	100	0.72
2	P11	41	30	1230	360.3	240.2	100	0.7
3	P12	40.5	20	810	237.3	158.2	700	0.81
4	P18	40.5	19	770	225.6	150.4	550	0.82
5	P17	41.5	20	830	243.2	182.3	60	0.8
6	P19	40.3	21	846	247.8	162.2	250	0.8
7	P20	41.2	27	1112	325.8	217.2	700	0.73
8	P14	41.6	17	707.2	207.2	155.4	80	0.83
9	P13	41	30	1230	360.3	270.2	250	0.7

Table 7. Half-decay time and shielding factor for textile samples obtained with induction charging method [28].

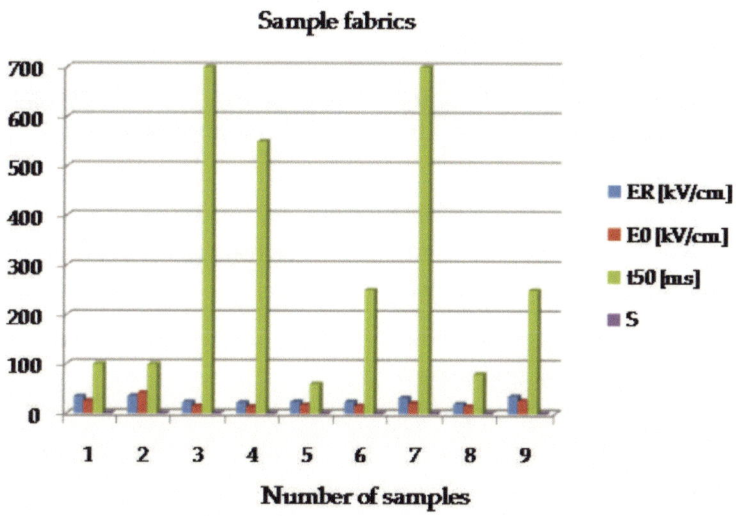

Figure 18. Graphical representation of parameters E_R, E_0, and t_{50}.

Half-decay time t_{50} and the corresponding value of electric field intensity E_0 are obtained from wave form signal on the oscilloscope. The shielding factor S for textile samples is determined based on relationship (10).

4.2.1. Discussions

Samples *5* (P17) and 8 (P14) presents the best values for time decay of electric charge and have shielding factors between 0.8 and 0.7. Samples *1* (P10) and 2 (P11) have values for time decay of electric charge of 100 ms and shielding factor 0.72 and 0.7, respectively. Samples *6* (P19) and *9* (P13) have time decay of electric charge of 250 ms and shielding factor of 0.8 and 0.7, respectively. Sample 4 (P18) have time decay of electric charge 550 ms and shielding factor 0.82. Samples 3 (P12) and 7 (P20) have time decay of electric charge 700 ms and a shielding factor of 0.81 and 0.73, respectively. In terms of mitigating the electric field intensity, samples 3 (P12) and 7 (P20) have the lowest attenuation compared to other samples.

According to standard SR EN-1149-5:2008, all analyzed samples can be used to manufacture protective clothing, with the following observations: (1) for materials that do not have electrostatic shielding effect, $E_R = E_{max}$; (2) for materials with electrostatic shielding effect, $E_R < E_{max}$; (3) if $E_R < E_{max}/2$ is considered that $t_{50} < 0.01$ s; and (4) if value $E_{max}/2$ is reached in 30 s, $t_{50} > 30$ s is considered.

Samples 5 and 8 show the best times of time decay of electric charge <100 ms and have shielding factor 0.8 and 0.83, respectively. The composition of sample 5 (P17) is one wool yarn + one yarn type 2 for the outer layer/front side of the fabric and one wool yarn + two yarns type 5 for the inner layer/back side of the fabric. The composition of sample 8 (P14) is one cotton yarn + one yarn type 2 for the outer layer/front side of the fabric and one cotton yarn + two yarns type 5 for the inner layer/back side of the fabric. Yarn type 5 is made of nylon filament at the surface saturated with carbon particles.

Both samples were obtained by plaited rib knitted structures, have 6% percentage of conductive yarn, and have the same number of yarns type 5 in composition which distinguishes them as the base yarn.

The authors consider that the fabric with plaited rib-knitted structures together with the type of conductive yarn and the number of conductive yarns used (max. 2) offers the most efficient protection for time decay of electric charge.

The difference of value for electric field intensity (with sample) and time discharge of electric charge may be due to natural yarn structure.

5. Conclusions

The ESD equipment development was supposed to choose a knitted bilayer structure that can provide good protection from accidental electrical discharges on the dielectric layer and a drainage of accumulated charges through conductive layer.

The presence of the yarn type 2 on the front side determines the performance improvement by using the yarns types 3 and 5 for all knitting configuration; knitting structure does not significantly influence the ability of electrostatic discharge.

The limits of variation for water vapor permeability are between 28.5 and 42.5%, statistical events that agglomerate into optimal zone indicate the presence of the cotton yarn in rib structure. Air permeability causes sensations of warm and cool of clothing products, the best value being obtained by the samples 10, 11, 12 (374–400 $l/m^2/s$), characterized by the presence of cotton yarn and conductive yarn element from nylon filament surface saturated with carbon particles.

Simple electrical measurements, like two-point and four-point DC measurements, reveal the good electrical behavior of the yarn and fabric of the analyzed samples, when containing carbon-covered yarns. The measured resistances are in the M□-range and are sufficient to avoid electrical charge buildup in the fabric.

Analysis of physical-mechanical and functional characteristics indicates appropriate values for chosen knitting structure designed for ESD protection equipment.

Technical textiles have multiple applications which make possible to conduct further researches to obtain smart textiles.

Determination procedures for evaluating specific performance of textile and their calcification as static or dissipative are surface resistivity, volume resistivity, speed decay of electric charge and shielding factor, and time decay of electric charge which are additional parameters to assess these materials.

From determinations conducted for surface resistivity to samples described in **Table 1**, two aspects stand out, namely (a) if the percentage of conductive yarns increases, these fabrics become more conductive reaching surface resistivity values less than materials with dissipative properties ($\rho_s < 10^6\ \Omega/\square$); (b) the textile structure (plain jersey and rib) influences the value of surface resistivity due to the presence of air in the textile structure, which makes the knitted samples with plaited rib structure to provide a higher conductivity than the samples with plaited plain jersey structure.

Also, when the voltage applied increases the value of current intensity, thus currents of the order of 10^{-3} A were highlighted. This aspect is not negligible in the case of higher voltages that may arise in case of real electrostatic discharges.

From determinations conducted for volume resistivity, it appears that the applied voltage values are lower and conduction processes in material are more intense than to surface. However, there are samples with volume resistivity of the order of 10^7–10^6 Ω·cm (P11, P12, P19, and P20).

From determinations conducted for time decay of electric charge and electrostatic shielding factor, three aspects stand out, namely (a) plaited rib structures offer the lowest times of decay of electric charge good and good electrostatic shielding coefficients; (b) the number of conductive yarns type 5 (filaments from nylon saturated at the surface with carbon) should not exceed two yarns (6% of conductive yarn) because at a greater number of conductive yarns intensive conducting processes can occur; (c) the aspects mentioned earlier must be correlated with the values obtained for surface and volume resistivity in order to make a complex analysis closer to actual needs and requirements of ESD protection systems.

Acknowledgements

This work was supported by a grant of the Romanian National Authority for Scientific Research, CNDI – UEFISCDI, project PCCA 179 – 2012 "ESD protective garments made with core conductive fibers knitted in double layer."

The authors also gratefully thank the staff of ELMAT Lab from the Polytechnic University of Bucharest for their support in the realization of some of the measurements, especially Prof. M. Dumitran.

Author details

Gabriela Telipan[1]*, Beatrice Moasa[2], Elena Helerea[2], Eftalea Carpus[3], Razvan Scarlat[3] and Gheorghe Enache[4]

*Address all correspondence to: gabriela.telipan@icpe-ca.ro

1 National Institute for Electrical Engineering ICPE CA, Bucharest, Romania

2 Transilvania University of Braşov, Braşov, Romania

3 The National Research & Development Institute for Textiles and Leather, Bucharest, Romania

4 TANEX SRL, Bucharest, Romania

References

[1] Agarwal BJ, Agarwal S. Integrated performance textiles designed for biomedical applications. In: Proceedings of International Conference on Biomedical Engineering and Technology IPCBEE. Vol. 11. Singapore: IACSIT Press; 2011. pp. 114-119

[2] Bahadir SK, Jevsnik S. Mint: Optimization of hot air welding process parameters for manufacturing textile transmission lines for e-textiles applications: Part I: Electroconductive properties. Textile Research Journal. 2017;**87**(2):232-243. DOI: https://doi.org/10.1177/0040517516629140

[3] Paasi J, Nurmi S, Kalliohaka T, Coletti G, Gustavino F, et al. Electrostatic testing of ESD protective clothing for electronics industry. In: Proceedings of Electrostatics Conference; 23–27 March; Edinburg. 2003. pp. 239-246

[4] Ardeleanu AS, Verejan A, Donciu C. Study regarding the knitting parameters in 3D textiles shields implementation. Mint: Acta Electrotehnica. 2010;**51**(2):128-131

[5] Helerea E, Moasa B, Ciobanu A. Analysis of electrostatic environment and discharge models. In: Proceedings of the IEEE International Conferences Applied Theoretical Electricity; 25–27 October 2012. Craiova: IEEE; 2012. pp. 1-4

[6] Baumgartner G. Consideration for developing ESD garment specifications. ESD TR 05-00 Report. ESD Association; 2000

[7] Chen HC, Lee KC, Lin JH, Koch M. Comparison of electromagnetic shielding effectiveness properties of diverse conductive textiles via various measurements techniques. Mint: Journal of Materials Processing Technology. 2007;**192–193**:549-554. DOI: https://doi.org/10.1016/j.jmatprotec.2007.04.023

[8] Ceken F, Pamuk G, Kayacan O, Ozkurt A, Ugurlu SS. Electromagnetic shielding properties of plain knitted fabrics containing conductive yarns. Mint: Journal of Engineered Fibers and Fabrics. 2012;**7**(4):81-87

[9] El-Newashy RF, Saad MA, Turky GM. Integration of conductive yarns into fabric by stitching. Mint: Journal of American Science. 2012;**8**(3):213-217

[10] Lemaire P. Forced Corona Method for the Evaluation of Fabrics with Conductive Fibres. Available from: http://estat.vtt.fi/publications/lemaire_poster.pdf

[11] Thilagavathi G, Raja ASM, Kannaian T. Nanotechnology and protective clothing for defence personnel. Mint: Defence Science Journal. 2008;**58**(4):451-459

[12] Rubeziene V, Baltusnikait J, Zuravliova SV, Sankauskaitea A et al. Development and investigation of electromagnetic shielding fabrics with different electrically conductive additives. Mint: Journal of Electrostatics. 2015;**75**:90-98. DOI: https://doi.org/10.1016/j.elstat.2015.03.009

[13] Chena HC, Lee KC, Lin JH, Kochc M. Comparison of electromagnetic shielding effectiveness properties of diverse conductive textiles via various measurement techniques. Mint: Journal of Materials Processing Technology. 2007;**192–193**:549-554. DOI: https://doi.org/10.1016/j.jmatprotec.2007.04.023

[14] Varnaite S, Katunskis J. Influence of washing on electric charge decay of fabrics with conductive yarns. Mint: Fibres and Textiles in Eastern Europe. 2009;**17**(5):69-75

[15] Hecht DS, Hu L, Gruner G. Electronic properties of carbon nanotube/fabric composites. Mint: Current Applied Physics. 2007;**7**:60-63. DOI: https://doi.org/10.1016/j.cap.2005.09.001

[16] Knittel D, Schollmeyer E. Electrically high-conductive textiles. Mint: Synthetic Metals. 2009;**159**:1433-1437. DOI: https://doi.org/10.1016/j.synthmet.2009.03.021

[17] Neelakandan R, Madhusoothanan M. Electrical resistivity studies on polyaniline coated polyester fabrics. Mint: Journal of Engineered Fibers and Fabrics. 2010;**5**(3):25-29

[18] Moaşa B, Helerea E, Ignat M, Telipan G. Experimental research on dissipative textile structures. In: Proceedings of the IEEE International Conference on Optimization of Electrical and Electronic Equipment OPTIM; 22–24 May 2014; Cheile Gradistei. Brasov: IEEE; 2014. pp. 155-160

[19] Dumitran LM, Bădicu L, Noţingher PV, Tănăsescu G. Premize pentru elaborarea unei metode de evaluare a stării de degradare a sistemelor de izolaţie ale transformatoarelor de putere pe baza curenţilor de absorbţie/resorbţie [Internet]. 2009. Available from: http://midmit.elmat.pub.ro/media/sec/LUCRARE%20Bran%202010_6ro.pdf

[20] Noţingher PV, Stancu C, Dumitran LM, Noţingher PP, Rakowska A, Siodla K. Influence of the ageing state of insulation systems on absorption/resorption currents. Mint: Revue Roumaine des SciencesTechniques – Serie Électrotechnique et Énergétique. 2008;**53** (2):163-177

[21] Stancu C, Noţingher PV. Influence of the surface defects on the absorption/resorption currents in polyethylene insulations. Mint: Scientific Bulletin of the Polytechnic University of Bucharest, Series C. 2010;**72**(2):161-170

[22] Dumitran LM. Electrical Insulation Systems (Sisteme de izolaţie electrică). Printech; 2008. 266 p

[23] Al-Saleh MH, Sundararaj U. A review of vapor grown carbon nanofiber/polymer conductive composites. Mint: Carbon. 2009;**47**:2-22. DOI: https://doi.org/10.1016/j.carbon.2008.09.039

[24] ASTM 257-14. Standard Test Methods for DC Resistance or Conductance of Insulating Materials

[25] Drake N. Polymeric Materials for Electrostatic Applications. Rapra Technology Ltd.; 1996. 168 p

[26] ***SR HD 429 S1 CEI 60093. Methods of Test for Volume Resistivity and Surface Resistivity of Solid Electric Insulating Materials. 2002

[27] ***SR EN 1149-3:2004,. Protective clothing—Electrostatic properties—Part 3: Test methods for measurement of charge decay

[28] Telipan G, Ignat M, Cătănescu L, Moaşa B. Electrostatic discharge testing of several ESD protective textiles used in electronic industry. In: Proceedings of the IEEE Electrical and Power Engineering International Conference and Exposition on (EPE); 16–18 October 2014; Iaşi: IEEE; 2014. pp. 602-605

[29] Carpus E, Vizileanu E, Scarlat R, Donciu C, Bonfert D, Popa A et al. Investigation of two-layer knitted structures with conductive fibres content. Mint: Industria Textila. 2014;**65** (3):145-152

[30] ***Keithley. Manual for the Use of the Electrometer Keithley 6715 A

[31] Moasa B. Contributions on Immunity Assessment of Equipment to Electrostatic Discharges [Internet]. 2015. Available from: http://webbut.unitbv.ro/teze/rezumate/2015/rom/MoasaCostelaBeatrice.pdf

Section 6

Future of Advanced Textiles

Chapter 15

The Long Way of the Success: From Idea to the Market

Adriana Ungureanu

Additional information is available at the end of the chapter

http://dx.doi.org/10.5772/intechopen.69210

Abstract

Due to the changes in the textile field, effective solutions come from two directions: incorporating ideas in physical new and very modern products or giving value to new ideas that can impulse later discoveries and potential applications by identifying specific niches. New trends ask to cultivate a culture of entrepreneurship not only in terms of business but even at the level of research as the last chance for those who cannot produce traditionally in market conditions. The commercial or research (innovative) niches must be identified at the national and international levels and conducted to new advancements that should guide to commercial destination by responding to special human needs. The purpose of this chapter is to reveal five case studies as examples of integrating smart textiles into the value chain. This research shows that success is a consequence of intensive- and commercial-oriented company policy, an effect of updating to nowadays conditions, and an inspired movement to the right business way. The chapter also represents an invitation for creative people to focus on developing innovative value chains.

Keywords: smart fibers, smart textiles, fibrous structures, technical textiles

1. Introduction

The international trade liberalization had an important impact on modeling the map of the textile industry. Thus, the elimination of quantitative restrictions forced relocation to cheap labor zones able to respond to a higher demand. Generally, the global changes supposed business updates in terms of price, quality, and innovation, but flexibility was required in front of the market threats, and it created new development opportunities. The studies [1] show that during crisis, the companies tend to reduce the outsourcing and keep the activity within their national boarders as a help between the local actors inside their value chain. More than this, sophisticated products (technical or smart) become a feature of developed countries without being the subject of the outsourcing phenomenon.

Textile reports [2] show that the behavior of textile actors is reactive (including those involved in research activity), which means they react in accordance with the trends of the markets.

Due to all the last changes from the textile field, effective solutions come from two directions: first, incorporating ideas in physical new and very modern products or second, giving value to new ideas and incorporate them in patents that can impulse later discoveries and potential applications by identifying specific niches. In other words, new trends ask to cultivate a culture of entrepreneurship not only in terms of selling but even at the level of research as the last chance for those who cannot produce traditionally in market conditions any more. The commercial or research (innovative) niches must be identified at the national and international levels and conducted to new advancements that should guide to commercial destination by responding to special human needs. For a value chain integration, a huge effort is dedicated to make well known the idea or the concept in front of those who could be interested in.

The main weakness of this subject is that expensive discoveries once registered in a patent, the researchers lose curiosity to cross the edge of theory and searching if there is possibility to extend the application, if there is any efficiency from the economical point of view, if there is a cover for some human needs to respond. So, their attention will be dedicated to new ideas, with no connection to the last ones, concentrating on other opportunities.

Reading carefully the literature dedicated to smart textiles, it is easy to notice that they cross by far the limits of "fashion" and they are in fact a superior form of technical textiles bringing unexpected solutions for every industry and offering obvious competitive advantages to those who use them. Another feature noticed is that traditional textile industry migrates toward a smart level due to specific technologies. Concluding, it is easy to understand that every discovery connected to new technology could offer one step in front of competitors. But the key to success is kept by the final user who often is not aware of the advantage of using smart textiles.

Not too many years ago, few people knew the meaning of "smart textiles," and this was a good motivation to focus on collecting information, dedicate researches on this subject, and then disseminate through specific ways: conferences, articles, meetings with important actors from textile field, and creating dedicated platforms. Some critics and old-fashioned mentalities became another good reason for continuing this work and improving discoveries and awareness about this new trend. Maybe the best way to persuade is to identify examples of good practice, to analyze them, and to focus on their secrets of success, from idea to the market.

That is why the purpose of this chapter is to reveal some European examples of successful companies involved in smart textiles. Their story started from a genius idea, but their secret was materialized through intensive efforts focused on the step to find the next user. These case studies bring essential information regarding the smart textile potential in solving some important human needs, the receipt of the success in very new, competitive, and money-demanding field.

Using the same algorithm of thinking, this chapter covers five different situations for integrating ideas in value chains, divided in two main perspectives: commercial (three case studies) and innovative (two case studies). Four case studies represent good examples of finding the best way to success, and the last cast study is more a simulation, a good reason for an imaginative exercise to find the next step to the final user.

As methodology, this research collected information from the Internet (sites, press releases, and international databases), sometimes combined with data obtained during interviews or from materials supplied by companies in specific cases. Since the target was to identify competitive advantages, the Michael Porter's model, the value chain configuration, and a Strenghts, Weaknesses, Opportunities and Threats (SWOT) analysis were very helpful. Also, the literature dedicated to new technologies underlines the necessity to identify more examples and new products, and this was a good inspiration to simulate a patent analysis and a proposal for the next stage of the research as a step closer to the final consumer.

This chapter contains three main parts: introduction, the second part contains five sections dedicated to each case study, and the third part concentrates on the conclusions. All the references will be inserted at the end of this chapter in order of their appearance.

2. Commercial versus innovative value chain

2.1. Integrating smart textiles into commercial value chain

Before digging the secret of success for some companies involved in smart textiles field, theoretical economic perspectives are required to emphasize the systemic interdependencies of competitive advantages into the triangle company-business environment-market.

Thus, some researchers [3, 4] argue the same view regarding the competitive advantages when they come from two directions: from the market and from the companies. They consider that at the market level, the demand and the offer represent the start for advantages, and they are shaped in a specific environment, where industries and companies are connected and guided by regulations created to ensure a healthy cooperation. In the same time, they consider that at the company level, the competitiveness is a result of the best strategy selected. Compressing the information, a double influence is noticed, the market has an impact on the companies' behavior, and the companies can change the market through new directions by using proper strategies or unexpected technologies.

Different opinions [5] consider competitiveness as a feature of companies rather than that of market, due to their strategies.

Some economists [6] consider that all industries due their existence as individual decisions and their amplitude and quality depend on accumulated experiences and on local advantages where they operate.

A short resume of the economic literature [3] shows that the analysis of competitive advantages may be observed from three perspectives:

- From the direction of demand and offer—this represents the oldest way to analyze a market; this method is not recommended during crises.
- Focusing on the net value—as an optimization of all factors involved from the subsidiary offers, this method is recommended for mature markets or products.

- From the perspective of Porter's five forces—this method allows an analysis in real conditions, by considering the entrances on the market, competitors, and substitutions (horizontally) and suppliers and buyers (vertically).

From these three perspectives above, the Porter's method brought its contributions to develop and build the entire specialized literature covering all the five dimensions and enriching the strategy theory. This method is used the most when market analyzing is required since it is easy to concentrate and translate all the information collected.

This subchapter presents the landscape of three companies—Elmarco (Czech Republic), Holmenkol (Germany), and Siderma (Romania)—where Michael Porter's five forces are adopted to build the market configuration for their smart products and understand their positioning into the value chain. Then, a SWOT matrix-table will reveal where the advantage comes from, what is the key of the success, and how these companies turned their advantage to become leaders. These case studies are built [7] by using information from the Internet (sites, international databases, or press releases). In case of Siderma, besides information supplied by the Internet, an in-depth interview and further brochures and documents [7] fulfilled the research. The main purpose is to understand whether they have some similar features by following the same rules guided by the market or do they discover their own roles?

2.1.1. Case study I: nanofibers from Elmarco S.R.O.

Elmarco [8] is a private company from Czech Republic specialized in producing nanofibers equipment with a lot of applications in many industries. The headquarter is in Liberec, and there are two more branches in the USA (Morrisville) and Japan (Tokyo). In 2013, the company had 35 researchers from a total of 80 employees. The activity started in 2000 as a semiconductor producer, but a partnership signed in 2004 with Technical University from Liberec changed the business way. In 2005, the Nanospider technology was created, and this became the key patent the entire company concentrated the efforts on. Nanospider represents an organic or inorganic electrospinning process, a versatile technology that allows production of fibers with a capillarity about 500 nm which means 1000 times thinner than the human hair. The nanofibers production using the Nanospider technology has the following advantages:

- High scalability—allowing a high productivity and a large dimensional demand.
- Uniformity of the fibers, and as consequences, uniformity of the tissues produced with these fibers.
- Good production conditions from the economical point of view and maintenance.
- Flexibility in raw materials used such as polymers and others.

The main characteristic of the nanofibers and nanomembranes is the possibility to use them in different fields such as: health industry—supplying the drugs *in vivo* system (dissolution into the human body) and *in vitro* system, wound patches, artificial tissues, organic tissues, or antiseptic textiles barriers for surgery; technical industry—separators for batteries, solar panels made up of titan dioxide nanofibers as a silicon substitute, air filtration systems, or water filtration as well as acoustic barriers; equipment for protection and high-performance clothing.

For 2017, the international market potential for nanofibers was estimated [9] at 852.3 million of dollars, and the perspective [10] of 2020 shows a market estimation of 2 billion dollars. This represents a chance for developing Elmarco business and keeping the international leader positioning.

During 2007–2010, Elmarco invested an important amount for building a research center with a production division. Most of its activity consists in nanofiber producing and selling (98% of its activity is dedicated to exports).

The business landscape of Elmarco is represented in two ways: first by using the five forces of Michael Porter for analyzing the market and second by configuring the value chain to understand the position where the company is integrated.

Entrance: Elmarco entered aggressively into the market since its Nanospider patent was unique, and it could be a leader from the start. To gain this, the company made a successful movement by signing a contract with academic environment and then obtaining strong funds from governments and European Union (EU) to sustain an intensive research activity. If another company wants to enter in the same field, it requires succeeding the patent barrier and considerable investments. Thus, the position of Elmarco is not easily threatened. Besides, the company signed a lot of international partnerships with famous universities and research institutes, which ensure a strong notoriety and a quick development potential. The position of Elmarco can be mined only by those companies that could bring new discoveries with low-cost implications and higher performances or large possibilities of scaling.

Competitors: The main nanofiber competitors [11] are as follows: Elmarco LTD., CONTIPRO, NAFIGATE CORPORATION, Nanopharma, and Nanovia LTD. From all these companies, the main actors are Elmarco, CONTIPRO, and NAFIGATE. The year 2010 was an important moment when ex-manager Ladislau Mares—the founder of Elmarco—created NAFIGATE, a real competitor since he knew all the vulnerabilities of this company. Starting from then, Elmarco met a regression at least at the personnel level when the number of employees decreased from 250 to 80. Meanwhile, NAFIGATE had an explosive development by signing partnerships with China.

Substitutes: Most of the products obtained through nanofibers can be replaced by traditional products, but Elmarco brought properties difficult to be replaced. For example, air filters or water filters can be obtained in better economic conditions and with better quality. Photovoltaic cells realized with nanofibers can successfully replace those made up of silicon. Classical impermeable fabrics are enriched with nanofibers, and new parameters such as air circulation are improved. In health field, new membranes are more soluble. In acoustic field, the porosity of fibers allows to improve isolation, ensuring a protection of 100%. New discoveries can pose a threat for the present ones. For instance, the researches concentrated on graphene could guide to surprising directions, and this could affect Elmarco in the future.

Suppliers: Poly Nano Tec is the main partner from the TransMIT GmbH (Germany), the largest raw materials producer for nanofibers, supplying polymers for extrusion processes and melting—blowing organic antibacterial agents. At the same time, the partnerships are an important source of information that could push Elmarco to new research developments. There are some spare parts suppliers, but the information about them is not available. Regarding the labor, Elmarco has an access to extremely qualified and cheap workforce in comparison with other states.

Customers: Until 2013, Elmarco produced 130 Nanospider equipment, and it is considered one of the most important international producers. The partnership policy was concentrated in two directions.

- Universities and Research institutes

To facilitate the research activity, Nanospider equipment was supplied to the Research Triangle Institute (USA), Kyoto Institute of Technology and Shinshu University (Japan), National Institute of Singapore, King Saud University from South Arabia, and Moscow State Textile University (Russia). A contract concerning the filtration with nanofabrics was signed with Akron University. Other partners are as follows: Nonwovens Institute (Raleigh, North Carolina, USA), Industrial Technology Research Institute (ITRI) from Taiwan, National Textile University of Faisalabad (Pakistan), and Ahmedabad Textile Industry Research Association—the biggest textile center from India with 98 units in the country. In 2011, the National Tissue Centre was inaugurated in Brno to develop modern methods for producing tissues and cells based on nanofibers. This project helped Czech Republic to be at the same level with the most famous institutes such as, Fraunhover Institute (Germany), Bioheart (South of Korea), and Indiana Stemcell, Purdue Nanotechnology Centre, and Wake Forest University (USA), dedicated to nanotechnology research.

- Business to business—customers from business environment [12]

In 2005, when the Japan branch was inaugurated, a joint venture with Atracell was born (named Nanopeutics) and the main activity was producing wound patches. Another partnership dedicated to the same subject was signed with HemCon Medical Technologies Inc. (Portland, Oregon, USA). CEZ company became a new partner in the field of energy, and thus, the solar panels were produced by replacing silicon with nanofibers. These new cells allow a better efficiency (about 80%) in comparison with those made up of silicon. In this case, when large-scale production started, the cost decreased significantly. Oerlikon Neumag from Austria is Elmarco partner in acoustic barriers. Thus, the nanofibers create a protection of sound by absorbing 100% of it. The Middle East offered a great opportunity for air and water filtration cooperation, and thus an important number of equipment is sold here.

Shortly, the Elmarco market can be seen in **Figure 1**.

Collecting the entire information, the Elmarco value chain can be seen in **Figure 2**.

From these two figures, it can easily be noticed that Elmarco became a strategic supplier for those interested in obtaining products based on nanofibers since its strength was the Nanospider patent. The weakness of Elmarco was sometimes the final user, and the intermediary producers or dealers need to work on this subject. The opportunities are gained from the research partners and the threats come from competitors.

Concluding, the SWOT analysis can be synthetized as shown in **Table 1**.

It is obvious that Elmarco' advantages come from the way of using the patent created together with the University of Liberec. Concerning the value chain, its position is difficult to be replaced by other companies since the financial support was dedicated to get a very good image and consolidate international partnership that could help in future discoveries.

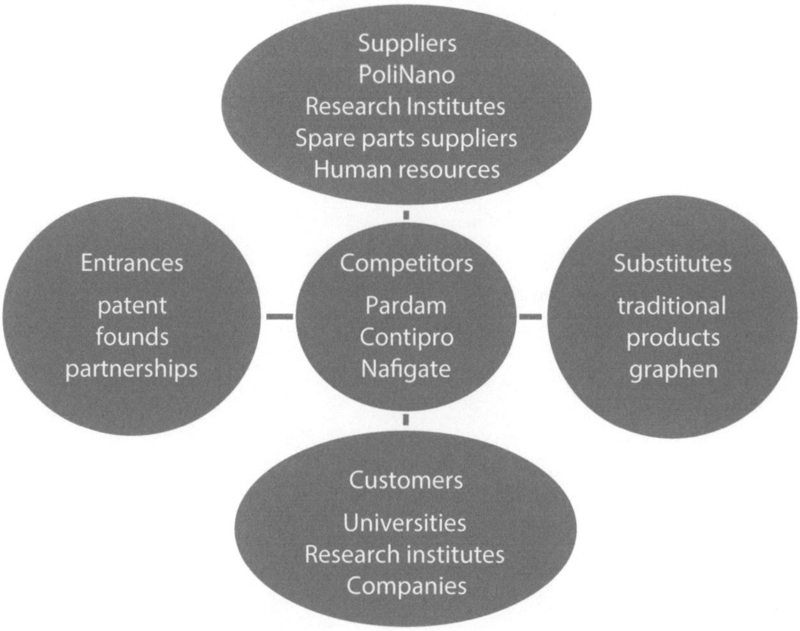

Figure 1. Elmarco market.

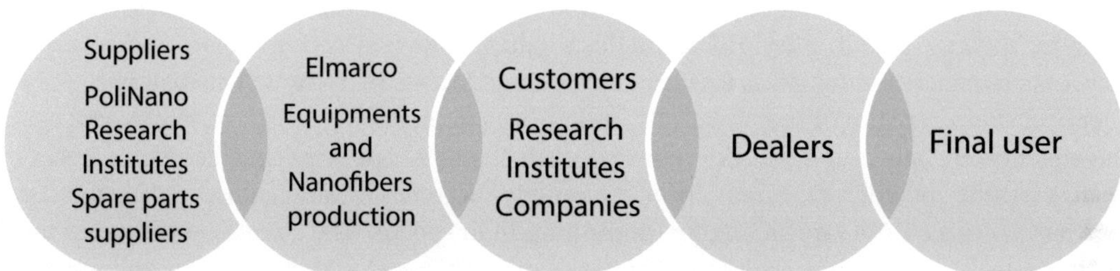

Figure 2. Elmarco value chain.

2.1.2. Case study II: nanoproducts dedicated to winter textiles from Holmenkol GmbH

Holmenkol GmbH [13] is one of the oldest companies in the world, whose history dates back to 1922, and it used to be a traditional wax producer. In 2001, the first liquid wax was realized through nanotechnology, and FORBES considered it the product of the year. Its headquarter is based in Heimerdingen, and there are branches in Germany, Norway, and Japan. Until 2008, Holmenkol worked as a division of LOBA company, the main shareholder from that moment, and due to its 18 patents, it became one of the most innovative companies in Germany. In 2006, 43 national teams' competitors attending the Winter Olympic Games used Holmenkol products.

In the start of 2008 [14], LOBA joined Holmekol and Nanogate—the most important investor dedicated to nanotechnologies—and all together built Holmenkol Sport-Technologies GmbH. In 2012, when 90 years of existence accomplished, the insolvency was declared, but at that

Strengths	Weaknesses
• Strategic equipment supplier • Patent • Financial support • High qualified workforce • International reputation • Strong national and international relationships	• Final user
Opportunities	**Threats**
• International research partnerships • New partners	• New discoveries • Graphen • Competitors (NAFIGATE)

Table 1. SWOT Elmarco market analysis.

moment, another company—Sporto-med GmbH, part of the Eimermarcher Group—became the main shareholder.

In the meantime, Holmenkol improved the traditional wax technology by creating the nano-CFC@technology patent, a hybrid technology which allows a certain cover and improves physical characteristics for skies, textiles, bicycles, and body care creams. This patent re-joined chemical characteristics of the materials used with nanotechnology discoveries, by changing the effect on the surface. A fluoride powder mixed with nanocomposites helps the water to change to a dry powder. During friction, the powder becomes liquid again and thus the abrasive resistance and slip rate are higher, permitting high speeds. The components are not used in their pure form, and they are integrated in a matrix. The products are sold as sprays, and the substances are not inhaled in lungs since the components are cemented on the product surface. The offer for textile winter clothes covers 16 nanoproducts. These were used to clean clothes at very low temperatures, conferring more water resistance, texture breathing at extreme temperatures, smell neutralizer, and extended textiles texture life.

Besides nano-CFC@technology, there are other types of technologies dedicated to textiles exclusively, such as Lotus-Hybrid Matrix (which stimulates a lotus water effect for different fabrics) and Hygiene Effect (which ensures a perfect cleaning during washing at very low temperatures by eliminating germs and bacteria).

Nowadays, Holmenkol is a Forumnano member and very active in public debates on Nanosafe platform regarding the risk exposure.

Collecting all the information, the market of Holmenkol can be designed by using the Porter five forces as follows:

Entrances: A company interested in joining this field is obliged to invest considerable amounts to produce similar products and to promote them since the notoriety of Holmenkol is difficult to be reached. Ninety years of commercial existence is quite hard to equal the trust of investors and traditional customers. Thus, the barriers are very high, and the direct competition is low.

Competitors: Regarding the Holmenkol offer dedicated to textiles, there are a lot of competitors since the modern products can meet the traditional ones, but the performances are difficult to be reached. The relationships created in 90 years shaped the customers to be a very high quality demanding, and from this point of view, the competitors meet disadvantages.

Substitutes: These can appear from new discoveries or some traditional products and could be taken into consideration, but the effect will not be the same. The price is a consequence of technology used and could separate the qualities and product characteristics.

Suppliers: All the products dedicated to textiles are made in Germany. This is the key of the best reputation offered to customers. Being a LOBA division for years, it was supposed that the raw materials came directly from inside the consortium for long time, as a guarantee of quality.

Customers: A wide range of customers use Holmenkol products—top athletes, international and national federations, and sports clubs—and they are applied on technical, high-tech, or thermal isolated fabrics such as Gore-Tex, Sympatex, Schoeller, eVent, Coolmax, Soft Shell, Thinsulate, Thermolite, and PrimaLoft.

Shortly, the market for Holmenkol products dedicated to textiles can be seen in **Figure 3**.

The value chain is configured in **Figure 4**.

Figures 3 and **4** show that the advantage comes from important investments in research activity, incorporating nanotech discoveries in successful products and updated to new trends, patent portfolio, and rich past, marked by the trust of investors and top customers. Despite its economic difficulties, Holmenkol is still a leader in its market. The weakness comes from its insolvency in 2012. The opportunities come from the future discoveries and new way of developments. The market trend can be mined by other discoveries that could oblige the owner to stop the research activity. Concluding, a SWOT analysis can be synthetized into **Table 2**.

Table 2 reveals that competitive advantages for Holmenkol were built in time, and it is interesting how this company could keep the customers and offer always innovative products. When the company met difficulties, the investors did not hesitate to come and offer their help to continue the research activity especially. Also, the collaboration with German suppliers consolidated the "made in Germany" image and helped Holmekol to be always the market leader.

2.1.3. *Case study III: SIDMAT 3, a revolutionary smart fabric from Siderma S.A.*

The particularity of this case study consists in analyzing the first Romanian smart fabric, and the entire information was collected from the Internet site, press releases, in-depth interviews, and documents supplied by the company, and it represents an important international reference for the specialized literature that claims lack of practical examples.

Siderma S.A. was created in 1971, and its main activity was the production of spare parts for shoes, automotive, constructions, and furniture industries. Now, it is a nonwoven leader, and the production goes to domestic or international market. The technology is a classic one, combining different fibers (cellulose, polyamides, wool, polypropylenes, and polyesters). The

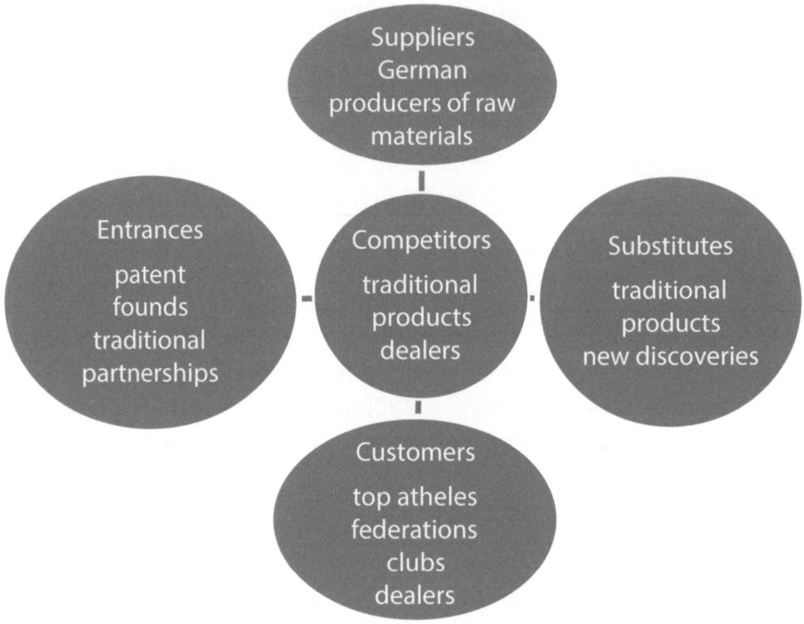

Figure 3. The Holmenkol market.

Figure 4. The Holmenkol value chain nanoproducts dedicated to winter textiles.

main technological process is carding-folding-punching, and the semi-finished fabrics have different widths, thickness, and density depending on applications.

Nowadays, Siderma has in its portfolio about 700 products, but 350 are available to be produced anytime. Generally, it produces shoes linings, knitted linings, thermobinding, interwoven in different colors, insoles, laminated assemblies, synthetic and wool furs, carpeting thermo-formable dedicated to automotive industry, geotextiles for construction industry, air and liquids filters, and nonwoven items for furniture industry.

Siderma offers a guarantee of its products by a high selection of the suppliers. Thus, the customers receive a REACH declaration (Registration, Evaluation, Authorization, and Restriction of Chemicals), which is a company statement, that the used substances are not in ECHA database (European Chemical Agency), where every 6 months, dangerous chemicals are listed.

http://dx.doi.org/10.5772/intechopen.69210

Strengths	Weaknesses
Research investments	Insolvency in 2012
Patents	
Rich history	
Trust of investors	
German raw material	
Top customers	
Example of good practice in nanotechnology	
National and international relationships	
Opportunities	**Threats**
New discoveries	New discoveries
New collaborations	Slowing research investments

Table 2. SWOT analysis for Holmenkol nanoproducts dedicated to winter textiles.

Siderma owns a collection of technical fabrics dedicated to military shoes as: Army, Shock Absorbent, Multistrat, Low Temperature Resistant Materials. Breathable, Special Insoles, Fireproof. The name of each collection represents the meaning.

The company also owns a research department where special customer demands or needs are covered from new discoveries and where new products are waiting to be proposed in new markets. The most representative product is SIDMAT 3, which helped the company to become an international leader, leaving behind famous brands from Western Europe. This product uses 78.3% from the production process of the entire factory, and it is required by top clients from Europe and Russia.

The technology of SIDMAT 3 is very simple: there are three layers, each produced separately and all together united by interweaving or sewing. The first layer is a polyamides nonwoven product consolidated under pressure at high temperatures, and it can be painted in different colors as in **Figure 5**. The second layer is made up of 80% wool by a mechanical interweaving. The mixture of wool and polyesters gives the final color. The last layer is made up of aluminized polyethylene, the only part not produced by Siderma, which is imported from abroad.

The principle of this product is that the wool maintains the foot temperature, the aluminum layer rejects the cold air and humidity that enter inside the leather military shoe, and the temperature remains constant due to the wool layer. These three layers are united by sewing and not by chemicals because in this way, the foot can "breathe."

SIDMAT 3 was tried in France twice, first at −50°C and later at −80°C. The result was a success, and the company obtained a certificate that proves its unicity.

SIMAT 3 (**Figure 5**) is a smart product since:

- it confers very low temperature and abrasive resistance.
- the technology combines classical production methods with modern ones, and the productivity is higher than that in the case of woven products.

- this product can be improved with other technologies, for instance, nanochemical substances for obtaining more particularities as fire protection, antistatic, odorizing, and many others. The product is supplied in rolls of 30 m or in plans, the width is 1.4 m, and the weight is about 650 g/sm.

The history of SIDMAT 3 is very interesting to be known. During communism time, the company had no idea about smart products since in that moment, it used to be the biggest synthetic leather producer from the South-East of Europe. In democratic times (after 1989), the economic situation of the company became worse and worse, and the conditions put pressure on the creative activity to activate the patents from the assets. SIDMAT 3 was born in this way, but it was different than nowadays.

During a fair in Turkey, Siderma met its future partner from Kazakhstan who owned a very modern technology, but was not able to produce. This was a great opportunity to recreate SIDMAT 3 for customers without being aware that this could be a smart product. During a business trip to ex-Russian countries, SIDMAT 3 was offered as a sample to a shoe factory from Azerbaijan. The owner of this factory made a test in a very cold room where the temperature was −40°C. During the test, ethyl alcohol was added to the shoes. It was an extremely dangerous test (simulating conditions of −65°C), and everything was filmed and posted on YouTube under the name of Daimoni Shoes (Test) [15].

Few weeks later after this film was posted, the first important customer of SIDMAT 3 appeared, who was already using a fabric from another competitor, but the physical properties of the fabric was changing at very low temperatures.

After this long introduction of SIDMAT 3, a market analysis was welcomed to emphasize how the company succeeded in creating competitive advantages and integrating this product into international value chain. The Porter model will be helpful again in constructing the whole commercial image of the product.

Figure 5. SIDMAT 3.

http://dx.doi.org/10.5772/intechopen.69210

Entrances: There is a relative high barrier for those interested in entering this market due to the patented technology, but new discoveries or geopolitical situation could afford anytime the access for similar products. That is why Siderma avoids advertising to not inspire the competitors.

Competitors: Even if SIDMAT 3 is difficult to compete with, similar products are supplied by GoreTex (USA), Tessile Toschi (Italy), and Lenzi Technology (Germany).

Substitutes: The partner from Azerbaijan tested GoreTex fabric, but it could not keep its proprieties at extremely low temperature.

Suppliers: They are selected depending on raw materials. Usually, they are from Italy, Russia, Germany, Romania, and Bulgaria.

Customers: They are from ex-Russian countries, but some of them are from Romania and Italy. The main reason of keeping this customers' portfolio is connected to payment conditions. Customers from the Russian Federation usually demand very high quantities and pay in advance in comparison with others who demand small quantities and ask for long terms of payment.

The SIDMAT 3 represents 30% of total Siderma income since it is a seasonal product, and it is produced every 6 months. Its market can be configured as shown in **Figure 6**.

The SIDMAT 3 value chain can be design as shown in **Figure 7**.

SIDMAT 3 includes raw materials with a cost of about 3.10 euros, the cost production goes up to 3.5–4 euros, but the list price is between 6 and 7.5 euros, and then the final price arrives at 150 euros per one military shoe pair.

A quite similar product can be produced with 40–50% higher price than Siderma, and this confers a leadership position into the market.

A SWOT analysis for SIDMAT 3 is presented in **Table 3**.

Concerning the value chain, Siderma controls the suppliers' selection to obtain best cost of raw materials, and it can decide the selling price with no pressure from competition. Another competitive advantage comes from the acceptance of the customers outside EU or North Atlantic Treaty Organization (NATO). Sometimes, the commercial relations are reduced due to the geopolitical situation, but partners coming from the Russian Federation are preferred because of payment agreements. Another important advantage of Siderma is its headquarter in an EU country, which allows an international standardization of its products from the technical point of view.

In conclusion, the Siderma example is very difficult to replicate since it requires a specific behavior in a certain niche where the commercial collaboration meets obstacles in a turbulent geopolitical environment.

All these three examples bring similarities and some differences. For instance, all companies analyzed started from an idea incorporated in a patent that created a successful product as

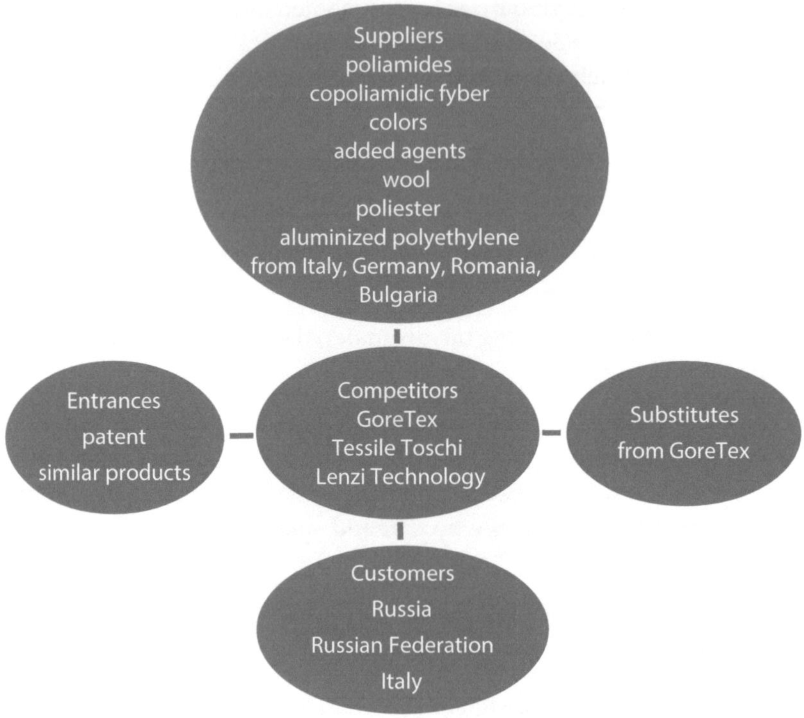

Figure 6. SIDMAT 3 market.

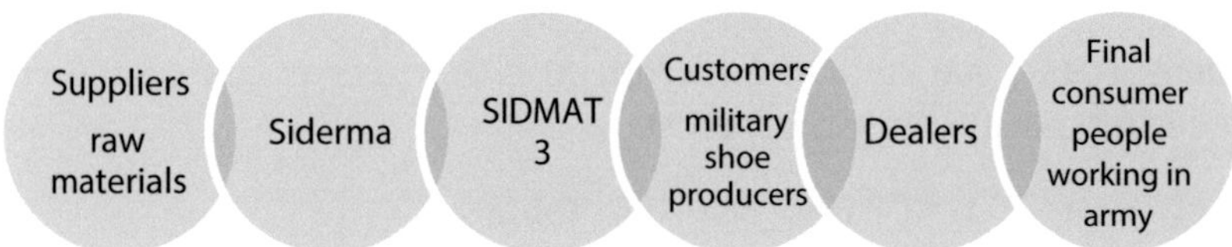

Figure 7. SIDMAT 3 value chain.

Strengths	**Weaknesses**
Patent	Imported raw materials
Partners	Payment conditions
Research activity	
Cheap price due to a cheap workforce	
Opportunities	**Threats**
New discoveries	Geopolitical situation
New markets	New competitors
New suppliers	New discoveries

Table 3. SWOT analysis for SIDMAT 3.

consequence of an intensive and costly research activity. Another common feature is represented by the international cooperation that brought strategic partners. New discoveries or new competitors could appear and undermining the position on the market. Opportunities are challenges toward new discoveries or new applications. Inside the value chain, the position is near the beginning and the suppliers are chosen carefully in terms of costs and quality. Also, all these examples offer unexpected approaches of textiles concepts in comparison with traditional ones.

The differences are in accordance with the field they act, the human needs they cover, and the partnerships they create. For instance, Elmarco is integrated into a commercial value chain, selling its products, but at the same time, it is a part of an innovative chain through the partnerships with universities and research institutes for spreading the potential discoveries. Also, Elmarco is a very open company, attending to all the nanotech international fairs. In the opposite way, Siderma plays very discreet, maybe because of the field where it is involved.

As a preliminary conclusion, all these three case studies showed that the company is looking to welcome human needs in the best conditions of price and quality. The market pressures them to do their best to let behind traditional products. But the financial effort is considerable, and they need to play big, at the international level.

2.2. Integration in an innovative value chain

Nowadays, business configures relationships depending on the way people and ideas interact. Allen et al. [16] show that people and companies create systems and the success supposed to be in the middle of a system. Relationships are created as consequence of needs or depending on market trends.

Lerner [17] considers that patents are essential for an innovative industry, and one of the most important inputs is the very high qualified workforce. He considers that usually companies cooperate directly with universities to start projects together, but the capacity is limited since a small company cannot play globally, and it can succeed with the help of a big one. Moreover, modern technologies are created by "hybrid models," where the research activity is concentrated in a laboratory, and this cooperates with a start-up financed by a venture capital. This helps small companies to rich a certain scalability of business since the researchers are the owners and their resources are constrained by tools, physical instruments, and equipment.

Finding different ways of financing a certain technology is the subject of a rich economic literature focused on. The main problem to find financial support [18] is that new technologies are difficult to estimate their potential, but it becomes attractive when competitive advantages are estimated.

Butaud-Stubb [19] underlines the 2020 European Strategy objectives, where research will be concentrated on energy efficiency, new materials and fabrics, IT integration, creating new products, and cross-fertilization. For a sustainable development, European companies will focus on ecological production processes, and the life cycle assessment will become a very important concept concerning recycling and carbon emissions. Thus, there it is a very important moment for the textile industry to reinvent and to respond to these European objectives. Starting from the first industrial revolution, when the textile industry had its historical role in human modernity, now other premises are created to permit a step to a new era by adopting technologies and finding solutions for new industries with no connection to the past. Nanotechnologies, microsystems, biotechnologies, photonics, advanced manufacturing systems, or advanced materials offer a wide range of discoveries and a huge opportunity for textiles researches.

That is why the concept of value chain is now adapted to the research activity, and it can be named "innovative value chain" keeping in view all the aspects of the commercial value chain theory.

The next sessions are focused on two case studies as examples of integration into an innovative value chain. The first case is about a company that reinvented itself turning from clothing production to research activity; the second one is a simulation of how an idea incorporated in a patent can be pushed to a step further into research value chain taking in consideration with ecological and economic aspects.

2.2.1. Case study IV: lesson from Davo Star S.R.L.

This case study cumulates information obtained during an in-depth interview with the company's owner.

The story of Davo Star Impex S.R.L started in 1990 when it used to be one of the most important Romanian exporter of women clothing, with more than 400 employees. Its production activity is still alive, but nowadays the company has only 25 employees, and the process has changed from Lohn production to full-product production.

A turning moment for Davo was in 2006 when the economic conditions obliged to make something else to survive. One day, in a middle of a meeting, one of its partners invited Davo to be a part of a research project. At the beginning, Davo accepted since it has no other way to do something more dynamic and was not prepared to enter to this unknown world of the research. The first project was LIGUIN covering a large wide of fields, including textile. The total value of this project was 1 million euro, and Davo brought its own contribution of 65,000 euro. Four years were enough to understand this new type of activity, and step by step, the shock was forgotten. The new activity offered the chance to meet an unexpectable and fascinating world, meeting very educated people, and learning so many new things from textile and other fields that a usual producer will never learn. At the beginning, the working style was difficult to understand. Business requires a dynamic style, asking results at once, and every day reception of final products. At the opposite site, the research activity asks for study, reading, and collecting information from all the known discoveries concerning a certain subject, and experiments are repeated many times for analyzing and comparing results. The first lesson Davo learned was that interesting results can be obtained even if the working style is slow.

After its first trial, the second chance became a desire to be involved in, and the story of the second project (SONO) with a total value of about 12,000,000 euros started in 2010. This new project followed the idea of the first one, but this time, the Davo' contribution was 750,000 euros. The result obtained in 2014 was an equipment for printing nanoparticles on the cotton fabric by using ultrasounds. It was a real success, and the European Commission evaluated it as a great impact on future technologies. Thus, Davo is now involved in the third project with a total value of 8,000,000 euros, and the company brought a contribution of 350,000 euros. The purpose of this new project is to continue the last project and to obtain nanoparticles from different active substances (and very expensive) that will be added on a wound patch. Davo has the responsibility of analyzing the inputs (quantity of energy used during experiences, quantity of substances, and fuel consumption) and the outputs (the quantities remaining inside the equipment, evaporation, and wastes). The main problem faced in these projects was waste management since Davo was in the situation to learn the process and to work carefully for collecting it. Another difficult moment was to identify the suppliers for these raw materials since very specific characteristics are required.

In this project, Davo is a part of a big team where three big universities (from Israel, Spain, and Portugal), a research institute (Austria), and small companies are involved (Germany, Great Britain, Austria, and Romania). When this project will be finished, the company from Great Britain will implement the product in its market.

Davo also learned also that during these projects, dissemination is very important. Since the owner is not a researcher, the information was propagated through commercial media (interviews, press releases, and textiles reviews), but its partners from research institutes or universities used to attend conferences and publish scientific articles in international journals.

The advantage of being a part of an international research project comes from reputation. A project attracts another one, and now, Davo is invited to be a part in other projects with new partners from Romania or abroad. The disadvantage of being part of an European project comes from EU long-term payments, and this forces Davo to find ways for financing the activity.

2.2.2. Case study V: patent integration into the innovative value chain

The idea of integrating a patent into an innovative value chain came after studying the methodology of potential sustainability [20] where the necessity to identify examples for new technologies was mentioned in very early stages. More than this, the production of a very new product considers not only economic aspects, but intangibles elements have an impact on the environment [21]. In the case of new discoveries, the knowledge is very important to be shared to gain the trust of potential customers.

These are enough reasons to consider an analysis of a patent potential since this is the earliest stage of a technology. The study will focus on a SWOT analysis that could reveal some directions to concentrate future research to improve its weaknesses and to spread the knowledge of problems that are solved. Thus, the weaknesses could be improved, and in this way, the next step could obtain a better position into the innovative value chain. Even this is an imaginative exercise closer to theory, and the steps proposed represent an algorithm of thinking available in every patent case and could invite patent owners to continue their research in the right direction to push the level of the discovery.

A deep analysis [7] shows a very large technical description of the entire process, but now a shorter explanation is concentrated in steps as a brainstorming of what should be done to follow the path from a theoretical idea incorporated in a patent, and then moving into the applicative research and following next steps in the market. The logic will be focused on a magnetic-coated yarn patent registered under the codes D01H13/30 D07B7/12 and RO128302 [22], and it was built with the help of its owner.[1]

2.2.2.1. Identification of advantages that could increase the possibility to integrate the patent in an innovative value chain

To identify advantages, it is very important to understand how the idea of this patent was born, what are the conditions to produce 1 km of magnetic-coated yarn in a laboratory condition and to identify sustainability indicators and their environment impact.

(a) Short idea history

The origin of this patent comes from the textronic field, by combining traditional textiles elements with magnetic ones, and the result could be multifunctional since the effect is magnetic flux, magnetic induction, magnetic permeability, and susceptibility [23, 24].

A method for making a magnetic yarn (**Figure 8**) includes coating the yarn [25] with one or more layers of magnetic mixture of ferromagnetic or ferrimagnetic powder, binders, plasticizers, and the like (deformer, maintaining in suspension) [26, 27].

(b) Producing magnetic coated yarns in laboratory conditions

To obtain magnetic yarns under the patent mixing formula, the following substances are required:

- Isotropic barium hexaferrite ($BaFe_{12}O_{19}$)—with BF symbol.
- Polyvinyl acetate ($(C_4H_6O_2)n$)—with symbol PVAc.
- Polyurethane adhesive—with symbol PUR.
- Glycol ($C_3H_8O_3$)—with symbol GlyC.

The process of magnetic yarns manufacturing requires certain input of raw materials, some of organic origin (polyvinyl acetate, polyurethane, glue, glycerol), other of inorganic origin (barium hexaferrite), which may be performed by a laboratory employee or by supplier. Actual production in the laboratory involves employee exposure during preparation of raw material, resulting in a suspension.

(c) Potential sustainability analysis of the magnetic coated yarn patent

Dedicated literature [20] offers a number of methodologies according to the specific subject under review, through which certain criteria related to environmental benefits (energy, water, raw materials, emissions of greenhouse gases, and pollution), consumers (product functionality, exposure to disease through product use, and cost-benefit), of those involved

[1]Grosu Marian Catalin – researcher at National Research and Development Institute for Textiles – Leather, Bucharest, Romania

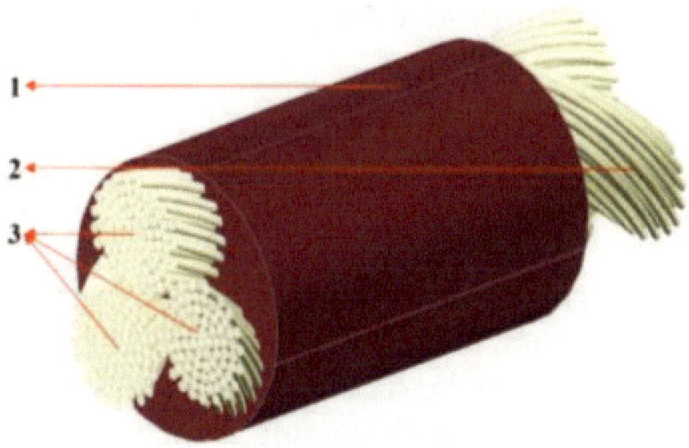

Figure 8. Representation of a magnetic yarn: (1) magnetic coating layer; (2) fiber; and (3) spun yarn within twisted yarn.

in the production (handling safety and health protection at work), and the company are to be observed (creation of markets, quality and performance at specific parameters, process optimizations, and efficiency). Relevant indicators for these criteria, which allow an analysis of magnetic yarns, include carbon footprint, energy efficiency, environmental risk, societal benefits, validity of resources, and risk perception.

Carbon footprint: This indicator regards emissions of greenhouse gases and usually is taken into consideration by studying similar products, but in this case, there is no reference. At a first glance, the barium hexaferrite has no carbon, and the rest of the raw materials, even if they have carbon, are not burned, so there is no CO_2 emission. From this point of view, the process is a clean one, but still there is a certain greenhouses emission of 0, 234 t of CO_2 because of the energy used in the process and the supply of the raw materials. Usually, it is taken into consideration that the emission during 1 year and the value should be around 2 t CO_2/year for an individual transport, even if our value is 9 t CO_2/year. Logistic should be treated carefully by the management team to be reduced.

Societal and users benefits: At this stage, it is required to develop applications within the field of sensors, electromagnetic protection of buildings, to respond to current societal needs. Recent researches have focused on the same area of interest. There were identified as research projects that generate the creation of unconventional technologies, to realize electromagnetic shields and multilayered conductive material shields. From the above-mentioned projects, the last one would approach to the magnetic wires as it uses polyester yarn, which attempts to be coated with aluminum nanoparticles and ferromagnetic material. Technological principle in this case is not similar, but the results can be compared. The project is ongoing, and the result is not yet known. But research in this direction emphasizes a focus area designed to bring solutions to pressing problems, confirming an attractive prospective for the future.

Incidents: During 3 years of experiments, there have been no recorded incidents such as disease inhalations and toxicity.

Environmental risk: At the experimental level, it results in a trace amount of magnetic waste, about 50% of the amount applied. An obvious problem identified is waste management, and in circumstances in which some of them are discharged by washing device, exhausts pipes deposits may be created.

Recyclability: For the moment, there is no possibility of recycling after use. It is premature to put the problem of recycling, as it is not yet a finished product. For now, no information on waste recycling is available, but future research may provide.

The validity of resources: Organic components (polyvinyl acetate, polyurethane adhesive, and glycerol) are easy to provide and relatively inexpensive. The wire is made up of imported cotton, and barium hexaferrite is the only element that is relatively difficult to find, still found in the country, but the prospect looks that the supply is imported (China) not always on great value given the small quantities required.

Risk perception: There may be concerns about the possible particles inhalation during processing and subsequent handling, but with micro-sized particles, there is no danger of inhalation. No information is known about changes in the magnetic properties of the textile yarn in time; the effect should be carefully followed.

Considering all these indicators, sustainability of the patent can be achieved by means of a SWOT analysis as the one below (**Table 4**).

Table 4 presents a clear potential for sustainability of magnetic yarns, but that does not necessarily translate into a competitiveness potential, since it is not a question of selling an app yet, although there are good prerequisites for both an economic process and an environmental impact.

2.2.2.2. Direction of orientation to integration into a value chain

Current research for applications of magnetic materials is mainly in the direction of nanotechnology, although certain studies [20] demonstrate the interest of researchers since 90s for barium hexaferrite nanoparticles. Historical evolution started from magnetic fibers has evolved in a direction of magnetic yarns to align new trends and research programs for the looming default to the nano (**Figure 9**).

EU clusters [28] dedicated to nanotechnology have developed strategies for shaping the value chains, which are now in training and are to be configured as follows:

Strengths	**Weaknesses**
Carbon footprint	Societal and users benefits
Low power consumption	Environmental risk
Lack of incidents	Cotton import
Low consumption of raw material	
Low costs	
Opportunities	**Threats**
Recyclability	Barium hexaferrite
Risk perception	Risk perception

Table 4. SWOT analysis of sustainability indicators.

http://dx.doi.org/10.5772/intechopen.69210

- Value chain 1 (VC1): multifunctional lightweight composite materials applicable for textile and sustainable sport, packaging, transport, energy, Information, Technology and Communication (IT&C), and construction;
- Value chain 2 (VC2): nanoenabled surfaces for multisectoral applications, such as humidity, water, plasma, and vacuum;
- Value chain 3 (VC3): structured surfaces in energy field, IT&C, textile, medicine, transport, and construction;
- Value chain 4 (VC4): functional ceramic and intermetallic alloys used in the field of energy accumulation and conversion, as well as IT&C;
- Value chain 5 (VC5): functional fluids for transport, construction, medicine and pharmaceuticals, consumer products, and IT&C;
- Value chain 6 (VC6): nano integrating for semi-finished and finished products, direct manufacturing, 3D structures for nano electronics and photonics, catalysis, and filtration;
- Value chain 7 (VC7): infrastructure for modeling and testing on full-scale, for complex adaptive systems for complete product design.

The variants mentioned above can be a source of inspiration to correctly identify future research orientation. Magnetic nanoscale wires would find utility in many of the areas targeted for funding of NANOfutures cluster.

Research could continue alongside for the development of magnetic fiber value chain by developing applications that will find their purpose both in Romania and Europe, but the first step requires the engagement of a proactive behavior that involves promoting the idea, and resources finding to activate the said patent into an active form. The mere fact is that there is a patent in a latent form in a complex database, but it does not make it visible. Good ideas, in any business, require to be promoted and applied. It is true that success does not always reap effort, but it is worth trying.

In conclusion, the patent analysis shows that it could be sustainable in terms of environmental indicators and resources. What determines a product to be successful or not is how it

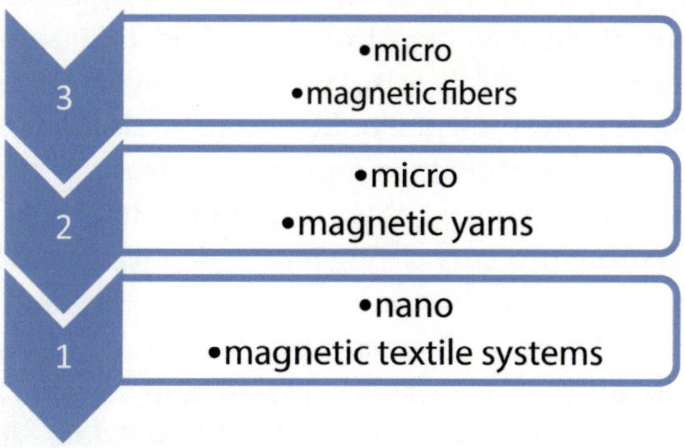

Figure 9. The innovative value chain of coated magnetic yarns.

manages to convince the user. Therefore, subsequent developments may be the key to success. Nanotechnology is the nowadays trend, and it coincides with development directions at European level and has the potential for integration into a value chain and likely financial support.

The message of this study is that any technical idea required to be pursued in relation to market conditions and access to resources for consumption and environmental impacts; otherwise, the effort may simply be an exercise in creativity, consuming time, and significant financial resources, which could be redirected to a more efficiently idea.

As it could be seen, in both case studies (Davo Star and patent analysis), the social and economic perspectives are taken in consideration. New technologies cannot be compared between them, and every step of the discovery is noticed carefully to control environment and health impact. Both require huge investments, and integration in international collaboration is *sine qua non*. When the discovery succeeds to cross the laboratory level, there are networks prepared and already configured at the European level to integrate these new ideas in planned value chains where specialists are available to develop later stages. Even if the success obtained in a laboratory is not a commercial one, there is the hope that every small step gets closer to the final process. As a conclusion, the best ideas could be implemented in the best way only inside the best variant of cooperation.

2.3. Final thoughts

Hopefully, this chapter revealed practical and theoretical examples of smart textiles from an economic perspective concerning their integration in different value chains, commercial or innovative. As it could be seen, even if it was an economic concept, an authentic interdisciplinary approach was required to get information from economic, textile, chemistry, physics, and business fields to have an impact in understanding the phenomenon of smart textiles.

This chapter covered a wide range of case studies focusing on equipment, a group of nanoproducts, a smart fabric, a solution to reinvent the activity, and a patent analysis. Three of these cases are from Romania, and they were luckier to be identified since smart textiles are quite nonexistent in this country, although it used to be one of the most famous textiles producer in the world. That is why these examples are welcome not only internationally, but even at the national level.

All these examples show that technology could help to reconfigure the international market, and the advantage seems invincible, but always for short time since new discoveries can undermine the entire company's effort. Also, even the investment required is huge, this is not covered always from the governmental or European funds (as in the case of Elmarco and Davo Star), and they can be supported from private sources (as in the case of Holmenkol and Siderma). The international context plays a very important role in every case, and the value chains represent a result of a very strong international cooperation. The integration in a commercial or innovative value chain confers a double opportunity for all the actors analyzed to be a supplier of a wide range of advantages due to their discoveries, and at the same time, they become opportunities users. This helps them to consolidate their position since they get reputation and trust, and thus they could open new doors for cost optimization or finding new ways of development.

The research activity is a specific feature for all of them, and this helps them to be in front of the others.

In these new markets, the technology diffusion is difficult to follow since they are not yet mass products or because the field where they act is not transparent at all.

The quality of raw materials supplied is very important. Investments are justified not only for great ideas or highly qualified workforce, but even for a healthy and economic sustainability.

If in Czech Republic or Germany smart textiles are well represented, Siderma seems an isolated and a surprising example from Romania.

All these approaches suppose specific niches, and they require an entrepreneurial culture [29] based on capabilities to create best environment for ideas, special equipment, unique applications, strategic partners, find ways of financing and sometimes special character to manage turbulent or geopolitical situations.

Finally, all these case studies demonstrate that smart textiles represent an alternative of what we live today, but surely, they will be the answer for the best future.

This research shows that success is a consequence of intensive and commercial-oriented company policy, an effect of updating to nowadays conditions, an inspired movement to the right business way. The chapter represents an invitation for creative people to focus on developing innovative value chains.

Author details

Adriana Ungureanu

Address all correspondence to: ungureanu.adriana@gmail.com

Faculty of International Business and Economics, "Dimitrie Cantemir" Christian University, Bucharest, Romania

References

[1] European Commission. European Competitiveness Report. Brussels: Directorate for Enterprises and Industries; 2012. p. 68

[2] Rosendo H, Scheffer M. In-depth assessment of the situation of the T&C sector in the EU and prospects task 6: Assessment of the evaluation of the research and innovation practices in five textile/clothing regions of the European Union. Describe prospects for research and innovation in the sector, final report entr/2010/16, prepared for European Commission Enterprise and Industry DG; 2012

[3] Gemawat P. Strategy and the Business Landscape. New Jersey: Pearson; 2010

[4] Morris M. Capturing value: A value chain approach to national export strategy development, the usefulness of value chain analysis as a policy intervention tool for developing countries, Executives Forum on National Export Strategies, Managing Competitive Advantages: The Values of National Strategy; 2002

[5] Hamel G, Prahalad CK. Competiția pentru viitor. București: Meteor Press; 2008

[6] Estall RC, Ogilvie B. Industrial Activity and Economic Geography. New York: Science Editions John Wiley & Sons, Inc.; 1967. p. 11

[7] Ungureanu R. A. The competitive advantages in smart textiles industry. Case of Romania [thesis]. Bucharest University of Economic Studies, Department of International Business and Economics; 2015. http://opac.biblioteca.ase.ro/opac/bibliographic_view/203208;jsessionid=EABBA40A0D114131B133E73558D3B03A

[8] Elmarco. 2015. Available from: www.elmarco.com

[9] Elmarco. 2012. Available from: http://www.elmarco.com/news/report-nanofibres-2012/

[10] Elmarco. 2016. Available from: http://www.elmarco.com/news/bcc-report-2016/

[11] Czechinvest. 2014. Available from: http://www.czechinvest.org/data/files/nano-brozura-obalka-nahled-final-1232.pdf

[12] Azonano. 2014. Available from: http://www.azonano.com/suppliers.aspx?SupplierID=1753

[13] Holmenkol. 2015. Available from: http://www.holmenkol.com

[14] Tesmasport. 2015. Available from: http://www.tesmasport.si/agodovina-podjetja-holmenkol

[15] Daimoni shoes (Test) [Internet]. Available from: http://www.youtube.com/watch?v=NTOD0kj-ICM

[16] Allen S, Deragon JT, Smith FC. The Emergence of the Relationship Economy: The New Order of Things to Come. Silicon Valley, California: Link to Tour World LLC; 2008

[17] Lerner J. The Architecture of Innovation. USA: Harvard Business School Publishing Corporation; 2012

[18] Jolly D. Development of two-dimensional scale for evaluating technologies in high-tech companies: An empirical examination. Journal of Engineering and Technology Management. 2012;**29**(2):307-329

[19] Butaud-Stubb N. Opinion of the European Economic and Social Committee on Growth Driver Technical Textiles. Bruxelles/Belgie: European Economic and Social Committee; 2013

[20] Möller M, Groβ R, Moch K, Prakash S, Hermann A, Pistner C, Küppers P, Spieth-Achtnich A. Analysis and Strategic Management of Nanoproducts with Regard to their Sustainability Potential. Nano-Sustainability Check. Umweltbundesamt; 2012

[21] Wong H K. Knowledge Value chain: Implementation of new product development system in a Winery. Electronic Journal of Knowledge Management. 2014;**2**(1):109-122

[22] Patentobserver. Available from: http://patentobserver.com/public/search/show.action?searchString=GROSU%2cMARIAN%2CCATALIN&applicants=HOSSU+IOAN&pageSize=30

[23] Bashir T. Conjugated Polymer-based Conductive Fibers for Smart Textile Applications. Goteborg, Sweden: Chalmers Reproservice; 2013. p. 14

[24] Rubacha M, Zięba J. Magnetic textile elements. Fibres & Textiles in Eastern Europe. 2006;**5**(59):48-52

[25] Smith WC. Smart Textile, Coatings and Laminates. Woodhead Publishing Limited and CRC Press LLC; 2010. ISBN: 978-1-84569-379-4

[26] Grosu MC, Lupu IG, Cramariuc OT, Hristian L. Magnetic cotton yarns-optimization of magnetic properties. Journal of The Textile Institute. 2015;**107**:757-765. DOI: 10.1080/00405000.2015.1061761

[27] Zięba J, Grosu MC, Frydrysiak M. Simulation of yarns textile magnetic core. In: 18th International Conference STRUTEX, Structure and Structural Mechanics Textiles-Conference book; Technical University of Liberec, Czech Republic; 7th – 8th December 2011. p. 375-378. ISBN 978-80-7372-786-4

[28] NANOfutures. European Technology Integrating and Innovation Platform on Nanotechnology; Nanofutures. European Technology Integrating and Innovation Platform on Nanotechnology, European Commission 2012. http://www.nanofutures.ro/files/NANOfutures_Roadmap_july_2012.pdf

[29] Ungureanu R. A. Entrepreneurship in the research field. SEA Practical Applications of Science. 2014:655-662